Praise for *The Secre*

Secret Life is a deep dive in the rela... ...d disease, the unseen microbial world and biodiversity—the essential determinants of all life. Although thoroughly scientific and easy to read, this book is one of its kind and a revelation for many. It will guide us to understand the health risks of cultural evolution, changes that may affect our survival and options for good life. Highly recommended!

—Tari Haahtela, professor of medicine, University of Helsinki, Finland

Brilliant read, a fascinating understanding of dysbiosis and encouraging all of us to look after our microbiome.

—Professor Desiree Silva, pediatrician

The Secret Life of your Microbiome skillfully repackages the intense science validating the fundamental role that our gut organisms play in virtually every aspect of our health. Not only do Drs. Prescott and Logan make this foundational information available to everyone, but in addition, *Secret Life* enables the reader to leverage these empowering discoveries and implement truly life-changing lifestyle modifications that will undoubtedly change their health destiny.

—David Perlmutter, MD, author, #1 New York Times bestseller *Grain Brain* and *Brain Maker*

In our age of information overload, reading and reflection are rare. With over 30,000 books published on related topics in the past ten years, I recommend strongly that you first read *The Secret Life of your Microbiome*. This book interprets our survival in the world, the immune system as our ambassador and negotiator in embracing the world (which we are not doing very well), and how we should engage holistically in attitude and action to thrive in our world.

—Professor John Hearn, Executive Director, The Worldwide Universities Network

Drs Prescott and Logan provide an alternative approach as they highlight the relationships between all life and human lifestyle in the biosphere.... written for the interested lay reader but an excellent starter for those of us in the health professions that realise there is a growing need to explore alternative visions and approaches to health and wellbeing. I really enjoyed this book!

—Julian Crane, MD, Research Professor, Department of Medicine, University of Otago, New Zealand

Drs. Prescott and Logan sketch an intriguing, optimistic narrative of microbes, the microbiome they constitute and the emerging microbiome revolution as driving forces in creating, sustaining and reclaiming health wellness and wellbeing. They provide evidence-based and inspiring insight on how to survive the "dysbiosphere," while "savoring the biosphere." As such, *The Secret Life* is a must read for health professionals and laypersons alike.

—Dr. Michael Garko, Ph.D., nationally syndicated host/producer, Let's Talk Nutrition

This delightful book makes very complicated subjects digestible for the lay reader, but is also a good read for the scientist. It beautifully ties every level of biological organization together, from the earth as a whole to the tiniest of organisms, and shows how they are all inter-related and even more importantly, interdependent. Written with humor and compassion, this work demonstrates how humanity has done a very poor job of stewardship of Planet Earth, especially over the last 100+ years, but provides many suggestions for reversing that direction.

—Christine Cole Johnson, PhD, MPH,
Chair, Department of Public Health Sciences,
Annetta R. Kelly Endowed Chair, Henry Ford Hospital and Health System

Drs. Prescott and Logan and have encapsulated an answer that allows us to understand why we live as we do, how our lifestyle choices connect with our immune system and neurobiology. By explaining how we got here, they also provide an answer as to how we can change the ways we engage with the hustle and bustle in order to thrive. This book is a must read for anyone hoping to live in an authentic, and grounded life.

—Martin Katzman (waiting for cretdit line)

Sprinkled throughout *The Secret Life of Your Microbiome* are delightful historical quotes from medical and science scholars that make the book a joy to read. This book paves the way to better understanding the connection between our internal and external environment through understanding the necessity of promoting diversity of the microbiome from inside to out—from our gut, to our skin and the environment that we live in. An excellent read!

—Alison C. Bested MD FRCPC, Director, Student Research Development,
Chair, Integrative Medicine, Associate Professor, College of Osteopathic Medicine,
Institute for Neuro-Immune Medicine

This beautifully written and important book needs to be widely read and studied. The *Secret Life of Your Microbiome* is ultimately giving us the key to achieving health and longevity. I am really excited about this book. A must read!

—Eva Selhub, MD, Tufts University, Department of Nutrition

The recipes Marlies contributed are essential in any gut-health journey. She's distilled down the most important foods to feed not only the organs and cells of the digestive tract, but also the symbiotic bacterial inhabitants that most people are missing. Probiotics are a requirement for great health, but it's even better to get them through your food, as Marlies teaches in her delicious concoctions.

—Summer Bock, Herbalist & Fermentationist, summerbock.com

THE SECRET LIFE OF YOUR MICROBIOME

WHY NATURE *and* BIODIVERSITY ARE ESSENTIAL TO HEALTH *and* HAPPINESS

Susan L. Prescott, MD, PhD
Alan C. Logan, ND

Cover design by Diane McIntosh.
Cover images: background © iStock; central © iStock.

Printed in Canada. First printing September, 2017.

New Society Publishers
P.O. Box 189, Gabriola Island, BC V0R 1X0, Canada
(250) 247-9737

LIBRARY AND ARCHIVES CANADA CATALOGUING IN PUBLICATION

Prescott, Susan L., author
The secret life of your microbiome : why nature and biodiversity
are essential to health and happiness / Susan L. Prescott, MD, PhD,
Alan C. Logan, ND.

Includes bibliographical references and index.
Issued in print and electronic formats.
ISBN 978-0-86571-851-7 (softcover).—ISBN 978-1-55092-646-0 (PDF).—
ISBN 978-1-77142-241-3 (EPUB)

1. Environmental health. 2. Biodiversity. I. Logan, Alan C., 1967–,
author II. Title.

RA565.P72 2017 362.1 C2017-903376-X
C2017-903377-8

Funded by the Government of Canada | Financé par le gouvernement du Canada | Canada

New Society Publishers' mission is to publish books that contribute in fundamental ways to building an ecologically sustainable and just society, and to do so with the least possible impact on the environment, in a manner that models this vision.

Contents

Introduction: A Meeting of Hearts, Minds, and Microbiomes

At its core, this book is a treatise on the interconnectivity of all life, so before we begin we want to share how we have become interconnected in our purpose and our life together. We also take this as an opportunity for each of us to introduce the other.

The cold edges of winter were slowly setting underfoot. The magnificence of Mount Fuji rose above us, a small group of strangers standing on a lakeshore at the side of a Japanese highway. Lost in translation. Waiting to reboard our bus. Not knowing where life was about to take us. Not knowing we would be changed for ever. Anticipation was in the air, our awareness heightened by nature. A strangeness of knowing. But not of what. We were on the edge of a journey that would take decades to unfold. This moment was one we would never forget. As our eyes locked that first time, our destinies became entwined. We felt the future before any words were spoken. A strong and powerful force was at play, its shape, yet to be revealed, one of deep connection that would later shape the rest of our lives. That was the beginning of our calling together. One that had started for each of us long before, but that would one day become united.

From that moment in November, 2003, our purpose together was set. We may have been part of a small delegation that arrived in Japan to discuss probiotics and therapeutic opportunities for bacteria in human health, but our first conversations were shaped by a much wider, much broader shared vision. We had come from opposite sides of the planet to a strange and foreign land. Our lives to that point could not have been more different. And yet we saw the world in the same way, framed by the deep and beautiful connections between all things. A beauty to be found in both the wonder of the natural world and in human nature. Even in the human feats that have created the great global challenges of our time, there is a beauty deepened by the quest to overcome these problems. Though our ideas were still forming, we both sensed the solutions to many planetary dilemmas would lie in understanding the complex symbiotic interconnections between all things—from the level of microscopic microbial ecosystems that reside *within us*, to the myriad macroscale environmental ecosystems that *we reside in* and completely *depend on* for our survival.

And so we spoke about health, we spoke about the future. How both human and planetary health depend on the vitality, diversity, and balance of all ecosystems. And how the same natural laws of interdependence, mutualism, and interconnectivity underpin life in all forms. None are truly separate. Diversity and mutualism buffer and protect us. They stabilise environments and make them more resilient to changes and threats and more robust when facing challenges. We could see the same patterns repeated on every level. Fractals of life. Recurring at progressively larger and smaller scales alike. Governing both biology and behaviour. What we might learn from nature, large and small, we might apply to society. The story of microbes, which are everywhere, is such a good example of how everything is interconnected, in ways that were once beyond our awareness. Our conversations soon drifted to how humans might work together—*with* nature. Since microbes *are* nature, that would also mean working *with* microbes to overcome the many modern challenges facing our planet today.

For centuries, humans have seen nature as something we must dominate, conquer, and tame, progressively eroding the natural resilience of ecosystems, with the loss of many species more vulnerable than our own. This is not sustainable. This must change. It is time to restore balance in our human social ecosystems, and in the many other ecological systems that we are interconnected with.

Our thoughts and feelings began reaching into the future, inspired by possibility and hope. But we knew so much needed to change in the world. We began to imagine what we might do, separately and together, to play our part in this. It would be 12 years before we saw each other again. But we never stopped thinking of each other. And of possibilities. The seeds were sown. In the winter that followed, they remained alive and waiting.

When Alan Met Susan

The obvious place to start in describing Susan, especially in a book about the microbiome, could be to underscore her international reputation as an acclaimed pediatrician and immunologist, whose highly cited research shines the spotlight on the importance of microbes in immune development and the detrimental effects of declining microbial biodiversity. Yet, by focusing only on her paradigm-shifting research in *The Lancet* and other journals, I wouldn't fully capture the much wider perspective she brings to the complexities of this story, which reaches far beyond ecology to the broader social and economic issues that drive environmental degradation, and with it, human disease.

In fact, it is her more holistic philosophies that I have always most admired—the way she recognizes that the health of humans, the environment, our social fabric, and our economy are interdependent. It takes guts to stand up in front of audiences in science and medicine and plead for the collective body not to lose sight of the holistic purpose. That there must be a stronger focus on finding common ground with more mutually beneficial cross-sectoral approaches that transcend competing interests. Messages of unity and collaboration must come from science and medicine as strongly as other sectors. Her ideas, her passion, and her optimism all struck a deep and profound chord within me.

Susan has smashed through the male-dominated glass ceiling of science. She has made extraordinary discoveries involving the most minute details of the immune system. However, in her writing and talks she leverages this knowledge, bypassing the often-labyrinthine details of immunology, and instead uses the immune system as metaphor for the way that we might all see life. Susan brings clarity to confusion. She reminds us that the immune system has a critical influence on the development and function of virtually every tissue and organ in the body. And therefore, so do microbes.

Seeing the immune system as an intelligent, mobile mind encourages us to look at everything differently—in a much more integrated way. But this goes far beyond the usual confines of human health, to the health of society and the health of the planet itself. Susan uses her credentials as an immunologist to engage people across many fields and to join almost any conversation, because everything becomes relevant from a holistic perspective. It's all interconnected. That's the theme of her research and her broader philosophy on life. These are among the many things that floored me when I met her almost a decade-and-a-half ago.

True to her philosophy, a large focus of Susan's career has been on disease prevention. In this, she encourages long-range thinking and advocates strategies that might not have immediate impact, with a view to long-term benefits for individuals and communities. She is a passionate advocate for how improving our environment, especially in early life, can benefit all aspects of our long-term physical and mental health—and how this can also mean better health and longevity for our children. Again, emphasizing interconnectedness, she argues that we need to apply the same long-range vision to addressing the adverse impact of rapid industrial growth on human health and the environment. Most of all, I admire how she works to build a collaborative mindset and strives to bring people together around any issue. She empowers and inspires others to act and to find opportunity for common ground. She reminds us what we can do—and that every choice we make can make a difference.

It is thanks to these perspectives that we were able to write this book with a wider vantage, one that frames the microbiome revolution in a much broader context, and delivers a more meaningful message as a result. For me personally, knowing Susan has profoundly changed how I see the world. She changed my life. For all the years that we have been apart, these ideas have been growing in me. And it is now my greatest joy that we are together again.

When Susan Met Alan

There is just something about Alan that stopped me dead in my tracks. I liked him immediately. It's hard *not* to like him. And I felt like I knew him, though we had never met before. This only got stronger when we started talking. Not only did this clever man have encyclopedic knowledge of the scientific evidence in his areas of interest, he was incredibly passionate and deeply caring, with a unique knack for applying his exhaustive knowledge in new ways, making connections between various strands of research that might not be apparent at first glance.

I was immediately struck by his many marvelous incredible and original ideas, with perspectives I had never encountered before. He was the first person to propose that altering the gut microbes with probiotics might have a role in improving mood and mental health. That might be commonly accepted now, with clear evidence, but in 2003 it was pretty radical. Few took it seriously. But the group of Japanese scientists from Yakult did, that's why they invited him halfway around the world. After listening to his ideas they went on to prove this connection, funding other Japanese scientists to undertake novel animal studies. Very few know about his catalytic role in spurring on research that would change the world. But I was there, listening to a man ahead of his time in so many ways. Within moments of our meeting I knew he was an orator like no other, with such a rich life history to draw on. A life experience that has crafted his strong sense of community and social justice.

Born in Belfast, his early years spanned the height of the sectarian strife of the Troubles in Northern Ireland. Even as a youth he was a passionate defender of the truth, always ready to stand up for an out-group who might be bullied. He already had a strong desire to help others, and was poised to make a difference in a troubled world in whatever way he could. He lived in New York City from the mid-1980s to the early 1990s during its darkest years of extreme poverty, hardship, and violence. Crime and desperation were the products of a failed system, as were alcohol, drugs, and disease. Despair and mental illness permeated the streets. Homeless communities were retreating underground for shelter and respite. His sympathies were always with the people most injured by the broken system. His work and travel through the streets of disadvantaged neighborhoods is a driving force behind his passion and advocacy for policies to address social injustice, health disparities, and economic inequalities. He is always fierce advocate for truth and justice.

He chose a broad liberal arts undergraduate degree—he loves history and the arts—but also completed his pre-med sciences and graduated *magna cum laude*. He already had ideas that the role of diet and nature were greatly underestimated in a range of medical conditions including depression and pursued the doctoral program in naturopathic medicine. Between pre-med and four years of his doctorate, he spent easily as much time studying to become an ND as I did in obtaining my six-year MD degree.

He understands that naturopathic medicine has many limitations and shuns its pseudoscientific associations. He knows that the experience and practice of MDs is distinct and acknowledges that he was able to side-step the required aspects of MD training that would have been overwhelming to him—emergency room rotations, trauma, delivering news to family members that no doctor ever wants to deliver, surgery rotations involving amputations, to name a few. But unlike my medical training, the naturopathic emphasis on nutrition and a more holistic approach to health plays an important part in serving humanity. I studied the body as a series of separate organs, with little integration or holistic understanding; he studied the connections between them and how they are connected to the external environment and nature as a whole.

I completed my training with virtually no knowledge of important things like nutrition and the effect of the whole environment on the health of the whole person. In a world of lifestyle diseases, the importance of nutritional and environmental health cannot be emphasized enough, yet orthodox medicine still places little emphasis on them. Alan was not only an outstanding student, he was greatly admired by his peers and lecturers for his ethics and his attitudes, and with stellar academic performance in place, he was voted valedictorian by his 130-member graduating class. As mentioned, people like him.

Alan has never been a clinician. He decided to focus his efforts on sifting scientific evidence and sharing knowledge and its practical applications with healthcare professionals, industry, and the public at large. He has battled anxiety most of his adult life, and his passage through naturopathic medicine into research and writing is as much for his own healing journey as it is for others. I've never met anyone with more respect for medical doctors and scientists. He thinks of us as rock stars. Always a collaborator, Alan has worked extensively with scientists and medical doctors on a number of projects including several books. For a decade, he was an invited faculty member within the Mind-Body medicine courses in Harvard's School of Continuing Medical Education. In fact, it was there in 2012 that he came across my book. A physician in the front row of one of his lectures had a copy of *The Allergy Epidemic*, and recommended it to him!

Working with Alan, on this project, and in life, has been the most joyous experience imaginable. Living in separate hemispheres, we wrote scientific papers together and shared research notes, but hadn't seen each other since 2003. Neither of us realised the depths of how the other still felt. Our reunion, at yet another microbiome meeting in 2015, was much anticipated on so many levels. When it finally occurred, the experience was nothing less than an atomic reaction, cascading across every aspect of our lives, bringing large-scale transformation across our multiple realities. That is a story for another day. But marrying Alan in 2016 was the happiest day of my life. A shared calling has now become a shared destiny.

Literally within hours of meeting each other again, we drew up the plans for this book—to say things that needed to be said. As we began the journey of this book, and of life, we soon saw that, together, the alchemy of our experiences provided a perfect platform for this work. Susan—schooled in orthodox medicine, but with holistic perspectives still shining through from her upbringing in a spiritually minded family. Alan—schooled in holistic medicine, but with a strong detail-oriented, evidence-based perspective. Like opposite sides of the same coin.

The microbiome is defined as the microbes (and their genetic material) found in various ecological niches, such as the human gut or skin. Early on in *The Secret Life of Your Microbiome*, you will see that this isn't another pop microbiome book that either claims some probiotic pills are going to cure all that ails us, or 300 pages of fluff and on page 299 you are told to eat your greens. We are, however, deeply grateful to holistic nutritionist Marlies Venier of British Columbia for her inspiring and elegantly designed recipes. A true artisan, Marlies was incredibly generous to loan her fermentation and microbiome-oriented recipes designs for inclusion in the pages of *The Secret Life*. Greens, yes, plus a whole lot more.

There are news stories galore about the microbiome, and how targeting it will transform medicine. It might. We will discuss some of those possibilities. But when you start turning all the news stories into a collage, pinning them on the wall like it's *CSI Miami*, you see they are parroting the same thing—just rinse, repeat, and add a chronic non-communicable disease (NCD). Microbiome linked to Depression! Microbiome linked to Obesity! Microbiome linked to Parkinson's Disease! Microbiome linked to heart disease, allergies, autism, type 2 diabetes, autoimmune conditions, cancer…

Who is asking *why*? Almost all of these news stories end with "*Researchers hope to be able to develop new treatments for…*" We are excited about that, too. But we need to start swimming upstream toward the wellspring, the origins of why *on Earth*, both literally and figuratively, we are seeing virtually every chronic, non-communicable disease being linked to a disordered microbiome. It would seem that something about the way we are living is disordered.

Secret Life is more about the *why*. It is an odyssey in ecology. Just like vaccination, your personal connection to nature and the steps you take to nourish the diversity of gut microbes is good for the herd. For all of us. Understanding these connections that escape the visual senses is critical to conservation. *Secret Life* is a book about empowerment, and a new way to see the inequalities that surround us. It's okay to ask what your microbiome can do for you, but much better to ask what you can do for your microbiome. Better yet, ask who is manipulating your health by pushing an unhealthy lifestyle upon you.

Ultimately, a crack in a liquid-filled container presents a problem that will not be solved with more buckets or a better bilge pump. The same lifestyle cracks that are warming our planet, causing environmental degradation, and provoking biodiversity losses are of vital interest to your microbiome. Our microbiome. The planet's microbiome. But what mediates microbes? Lifestyle. But what mediates lifestyle? Not just a healthy diet, but so much more.

One of the wonderful aspects of this evolving microbiome revolution is that it is uniting physicians and scientists from many distinct fields. At this point we don't know what the utopian "ideal" microbiome is, or what it might have looked like once upon a time. Just like Atlantis, this fantastical place may be lost forever! But we have learned plenty about the microbiome of our still-living, but increasingly small communities of brethren who eke out an existence not dissimilar to our Paleolithic ancestors. These are our relatives, too. They may have higher rates of mortality in early-life, but seem particularly resistant to chronic non-communicable diseases. What can we learn? All indications suggest that their lifestyles are crystallizing in the form of essential interactions between the immune system and the microbiome, which, in turn, flow outward toward health. We will expand on this in detail.

As the microbiome gives up details about its once-secretive life, it has become clear that it can never be separated from its environment. In other words, every single aspect of the modern environment can find its way to both the immune system and the microbiome. An app luring you to eat fast food? Yes, it is of relevance to your gut microbes. Your mindfulness in everyday life and total time immersed in nature? Also relevant to skin and gut microbes. Which are relevant to health.

We underscore the desperate need for a science that is in our best interests—what we call *biophilic science*. Science, that is, that concerns itself with the promotion of quality of life and sustainability. We may share our frustrations at some of the gamesmanship that goes on, the manipulation of scientific currency and its use for various agendas or stock options. But ours is a hopeful message of unity. Along the way we will provide quotes from those who were waving warning flags many years ago. Scientific soothsayers who were trying to wake us up. Many of them wrote at a critical turning point in human history, the second wave of the Industrial Revolution. Their writings take on so much more meaning now that brighter light is being shed on the importance of microbes, natural environments, and traditional lifestyle habits in the promotion of human health.

Ours isn't a back to nature call. There is no going back. Plus, we kind of like refrigerators, drains, medical advances, Google Books, and while keeping screen time down, at least one of us loves *Australian Survivor*. Nah, we both do. Rather, we provide a "forward *with* nature" message. The PhD scientist perspective. The MD perspective. The ND perspective. Each line of this book was written and agreed upon by both of us.

A better, more connected world awaits. One where we sustain the planet and the life that sustains us. The secret life of your microbiome has much to teach us…

Yours in Health!
— Susan and Alan

1

Savoring the Biosphere

It is, I submit, a condition of sanity to know the country and the seasons, the hills and the sunrise, the birds and the flowers; to know—not merely to read about—the sting of the wind-driven snow, and the changeful music of the sea. There would be less psychopathology of everyday life if we kept up our acquaintance with the bonnie briar and the cry of the moorland.

— Sir John Arthur Thomson, *British Medical Journal*, 1914

The Thomson Forewarning: Savor the Bliss-Biosphere

It was a dawn of a new era. The Age of Reason had ushered in a renaissance of knowledge, science, and technology, and a new philosophy. Explosions of thought and discovery manifested as hope and faith in a newly mechanized world. Aurora—the mythical heroine casting dawn light—stretching her arms out to many. Change hung expectantly in the air, ready to crystallize in whatever form we would call forth. But what kind of change? Would we move ever more closer to the gluttonous side of industrialization? Or would we heed the call to move forward, utilizing technology *with* nature to restore balance lost? Would we focus on the inherent natural harmony and interconnections of ecosystems, including our own communities, or would we belittle the importance of these things in favor of a far less balanced kind of progress?

There were many who hoped we might find a peaceful road to progress through equality and justice—for humanity and our environment. But we chose a much more difficult road. A road very dark at times. Yet it has been a century of great discovery, vastly so, and there has also been much light. We have seen advances in public health, surgery, and life-saving medications. Progress toward equality. Even so, progress towards enlightenment has struggled against a relentless culture of immoderate consumption, intent on holding power by building division, focusing on differences and reinforcing fear. It is from within this culture, with eyes conditioned by it, that we first saw the microscopic landscape of the microbial world as a dangerous frightening world filled with countless hordes of tiny enemies, all bent on human destruction. We began waging war.

Through the polarized lens of black-and-white thinking we saw neither complexity nor beauty. Microbes caused infections, so the simple answer was to disinfect the lot of them. They weren't considered to be a part of nature, or at least not one that we needed. We mistook for sinister foes our ancient friends—friends who had been sustaining us since the very dawn of our time on Earth.

Microbes are at the origins of all life on our planet, silent partners to the evolution of every form of life and long, long before *Homo sapiens* arrived on the scene. We are only now learning of their true nature, and, with that, the depths and complexities of our relationship with nature itself. Microbes are fundamental to the core of all ecosystems, large and small, including those in our own bodies. We are even starting to see that microbes may hold answers to the environmental and health catastrophes we have created, oblivious to the vital balance of all ecosystem. The story of microbes is a story of collaboration, of mutual benefit, of balance, and of the inherent interconnectedness of every single thing on this planet. These are lessons that our global society needs now, more than ever.

As in 1914, we stand on a threshold. At a crossroads. Our getting of wisdom in the last century has been a painful process for so many: think of how the suffering of wars have led to medical breakthroughs or the ways in which drug side effects have been brought to light. But it does not have to continue that way. If history has taught us anything it has shown that we need to temper great discoveries with greater wisdom. History is important because it contains many lessons overlooked in the haste for narrowly focused progress. It shows us how we might choose to do things differently this time. In the pages of this book we will often revisit the thoughts of many who have gone before. Many of them swam upstream towards the origins of humanness. Wisdom is built on knowledge. It is also built on awareness. And the ability to reflect—to see things anew and make different choices in this, our new renaissance.

Three Days of Peace and Medicine

The second wave of the Industrial Revolution (1870–1914) was an incredible period of human history defined by a galloping pace of urbanization, scientific innovation, and technological expansion. Exciting, certainly. Culturally, the Earth seemed to be spinning faster, just as it is now in our own rapidly changing period of human history. At its height in 1914, a 1,500-strong crowd of influential physicians and scientists gathered in Aberdeen, Scotland. The Annual Meeting of the British Medical Association was the Woodstock of medicine at that time. The who's who attendees included medical giants such as Sir William Osler and Carl Jung. Science reporters described the meeting's atmosphere as charged with electricity.

Mostly, the energy at the gathering was manifest in the form of high-voltage optimism. Based on the massive improvements in public health already realized, it was possible to dream of scientific advances and the application of technology in ways that would continue to transform medicine and conquer disease. Well, at least infectious disease. Much like at contemporary medical conferences, commercial exhibitors were on hand to show off their wares: electrotherapeutic devices, surgical instruments, hygiene appliances and chemicals, medicinal foods, mineral waters, therapeutic beverages and, of course, an expanding display of pharmaceutical preparations. Neuropsychopharmacology was taking its first steps, knocking on the door of chemical Zen. Anti-stress, hypnotic agents such as the drug Aponal were unveiled and positioned for sleep disturbances, tension, and anxiety.

On the evening of July 31st, 1914, John Arthur Thomson (1861–1933) strode toward the central podium of the Aberdeen meeting. Eyes were fixed on this internationally respected biologist, a Scottish native who could easily be described as a celebrity. He was invited to give the highly anticipated keynote address known as the Popular Lecture. The annual Popular Lectures had begun a decade earlier and were tremendously successful. The purpose was to disseminate technical aspects of science and medicine through the prism of practicality.

The lectures were a perfect match for Thomson and his ability to translate theories into practical experiences that could be visualized by wide audiences. He had made a career out of delivering complex matters of evolution, science, and health to fellow scientists, medical professionals, educators, and the lay public. The Popular Lectures were the TED Talks of the day. Something to savor. That is, before the current Epoch of Excess.

Amid the wining and dining, and the drinking from golden chalices of technological utopias, Thomson made his way to center stage with some concerns on his mind. Concerns about the future of humans in a world increasingly disconnected from nature. About urbanization and its contribution to mental ill-health. About viewing the environment through anthropocentric lenses of modernity, goggles that obscure biological truths such as the interconnectedness of all life. Each stride toward the podium brought him one step closer to the delivery of a dire forewarning.

Anyone in the audience that knew even a little about Thomson would have understood that he fully embraced scientific exploration as the path to a better world. He wrote extensively about the historical value of science and its application to human progress and quality of life. He referred to science as the torch to wellbeing and emphasized that further understanding of the environments in which humans reside, work, and play could be leveraged to promote health. And in the process, as we apply science to improve the human condition, we should aim to do so with, in his words, "less wasteful exploitation of the Earth." For Thomson, health of the Earth and living creatures within natural environments couldn't be uncoupled from human health.

Months earlier, the scientific and medical community had been alerted through press releases that the title of Thomson's Popular Lecture would be *Vis Medicatrix Naturae*. This expression had long-since been attributed to Hippocrates and translates from Latin as "the healing force of nature." It had become an idiom of sorts; many saw a literal meaning, others a figurative interpretation. With increasing focus on technology, surgery, and pharmaceutical drugs, early 20th century physicians were already distancing themselves from the phrase, leaving it for use by naturopaths, pseudoscientists, and shysters selling useless blood purifiers, colon disinfectants, and the like.

Why would Thomson choose a controversial term as the title of his talk? Only a few years before, one of England's most famous physicians, Philip Henry Pye-Smith, had informed an inaugural class of medical students that *vis medicatrix naturae* was a mere "figment which owes its prevalence to its Latin dress." The idea that nature, however it might be defined, provides some sort of vital healing force to humans was incompatible with the dawning of the new medical era. As such, it should be expunged. From the opening lines of his lecture, it was clear that Thomson was well aware that the title would stir interest, and that he would have to cast *vis medicatrix naturae* in a new light.

About a century before it became trendy to discuss so-called positive psychology, Thomson already had a name for what he hoped would be a burgeoning scientific field—"the psychobiology of joy." He drew attention to fledgling research demonstrating that positive emotions could influence physiology, and, theoretically, risk of disease and discomfort. He focused especially on the ways in which stress vs. joy could hinder or help operations within the digestive tract, which in turn can strongly influence health.

As he said: "It may be well for us, on our own behalf and for our children, to ask whether we are making what we might of the wellsprings of joy in the world; and whether we have begun to know what we ought to know regarding the biology or psychobiology of joy." Thomson wasn't secretive about what he felt was the primary wellspring of joy, a source that was being obscured by brick, mortar, and indoor life within an increasingly urbanized world. It was nature.

Contact with nature was the fountainhead of joy. The increasing rates of depression coincident with urbanization, could be a byproduct "of neglect of the well-springs of natural joy—of delight in nature among the foremost."

Thomson took the podium and plunged into the deep end of *Vis Medicatrix Naturae*; at the outset he acknowledged there are many ways to interpret the phrase, and in quick succession he dealt with the most obvious. First, he made reference to its interpretation as the traditional use of medicines derived from natural sources. He rebuked the modern medicine men who "scoff too loudly at the old prescriptions," but he wasn't there to talk about that aspect of the phrase. Then he took another major interpretation off the table, that is, the extraordinary capacity of living creatures to heal their own wounds. He made reference to a giant sequoia that spent over 100 years (of its over 2,000-year lifespan) folding its tissue over a deep wound sustained in a fire.

He highlighted another key way in which *vis medicatrix naturae* was explained: the ability of aspects of nature to create conditions for life (e.g., worms and their ability to enhance soil quality). This will be of particular focus in this book, but it wasn't Thomson's primary interpretation of the Hippocratic phrase on that July night. However, he made special note of the ability of "bacteria and infusorians make a clean thing out of an unclean...over the world—on land and sea—a body of scavengers of all sorts and sizes who clean up while we sleep." Throughout his career Thomson was quick to point out that much of nature's beauty, let alone ability to promote life, is a product of unseen, microscopic creatures.

Finally, he removed one more interpretation of *vis medicatrix naturae,* the view that it means the relative absence of chronic disease within wild nature compared to that encountered by domesticated animals and cultivated plants. Since Thomson had written extensively on that topic, especially from an evolutionary standpoint, the crowd may have been anticipating a lecture on the fundamental healthfulness of wild nature. Then, his more pressing concerns and his interpretation was wholly revealed:

> What, then, do I mean tonight by the healing power of nature? I mean to refer to the way in which Nature ministers to our minds, all more or less diseased by the rush and racket of civilization, and helps to steady and enrich our lives.... There are deeply-rooted, old-established, far-reaching relations between man and Nature that we cannot ignore without loss.... In a period of evolution that has been mostly urban we miss our contact with Nature; most of all, perhaps, in youth, for it remains true of the child who goes forth every day, that what he sees becomes part of him for a day or for a year, or for stretching cycles of years...

With surgeon-like precision, Thomson spent the remainder of his time at the podium making his argument for the medical importance of contact with nature in an urbanizing world. He made the claim that awareness of nature promotes attention, creativity, and awe. Observation of nature facilitates rational progress; nature heals because it speaks to both our moral and intellectual ear, illuminating the importance of concern for others above the self. Ultimately, his message was that contact with nature—soaking it up in a mindful way through all senses, not only the visual—was an essential means to quality of life and a vaccine against psychological problems. It was a serum that was being overlooked.

Thomson foretold the world in which we are now deeply entrenched. His focus on the perils of a childhood disconnected from nature was an effort to have leaders in medicine understand that it would have multigenerational effects. Contact with nature in childhood fosters a realization of the importance of nature, in his words, for stretching cycles of years. Thus, lack of contact with nature in early life begets lack of contact with nature in the early life of the next generation. He emphasized the word *loss*. The diminishing relationship between humans and nature, as he said, cannot be ignored without loss.

Since his task was to deliver the Popular Lecture, Thomson didn't encrypt his words with technical talk and scientific jargon. He was firm in his conviction that a prerequisite of true mental health was a knowing of nature. Long before the World Health Organization formally stated in 1948 that health is not merely the absence of disease, Thomson said this: "Just as peace is more than the absence of war, so positive health is more than the absence of disease."

Thus, he was speaking to the Aberdeen audience from the point of view that mental health, or sanity, meant flourishing. Less psychopathology of everyday life really meant quality of life and the ability to learn, grow, and thrive. It was clear that one of the impending societal consequences of a disconnect from nature would be the erosion of positive health. But who was tallying the losses? Was anybody? He hoped the physicians in the audience would be moved by his plea.

Following his verbal argument for spending time in mindful contact with nature as a medicinal agent, he concluded the session with some visuals. Even without a black turtleneck or having read "The Presentation Secrets of Steve Jobs"—which wouldn't be published for another 95 years—Thomson wowed the crowd with nature images from his personal collection. They were projected on lantern slides while the auditorium was placed under dim light.

Soon after, the crowd dispersed and the hall was in darkness. And so was Europe.

About 72 hours after Thomson wrapped up, Britain declared war on Germany and the British Medical Association wouldn't have a grand meeting for another six years. The war machine transformed medical priorities and undoubtedly advanced medicine in ways that help all of us today. But save for some who applied its principles to heal returning veterans suffering from shell shock, *vis medicatrix naturae* wasn't exactly rolling off the tongues of physicians. Despite the elegant articulation, Thomson's premonitions went largely unnoticed. Only in the first decade of the 21st century, and especially the last few years, have scientists assessed more precisely the toll modern societies pay for living with a relative deprivation of nature.

Additional Points of Reflection

In our own journey here in *Secret Life* we will reflect upon the many ways in which ignoring *vis medicatrix naturae* manifests itself in modern society. Unlike Thomson, we have the advantage of accumulating research to draw from. Lots of it. We will make our own efforts to illustrate the abundant research and bring it to life. However, before we follow those scientific tributaries toward their ultimate destination—the argument that all our connections to swaths of visible and unseen forms of life influence us in ways largely unrecognized—some additional nuggets from Thomson can provide a modern point of reflection. Especially important was his introduction of the word *biosphere* into modern language.

Throughout the war that was supposed to end all wars, and right up to the end of his life in 1933, Thomson continued to write extensively on health, nature, and the interconnectivity of life. In a 1902 lecture at the Royal Botanic Gardens Nature Study Conference, Thomson was the first scientist to use the term biosphere in the English language. He was referring to the way life (Greek: *bios*) surrounds us like a living globe (Latin: *sphaera*). To the audience he underscored the ability of this living, seasonally rhythmic sphere to influence our own daily life. The biosphere, according to Thomson, had a grip on our physiology and psychology—our cognition, perceptions, and mood. In his famous Gifford Lectures (1915–16), he expanded his thoughts, stating the following:

> There are in our world three spheres which overlap one another. There is the cosmosphere—from the solar system to the dew drop, from the moon to the moonstone, from the sea to the snow-crystal—the Domain of the Inorganic.... Secondly, there is the biosphere, the Realm of Organisms, where the laws of matter and motion still hold, but are no longer exhaustive, since another aspect of reality has sprung up, which we call life.... Thirdly, there is the sociosphere, the Kingdom of Man.

We cannot overestimate the importance of these spheres and the key word *overlap*. For the sociosphere—our cultures and the collective actions we take at given points in history—determines the extent to which we humans appreciate the importance of the biosphere. Not just the importance of the biosphere in sustaining life in obvious ways, such as the ability of green plants to help provide the air we breathe, the water we drink, and the food we eat. The societal recognition, or lack thereof, that this enveloping realm of all living organisms is actually a contributor to our own vitality and wellbeing in less obvious ways as well—that was the bit that Thomson was concerned about. There was a bliss to be found in the biosphere, and he feared we were running away from it, hurtling toward an abyss.

Even though it was inanimate, Thomson viewed the cosmosphere as full of life. Its matter and energy made their contributions to the theater of life. The moon, moonstones, snowflakes, ice, flowing water and all other "structural" aspects of nature's built environment can influence the vibrancy of the biosphere in various ways including climate, weather, air, and water quality, to name but a few. We, of course, contribute to these aspects of the cosmosphere because of the way in which we live our lives in the sociosphere. Thus, the depth of appreciation for the biosphere from a cultural perspective also dictates the extent to which we will pillage or, conversely, protect the biosphere that feeds us both literally and emotionally. Culture and modernity are shaping us toward diminished societal connectivity with natural environments. By extension, what if cultural shifts lean us toward more narcissistic tendencies? This is where we, as stewards of the biosphere, could get ourselves into a place of more pillage and plunder, less protection and preservation.

Maybe Thomson couldn't have imagined huge mechanical arthropods in the form of American cars traversing the Earth by the millions while averaging ten miles per gallon of fuel. He didn't use the term "greenhouse gases" but he surely knew by witnessing the Industrial Revolution first hand that humans could impact air and water quality within the cosmosphere. Stunningly, in his final writings he had this to say: "Many other examples might be given of man's direct influence on the cosmosphere. Let us take one more. Man's production of carbon dioxide by burning fuel has been increasing at a prodigious rate. What if its continued increase leads to an overloading of the atmosphere with this poisonous gas—an overloading which the vegetation, hungry as it is for carbon dioxide, will be unable to lighten! What a hurry there would then be to plant trees!"

In his writings he explained that when humans conquer part of the cosmosphere by building dams and altering the course of rivers this, of course, can alter the biosphere. He knew full well that humans could also directly compromise the biosphere, tearing down trees and contributing to animal extinction: "A minus intersection is illustrated when man stupefies an animal or impoverishes the biosphere by his ruthless exterminations." He claimed that each sphere—cosmosphere, biosphere, and sociosphere—was "enveloping and interpenetrating" each other. It was all connected.

Time Sink

In making his treatise, Thomson underscored another critical point concerning health in an urbanizing world and our diminishing contact with and knowledge of nature. Since there are only 24 hours in a given day, he was concerned about time use. In the technological world we are a product of what we do, but also what we are not doing. In his words, "Mechanization of man leads to depression and fatigue, and consequent artificial short-cuts out of both." Think energy drinks, stimulants, and caffeine. He stated further: "Along with improved functioning must be included improved use of leisure time—more play and less mere looking at it... we are molded not only by what we do, but also by what we do not do." One of things we were increasingly not doing when he wrote, and are currently not doing as if on steroids, is making contact with nature.

Again, Thomson didn't live in the world in which we do now, where viewing a screen is the dominant use of our leisure time, but he certainly had concerns about the creep of perpetual spectatorship in modernity, an encroachment of non-play into a formerly active life. On the other side of the activity coin, there was often not enough time for mindful reflection. Said Thomson: "Sometimes also there should be mental fasting, there is so little time for reflection between our intellectual meals and almost pathological devotion to print." Both of these devotions—to spectatorship vs. play and salivation to the Pavlovian media bell—could compromise mental health. But they could also minimize an understanding of the web of life. In a non-active relationship with nature, even assuming any relationship at all, the senses are not fully engaged in the realization that the biosphere needs to be protected.

Later we will discuss global inequity and matters of social and environmental injustice. We will focus especially on how these factors impact equitable access to natural environments and equitable ability to reach individual human potential. For now, we will point out that Thomson viewed access to open air, gardens, and sunlight—nature—as a matter of equity. When it came to inequality, urbanization, and health, he recognized the advantage held by the elite, stating: "In urban conditions especially, all but a few men and women—the elect or elite—are apt to acquiesce in a mediocre standard of health." He explained that "pathologically [money] means power over the bodies and souls of other people; normally it means securing increased wellbeing." He wanted greater numbers of people to attain greater equality of wealth for its non-pathological purpose—wellbeing.

But money wasn't everything. He turned out to be correct in his assertion that beyond a certain level of decent income—that which supports quality of "food, space, fresh air, sunlight and rest" or what we might now call a "living wage"—wealth does not produce happiness. He referred to United Kingdom mortality statistics showing that although certain factory workers were making higher wages, they had a shorter lifespan compared to those working in natural environments.

On the other hand, as we will discuss later, it would be difficult to obtain time for contact with nature *and* rest while feeling obligated to pull two full-time, minimum-wage jobs at the likes of big-box stores and coffee chains. A living wage at least allows the recipient a *choice* to participate in bliss, or *biosis*, a joyful mode of living. How can we have conversations about *vis medicatrix naturae*, or discuss the importance of active engagement with nature or allowing microbes to work toward our health, while pretending that these broader issues are unrelated?

When speaking of mental health, Thomson recognized that it was not defined by the absence of some sort of clear disease-oriented criteria. There was no distinction between health and mental health. Health was defined as the realization of a life of fullness and freedom. In his words:

> It is impossible to look round today without being impressed—sometimes dismayed—by the frequency of mental and nervous disorders.... It must never be supposed that normal and abnormal [speaking of mental disorder] can be distinguished like sheep from goats...health means a harmonious correlation of the parts of the body and of the inner life with the outer; a quality of vigor, difficult to define, which implies reserves of energy and a power of initiative; a fitness of adjustment to the conditions of life in so far as these make for conservation and enrichment; and a certain joyousness which acts as a tonic.

We will frame our unfolding discussions on the interconnectedness of life in relation to human health in the light of an absolute truth: there is no health without mental health. By extension, as we travel through the research surrounding natural environments and personal health, we will do so from the perspective that mental health is not the absence of some coding criteria in a four-inch thick diagnostic manual of mental disorders.

Defined mental disorders such as major depressive disorder are serious, indeed even life-threatening illnesses. However, our discussions within *Secret Life* will span the bridge of relevancy from the "walking wounded" (functioning, but not thriving) to those with bona fide disorders. In other words, the research-proven perspective is that someone with low-grade anxiety and depressive symptoms, feeling chronically tired and perceiving themselves to be in less than optimal health, is suffering, and they are on a trajectory toward diagnoses of many overlapping conditions. Major depression. Cardiovascular disease. Diabetes. Obesity and many others. We will make every effort to speak toward prevention as well as exploring the role of nature-based therapy as an adjunctive treatment in specific mental disorders.

Nature Play

Thomson wrote extensively on the importance to children of studying, learning, and growing within natural environments, especially in early life. But also of play. He maintained that the play-instinct in mammals had been developed over the ages because it worked toward a health advantage. Playfully, in his own words, he suggested that animals do not play simply because they are young, but rather they keep themselves young so that they may play. With a more serious tone he emphasized free play and games for children. He stated that allowing children to play at their own games was a natural safety valve, the closing of which would be a disaster: "They are opportunities for the free play of individuality, originality, idiosyncrasy—variations, in short, more or less sheltered from selection; they are necessary to the perfecting of powers—physical, emotional, intellectual—which are afterward of critical moment. Play is thus a rehearsal without responsibilities, a preliminary canter before the real race, a sham fight before the real battle, a joyous apprenticeship to the business of life."

In the early part of the 20th century there was an increasing uneasiness concerning the lack of opportunity for fulfilling what appeared to be innate needs for nature play within rapidly growing cities. It is interesting, given the current crisis of indoor life among our screen-fixated youth, and the continuing inequity of opportunity for play in green and safe, clean, nature-rich environments, to look back and read concerns that could easily apply today. For many in our "developed" societies, the same lack of opportunity has remained, unchanged. However, even when opportunity exists, the normal drive for nature play may now only smolder, smothered by highly structured activities set up for our children, and the omnipresent screen media that has changed the game.

> Oh! The pathos of the efforts of little children in some of the narrow, crowded alleys of our city, vainly trying to achieve this self-realization! There is no sufficient opportunity—the material that formed the opportunity and the need for the nervous reaction of the human race through thousands of generations of upward progress is wanting; no earth to dig in, no trees to climb, no animals to tame, no fruit to gather, no seeds to plant, no banks to jump from, no natural dangers to flee from, no pursuers to dodge. Like drowning men in a great sea

> of need, they catch at the miserable straws of opportunity, and sink—many of them never to rise.
>
> — George E. Johnson, *Why Teach a Child to Play?* 1909.

There are so many layers to the elegant plea made by Harvard's G.E. Johnson. He was pleading on behalf of all children and, more broadly, for generations to follow. Nature play provides opportunity to learn where food comes from, to understand animal behaviors, to witness growth and resilience of other forms of life. It is also an essential course in risk assessment. Within reason (obviously, he wasn't talking base jumping for kids!), Johnson was making the argument that children should be allowed to make their own inquiries and experience the pleasure and possible perils of contact with biodiversity. As we will discuss later, research is now backing up his century-old claims. Nature play not only facilitates joy and positive emotion. It does more. In the course of human history it has been an ideal path to the development of a healthy nervous system.

Three Billion Years and Five Miles

Life on Earth—green life—has existed for over three billion years. Our own genus, *Homo* is close to three million years old. Really, though, how can we even have a rational perspective on that kind of time? Think about what it's like when your flight is delayed and you have to hang out at the gate for three hours, or when you are in your third consecutive hour-long meeting of the day and you start to realize it's just a meeting to discuss having a meeting about other meetings. Three hours feel like forever.

The point is that it has literally taken ages to shape the Earth and us. For almost all of three million years our human experiences—the traumatic and triumphant, the painful and pleasurable—along with those of our *Homo* genus ancestors, have been within natural environments. Until we started propelling our fellow humans into space and down into the depths of oceans, all of our experiences occurred within a very thin terrestrial portion of the biosphere. Thomson was right in describing the biosphere as an envelope. It is technically the five-mile band from ocean floor to atmosphere where life (microbial at the extreme ends of the envelope) has been found to exist. It would be unrealistic to think that our ancestral time within the biosphere *isn't* presenting itself squarely and resonating deeply in the mind-body interface of the modern, iPhone-clasping man and woman.

Imagine trying to break those three million years into hours. Think of all of those hours our ancestors "waited at the gate," watching carefully for animals to eat and observing the patterns of those that could lead them to fresh water and edible plants. Learning through tragedy the difference between edible and inedible plants, and which animals can bite hard and strike back. All those hours our ancestors spent studying the natural environment—associating the animate and inanimate aspects of nature with sustenance, shelter, and danger—remain within us. Like a tuning fork that seems still, but upon closer inspection is vibrating elegantly, those hours resonate in you right now.

The Vibration of Vigilance: Biophobia

Psychologists have started to pull the onion layers back on the extent to which our ancestral experiences currently manifest themselves in daily life. Some of the ways our Stone Age experiences dictate modern cognition and behavior are not so subtle, while others appear quite stealthy and seem to show that we are marionettes pulled by the strings of our ancient, nature-based past. The resonance of the past influences the way we absorb the environment, that is, the way we scan sights and sounds that were highly relevant to us many millennia ago.

For example, when we are presented with photos that contain a number of different animate and inanimate objects, we are drawn immediately to any animals in the scene. This is what psychologists call attentional privilege. Rapid attention toward ancient threats from the animal kingdom—the sights and sounds of a slithery reptile for example—can be observed in infants long before they are influenced culturally by scary story books and other conditioning. When psychologists monitor the infant physiological responses to such creatures, they find a very rapid startle response. Not only that, the acquisition of a fear associated with reptiles and spiders is much easier to achieve, and much harder to extinguish, than other more contemporary threats.

Other examples of this resonance are found in the growing field of research called evolutionary consumption. Put simply, this area of research demonstrates that modern consumer behaviors are often a product of the same evolutionary forces that have made us who we are biologically and physically. For example, researchers have found that after we've taken a stroll through a crowded marketplace with lots of commercial offerings, we are more likely to have distinct memories of the locations hawking calorie-dense foods. This enhanced recall had nothing to do with personal taste and food choice preferences; researchers controlled for that. It is more likely a reflection of all those hours spent in the Stone Age foraging for the denser calories required to sustain life.

There are other examples, too. For instance, researchers have linked our ancient need for fresh, preferably flowing, water to some very modern preferences like our penchant for shiny, glossy things. It seems to be why we have a distinctive, generalized preference for the shininess of glossy objects over matte. Think sun reflecting off ripples of water. Sure, there are a few who buck the trend and matte down their high-end European cars, but the dealerships are hardly full of matte-painted vehicles.

The idea of resonance is important because if we can scan scenes and selectively prioritize the living from the nonliving, if we can develop biophobias (innate and immediate fearful reactions to ancestral dangers), and be led to high-calorie and high-gloss consumer products by way of our ancestral past, then it allows us to make the argument that we may also have built-in, or innate, biophilia. That is, at birth we may very well have an attraction to all living things. Although difficult to prove, is it really that far-fetched? Why wouldn't we have an extreme interest in the life that has sustained us? If we can have startle responses and rapid development of phobias toward coiled reptiles that caused many a death to us and our ancestors over our three-million-year *Homo*-genus history, then why wouldn't we have a philia toward other forms of life? The living beings that gave us life.

Scientists have shown that we seem to universally prefer images of nature scenes compared to photos of an urban built environment. And it's not like the researchers are stacking the deck by making sure there is garbage and graffiti in the pictures of the human-built environment. Typically, these studies use clean city streets without people or animals. The images of nature usually have lots of green vegetation and sometimes bodies of water, but mostly researchers avoid using extraordinary nature scenes like the Grand Canyon or Himalayan peaks because such scenes are known to provoke very high levels of awe. That would be stacking the deck.

But it's not just that we prefer nature scenes after we contemplate them as if at some wine and cheese-filled art exhibit. Rather, scientists can detect the preference for nature scenes even when they are presented to research subjects at an incredibly rapid rate—for a mere one-hundredth of a second, or what is known by stopwatch aficionados as a "jiffy." We also seem to like more realistic nature images. For example, when researchers display nature images in three- versus two-dimensional form, there are changes to brain oxygen use in ways that would typically reflect improved mental functioning.

In a series of recent brain imaging studies, Korean researchers have compared brain responses to urban built environments and rural nature scenes, using functional magnetic resonance imaging (fMRI). The results of their studies have been remarkably consistent—scenes of natural environments increase activity in brain regions associated with positive mood, emotional stability, sharing resources, empathy, and love. Again, this study was not set up as an act of contemplation; images of nature were presented in a two-minute block, but the researchers showed a new image every couple of seconds. By comparison, city scenes reliably increased activity in the amygdalae. These almond-shaped structures are known as the fear centers because they process threat, arousal, and risk assessment.

Obscuring the Roots

These types of studies lend support to Thomson's claim that "there are deeply-rooted, old-established, far-reaching relations between man and Nature," relations that might be pathologically suppressed by ways of life and cultural changes that have been shockingly rapid in the span of evolutionary change. These studies, and many others we will refer to in the following chapters, indicate that biodiversity matters to positive emotional health. We aren't just looking at "green space" or "the countryside" while in an MRI scanner or engaged in similar psychological studies. When we look at nature images we are typically seeing much more of the biosphere, Thomson's "Realm of Organisms," or a surrogate of their representation as opposed to what we find while viewing scenes of urban brick, steel, and glass.

Thomson focused on the psychopathology of everyday life because he knew that being in a place of positive emotion, or, put simply, a good mood, was an asset to our ancestors. *Joie de vivre* has helped us survive and thrive. Modern science backs up that assertion. In the harsh world of our ancestors, can you imagine how refreshing it would have been to taste fresh water? Can you imagine after a long hunt or forage how delicious food must have tasted, even though it wasn't laden with highly-processed sugars, fats, and sodium? Even though it wasn't wrapped up by the delicate hands of a sophisticated marketing machine that includes a character in a clown suit cajoling us and our children to eat fast food, it seems safe to assume they were still lovin' it.

The point is that even while sidestepping predators, our ancestors must surely have looked around themselves with fascination, amusement, and awe—cognitions and emotions that were all experienced in natural environments. By extension, the environment in which those experiences took place is an essential part of who we became as humans. The theater. Here, one of our prevailing arguments is that contemporary science often looks at human existence outside of its larger theater—outside of its functional connectivity to the past, and away from the unseen and seemingly unrelated forms of life that surround us.

But cultural forces are strong, and no matter what our brain responses may be in an MRI machine, we can easily overlook what might be good for us. The progressive loss of human interaction with biodiversity is something contemporary researchers refer to as "extinction of experience." As we will discuss, the extinction of our experience with biodiversity cannot be separated from our role in the ongoing extinction of biodiversity itself. Perhaps if we were more fully aware of Thomson's *vis medicatrix naturae*, we would be less likely to stand by idly and remain spectators to global biodiversity losses, climate change, environmental degradation, and dramatic increases in chronic, non-communicable diseases.

The Dysbiosphere

Our existence, and that of our ancestors, was without massive environmental consequence until we started to cultivate crops and domesticate animals some 12,000 years ago. But even in that time span, things changed at a snail's pace until the primary Industrial Revolution. Today, it isn't difficult to make the argument that the biosphere first described by Thomson is now a *dys*biosphere. Life in our sphere is in distress. We are in the midst of massive biodiversity losses, global environmental degradation, rapid urbanization, and, of course, the reality of climate change.

So, what is dysbiosis? This Greek-rooted term translates as "difficult living," or "life in distress." Often it is used in the context of disturbances in the life of microbes that inhabit our skin and gastrointestinal tract. In the microbiological sense it is more narrowly defined as marked changes to the structure of complex microbial communities that would otherwise be living in a peaceful manner with us. In other words, it is a disturbance in the world of our microbial friends, and since their world is our world, it really means there is a disturbance on us (skin) or within us (gut). Technically, it is a state of change that involves the loss of beneficial microorganisms, and/or the expansion of potentially harmful microbes, and/or the loss of overall microbial diversity and we will have much to say about that topic later.

Within *Secret Life*, we will use the term dysbiosis (or dysbiotic) in both its global meaning—life in distress—and its more narrow microbial meaning. Most often the distinction in use will be obvious and we have made efforts to specify microbial dysbiosis where appropriate. We will also use the term probiotic (which translates as "promotion of life") with a broad interpretation, and not always simply as a commercially available product sitting on a shelf. Although we will discuss them, *Secret Life* isn't a book about probiotic supplements; it is a far larger discussion of the ways in which lifestyle—and the forces that drive lifestyle—interact with unseen life, and, ultimately, the quality of all life on Earth.

Looking around at the state of affairs in global health, divisiveness in politics, and socioeconomic inequalities, it is easy to see that life is in distress. The rates of non-communicable diseases—type 2 diabetes, cardiovascular disease, asthma, allergies, autism, depression and other mental disorders, autoimmune conditions (e.g., Crohn's disease, ulcerative colitis, celiac disease), neurodegenerative diseases, and others—are now described as a global epidemic. Tellingly, many of these conditions are increasing in tandem with urbanization, westernization, and the lifestyle of modernity, including a drift from traditional diets and contact with nature. This is dysbiotic drift.

Climate Change and Dysbiosis

> **The concept of global warming was created by and for the Chinese in order to make US manufacturing non-competitive.**
>
> **— Donald J. Trump, 2012**

No, climate change isn't a hoax. The undeniable scientific facts aren't a product of some foreign regime in cahoots with 97 percent of the international scientific community. We can see the evidence in the air bubbles deep in the ice core throughout Antarctica. This long-trapped Pleistocene air provides a clear way to demonstrate that carbon dioxide and other greenhouse gases in the current atmosphere are the highest they have been in 650,000 years. The colossal increases in carbon dioxide in the atmosphere since the Industrial Revolution, and especially over the last half-century, are in line with separate research that has been tracking human-generated sources. This is real. Not a conspiracy against manufacturing and commerce. Quite the opposite. Lifestyle is influencing the cosmosphere. And the biosphere.

In 1933, Thomson visualized a day when there would be a rush to plant trees to offset atmospheric carbon dioxide. However, he was comforted in reminding readers, correctly, that "since carbon dioxide is soluble in water, the sea acts as a regulator of the proportion of this gas in the air." Remarkably, data published in 2014 by the World Meteorological Organization indicates that we are now past a point where human-manufactured carbon dioxide can be handled by the absorbency powers of oceans and plants. Even the best brands of paper towel have a limited capacity to clean up if the tap is left running.

It may seem silly to even discuss these realities and make the point that a collective of the foremost international scientists do not waver from their consensus statement: "Human influence has been the dominant cause of the observed warming since the mid-twentieth century." It would be tempting to just have a giggle about a tiny minority of misinformed (or worse, cleverly slick with a self-interested agenda) climate deniers that have a bit of a platform on the contemporary soapbox known as social media. But we can't laugh it off. Their ignorance spreads like the invisible greenhouse gases, and, sadly, these merchants of doubt have actually penetrated the minds of those who should know better.

Trust in Societal Teachers

Some of those who should know better are medical doctors, often held out to be the leading experts on lifestyle and the environment. The public generally sits up straight when a doctor speaks. Medical organizations cherish and encourage this sovereignty, rushing to the scene lest any intruders suggest that the Emperor might need a new Armani suit. The basis of this privileged position regarding knowledge of the environment—in all aspects of the term—and the way it presses upon the health of the person showing up in a clinic, is transparently, tissue-paper thin.

Numerous studies show that medical school education related to diet, exercise, and other specific lifestyle factors is paltry at best. Moreover, graduating medical students report lack of clinical preparedness to handle specific diet and exercise advice. Test scores on nutrition knowledge average about 50 percent—no different than they were in the 1970s and 1980s. Ecological medicine—teaching as if the environment matters—is essentially nonexistent. Which brings us back to climate change.

Shockingly, in a pair of 2016 studies that polled the members of the American Thoracic Society and the American Academy of Allergy, Asthma & Immunology (AAAAI), only seven to ten percent reported being very knowledgeable about the association between climate change and health, with as many as 47 percent indicating that climate change is either not happening, or is mostly a natural phenomenon, or that human activity is no more a factor than natural processes. Which means that 47 percent of physicians in specialties that deal with where the environment meets health—allergy, asthma, lung conditions—are completely at odds with science. These disheartening statistics aren't advertised in the abstract of the AAAAI study, which paints a rosy picture of physician awareness, but are readily available in the tables within the report itself.

Doctor means "one who teaches" (Latin: *docēre*). Yet it seems obvious that fitness for teaching matters of lifestyle, individual vs. social responsibilities, ecological influences on health, and how these variables intersect in the doctor's office is based on *docēre* self-teaching and political views. Research involving US middle- and high-school teachers also shows that personal political ideology predicts their classroom approach to the topic of climate change. We have a ways to go in educating our educators, and undoing the damage of the merchants of doubt.

Biodiversity losses aren't a hoax either. Scientists, if anything, have been ultra-conservative about their estimates of historical species losses over the millennia. Just like normal oscillations in global climate over tens of thousands of years, so too have there been lots of species that have had their day in the sun and are long gone. Extinction of animal and plant life has been part of the Earth's history for as long as there has been life upon it. However, just like ongoing climate change and CO_2 in the atmosphere, what we are now witnessing in terms of species loss is not a normal oscillation. Current species loss, by some estimates, is one thousand times higher than the normal oscillations.

Of course, there have been times in the Earth's history when massive extinctions have occurred—at least five big ones that we know of, including the one that ended the reign of kids' beloved dinosaurs. We are now in the midst of a sixth Big One. In fact, scientists are warning that what we are witnessing before our very eyes is the largest loss of species since the loss of dinosaurs 65 million years ago. In 2015, after careful analysis of the existing data on biodiversity loss, an international team of expert scientists led by Gerardo Ceballos had this to say in the pages of *Scientific Advances*: "If the currently elevated extinction pace is allowed to continue, humans will soon (in as little as three human lifetimes) be deprived of many biodiversity benefits. On human time scales, this loss would be effectively permanent because in the aftermath of past mass extinctions, the living world took hundreds of thousands to millions of years to re-diversify."

The researchers went on to a call for urgent action: "Avoiding a true sixth mass extinction will require rapid, greatly intensified efforts to conserve already threatened species and to alleviate pressures on their populations—notably habitat loss, overexploitation for economic gain, and climate change. All of these are related to human population size and growth, which increases consumption (especially among the rich), and economic inequity. However, the window of opportunity is rapidly closing."

Truer words were rarely spoken. Especially about the essential need to move quickly on the last embers of opportunity. It's so hard, though, for us humans to take action on things that appear so far removed from the present. Our dominant thoughts are about the here and now. Avoiding pain, and feeling pleasure. Today.

Perhaps if we weren't so disconnected from the natural environment, and a little more tuned in to the way in which *vis medicatrix naturae* can provide immediate benefits, we would protect the source of those benefits. The same way a dog protects the human companion who nourishes it with food and love. Psychologist Elizabeth Nisbet calls this awareness of nature's immediate benefits "the happy path to sustainability." The problem is, our good friend Dr. Nisbet has found that humans consistently undervalue the extent to which experience in natural environments can subsequently promote a healthy mental outlook and make us feel more vital. Changing that underestimation—through experience and education—is part of the solution.

Cats, Cows, Clover—It's All Connected

In the pages of *Secret Life* we will explore the ways in which life and lifestyle are interrelated—often through our microbes. To set the stage for this, a brief mention of Thomson's illustration of the web of life and lifestyle—a descriptive that combined the writings of Charles Darwin and the famed biologist Thomas H. Huxley—seems worthwhile.

Darwin conducted clever experiments demonstrating the essentiality of insect pollination. When patches of red clover growing in fields were shielded from visits by bees (Darwin covered a clover patch with gauze-like material, allowing plenty of light and water for growth) he found that seed production—providing opportunity for new red clover growth—was minimal. But there was more. The visits by bees were limited by higher numbers of field mice because mice were enemies to early beehive development. However, the number of mice was determined by the quantity of neighborhood cats. More cats meant less mice, more bees, and thus more clover.

This remarkable observation, which is really the heart of ecology, was built upon by Huxley. Although somewhat in jest, he added a social sphere to the clover. In lectures published in 1872, he noted that the quantity of neighborhood cats were a product of the kindly women who cared for them; thus, since the entire British Empire was secured by strong soldiers who fed upon beef, which was from cows that grazed upon clover, it was actually the animal-loving women, crudely referred to as "old maids," that were influencing the status of global affairs. Light-hearted as it may have been, it is a poignant illustration of the beautiful web of life and the many under-appreciated threads joining Thomson's cosmosphere, biosphere, and sociosphere.

In this modern era we are searching for something to save us from ourselves. In the last decade, stacks upon stacks of book titles have included climate change in the title. Over 7,000 by our count on Amazon.com. More than 17,000 books include happiness somewhere in the title, more than 7,000 use the term lifestyle, 1,700 include longevity, and 1,600 the word biodiversity. What is missing, in our view, is a detailed description of why these topics—or buzzwords—are often discussions that are one and the same. We will provide our own clover stories as we discuss *vis medicatrix naturae.*

We close out this chapter with the words of another famous Scotsman. Famed naturalist John Muir, known affectionately as the "Father of US National Parks" because of his influence on their late 19th century development, left us with a treasure trove of elegant prose. At the very time Thomson was writing about *vis medicatrix naturae* on the other side of the Atlantic Ocean, Muir was approaching the end of his life. They would have both walked the very same Scottish terrain in their youth. Remarkably, both men were born in East Lothian, an area just east of Edinburgh. As you can see, they had the same Big Picture ideas: "When we try to pick out anything by itself, we find it hitched to everything else in the universe." (John Muir, 1869.)

2

Surviving the Dysbiosphere

Survival is not enough. Seeing the Milky Way, experiencing the fragrance of spring and observing other forms of life, continue to play an immense role in the development of humanness.... Otherwise man may be doomed to survive as something less than human.

— René Dubos, microbiologist, environmentalist, *Life*, 1970

Shifting Psyche in the Dysbiosphere

At all hours of the day and night, just as surely as birds migrate and the Earth rotates on its axis, massive juggernauts traverse national highways and city streets. These semi-trailer trucks—40-ton transport vessels unimaginable just a century ago—are filled to the brim with cargo that is driving dysbiosis—life in distress. Flinging open the back doors for forensic examination, we soon see the stash that is eroding life and health. Sugar-rich foods and beverages, cigarettes, ultra-processed foods, high-calorie/low-nutrient foods far removed from nature, energy drinks, and the raw material for fast food that gets assembled by workers who don't receive a living wage are making their way to a city near you. These are all markers of a system of personal, public, and planetary health run amok. Yet, all the while, they are also our desperate attempts for a healing balm for all that ails us, along with literally truckloads of antidepressants. Distress in life serves only to increase the bloated haul.

Close to 30 million tons of sugar are hauled around the United States annually. The markers of our sedentary life—37 million new televisions, 160 million new smartphones—scurry over interstate highways. Many more are destined for households in westernized and developing nations alike.

Psychotropic medications are increasingly prescribed to children, teens, and adults. Although prescription drugs don't weight much, they represent one of the most valuable commodities transported on US highways and rail routes—$914 billion per year. Despite having a per-pill weight which is next to nothing, the haul still amounts to over 200 tons of antidepressants, anti-anxiety, attention deficit, and sleep-enhancing medications. We also turn to dietary supplements to soothe us and help fill in nutritional voids—$30 billion worth of approximately 30,000 different dietary supplements in North America alone. Just one small segment of this industry—omega-3 and other fatty acids—tells a story of the sheer volume of our desire to be fixed: 120 million tons of fatty acid supplements are moved around our global transportation systems.

Then we have energy drinks. Juggernauts deliver $43 billion worth of so-called energy drinks to keep up with global demand. Experts in business marketing describe it as an unquenchable thirst for more energy; they foresee a 40 percent growth in sales and profits galore by 2020. We, on the other hand, foresee an increasingly fatigued global population trying to prop themselves up and survive the demands of modernity via the contents of little plastic bottles or large tins of packaged stimulants. Each can and bottle is interconnected to so many other issues of our time. For example, Dr. Subin Park and colleagues have found that energy drink use is linked to sleep problems, depression, suicidal ideation, and stress. They found that consuming junk food magnifies these energy drink-neuro-behavioral-emotional links.

Visualize all that global locomotion for just a moment. The colossal movement of products, the energy it takes, the planetary fatigue it induces, the pressure it places on the biodiversity we are increasingly detached from. A core theme of *Secret Life* is the interconnectedness of life in promoting health, which therefore cannot be removed from the interconnected forces that threaten that vital force. One by one the semi-trailers pull up to the urban loading dock, which is essentially our own gullet. Heavily supported by prodigious promissory notes written by marketers who pledge us a better life, each load brings us further away from our ancestral past.

No Health Without Mental Health

In order to make our argument that all aspects of life—both seen and unseen—are connected to your health and vitality, and, by extension, the health of those around you, we first need to take a step back and take stock of where we are at, mentally, in westernized societies or so-called developed nations. We apologize up front if some of this chapter might seem, for lack of a better word, depressing. However, as the pages of *Secret Life* unfold, and the messages of positivity break through, they will do so from the oft-overlooked vantage of mental health. This chapter is an essential prism so that we can later crush the superficial dismissal of nature by many political leaders, institutions, and those in positions of power and authority. In many of these halls, nature, if on the radar at all, is viewed merely as a low-ranking variable in modern health. To illuminate our argument, we must first shine the optimistic, variegated light of our *vis medicatrix naturae* dialogue on the unsettling crises in our midst.

Although we will focus on health from the mental and emotional perspective, we do so with the understanding that emotional health is deeply connected to the epidemic of chronic, non-communicable diseases in our midst. Allergy, asthma, and autoimmunity are the barometers of change. These diseases are happening at increasing rates, especially early in life. Consider that pediatric inflammatory bowel disease has tripled in some countries over the last half-century. But careful study shows that rapid increases have occurred mostly over the last decade. Based on twin studies, genetics are not the issue, accounting for only about 20–30 percent of the causation. Also, although its tempting to dismiss this as simply increased awareness by doctors, the increases in pediatric type one diabetes are increasing along similar lines. Increases in type one diabetes aren't a matter of physician awareness. It is fairly obvious, life-threatening disease and has been since doctors rode in a horse and buggy. The only rational conclusion is that something, or some things, have changed in our environment, lifestyle, or both.

Onwards then, to the current psyche in developed nations. As defined, the term "developed" might infer that nations like Australia, Canada, the United Kingdom, and the United States are all grown up and fully mature. When it comes to dealing with infectious disease, early-life mortality, and the delivery of advanced healthcare, these and other G12 countries are certainly wise beyond their years. But not so much when looking at them through the filter of what *actually* defines health.

Health, as Thomson made clear and as the World Health Organization (WHO) has now formally defined, is not the absence of disease but rather a state of complete physical, mental, and social well-being. WHO has more specifically defined mental health as the ability to reach one's potential while coping with the normal stresses of life. Of course, defining what, exactly, are the "normal" stressors of life in our increasingly complex, urbanized, technological, nature-detached world, is a bit of a challenge. However, we can all understand that, just because someone doesn't meet the checklist criteria for a mental disorder, doesn't mean they are thriving. Nor does it mean they are healthy.

Beyond the Borders of Disorders

The international statistics on mental health in developed nations are troubling to say the least. In the United States 25 percent of the adult population reports having a mental illness in a given year, and 50 percent of adults over the course of a lifetime. Those are *actual* disorders. Primary interventions—pharmaceutical drugs and cognitive-behavioral therapy—can be powerfully effective. They just aren't effective enough for everyone. Plus, they are often used after months or even years of suffering. Prevention of mental disorders desperately needs to be prioritized.

Since our focus is on mental health, we won't enter the debate about whether or not there is a more recent epidemic of defined mental *disorders*. Although some studies have shown major increases in diagnoses of depression and anxiety in recent years, others have not. Whatever the accurate percentage of westerners that currently meet various criteria for depression and anxiety disorders may be, we can all agree that it is unacceptably high.

There are untold numbers of people who sit just below the threshold of diagnostic criteria for depression and anxiety disorders. In medical jargon the individuals who look good from afar, but are far from good, are labeled as having subsyndromal, subthreshold, or subclinical conditions. Psychological distress, fatigue, sleep difficulties, and other symptoms associated with subclinical mental health disorders are commonly reported in primary care settings. Only eight percent of people show up to their doctor to discuss mental health issues like depression, anxiety, alcohol problems; they are much more willing to talk about physical ailments. But when prodded just a bit, it turns out one in three are actually experiencing emotional symptoms that they were reluctant to discuss.

Only in recent years, have researchers started to get a handle on the true suffering, unrealized quality of life, and difficulty performing daily activities, experienced by those who sit just below diagnostic cutoffs. Historically, scientists and clinicians have been hyperfocused on individuals that meet highly-debated criteria for various mental disorders; the consequences of this are now clear—many suffering individuals have been overlooked. We think it is an important part of the discussion because living on the borderland around "nearly" having major depression and/or anxiety disorder(s) sets one on the path to a higher risk of other chronic diseases, and, most certainly, impaired quality of life. The physiological and metabolic changes—oxidative stress and inflammation—that lead to damage to cells are there on either side of the line.

In fact, scientists discovered recently that individuals with subthreshold depression experience loss of grey matter and increased mortality. That's code for the reality that the mind-body interface really isn't interested in diagnostic manuals: moving along the mental state continuum in the direction away from vitality damages cells and can shrink your brain and shorten your lifespan. We will describe in *Secret Life* how recognition of the interconnectedness of life and lifestyle can provide an antidote to the erosion of health. Provide a sea wall to protect your grey matter.

The Resonance of Stress

Shifting mental health can also be gleaned from perceptions of stress. The vast majority of international academic studies and surveys conducted by various mental health organizations show significantly higher reporting of psychological distress within the last few decades. In fact, never mind decades, they show significantly more distress in the last few years. If youth and adults of all ages are reporting higher levels of perceived stress—and *perception* is *reality* when it comes to thoughts and feelings—then we have a clear roadblock to the realization of mental health.

Very telling are the increases in cases involving symptoms of depression, anxiety, and stress fielded by the employee assistance departments of major global companies. For example, one survey involving 100,000 employees located in Europe, Asia, and the Americas reported a 27.4 percent increase in such cases from 2012 to 2014. Similar findings from an Australian employee assistance firm covering 600,000 employees has shown a 15 percent uptick in stress-related sickness and cases of flat-out AWOL from work due to stress in recent years. There are many legitimate reasons why this might be the case, often pinned on horrible bosses, modern workplace culture, and excessive demands. However, our children, teens, and university students are also experiencing the same trends.

New research from Dr. Jean Twenge shows that compared to their peers from 1980, North American teens are much more likely to report symptoms of distress such as trouble sleeping, shortness of breath, diminished cognitive focus, and memory recall. Also in the last few decades, young adults in university are more likely to report feeling overwhelmed. Recently, Canadian scientists from the Centre for Addiction and Mental Health found that the number of teens reporting moderate to serious psychological distress has increased since 2013. Meanwhile, rates of disability due to mental or neurodevelopmental conditions has increased by 21 percent over the last decade. It's an upward creep of stress, especially among our youth.

It seems safe to say that despite increased awareness of mental illnesses, refinements in psychotherapies, dramatic increases in psychotropic prescriptions, marginal gains in breaking associated stigma, an inflated sector known as the wellness industry (however loosely that may be defined, including coaches, gurus, and therapists of all sorts) there is little evidence to suggest that recent cohorts in westernized nations are in a better place psychologically. All indications point to the way we live, and how that clashes against our still underlying desire, as Dubos said, to see the Milky Way.

Empathy, Perspective Taking, Civility

René Dubos was worried that the deep technology of the modern world—the technosphere as he called it—was beginning to erode our humanness. Our values and vitality. Specifically, he wrote, "Technological civilization is threatening the elements of nature that are essential to human life and the values that make it worth living." Humanness has been forged over the millennia. As Dubos put it, "Humanness is the expression of the interplay between man's nature and the environment, an interplay which is as old as life itself."

Empathy—the ability to take another's perspective, to understand and share their feelings and emotions—is, among other things, a key part of what Dubos described as humanness. He feared it was being vacuumed up by a runaway technology, planning, and policy machine, one that was operating without forethought for unintended consequences. Operating, as he said, without imagining what kind of world we would all *want* to live in. In his last publication, *Education for the Celebration of Life: Optimism Despite it All* (1982), he even went so far as to say that in the future, those with empathy will be the most valuable members of global society: "The most valuable members of future technological societies may turn out to be not those with the greatest ability to produce and distribute material goods but rather those who, through empathy and happiness, have the gift of spreading a spirit of good will."

The science surrounding empathy has grown by leaps and bounds since 1982 and recent studies have confirmed Dubos's speculations, at least from the perspective of empathic skills having the potential to ripple outwards. Higher levels of empathy, in general, have been associated with overall wellbeing, life satisfaction, rich social networks, better workplace performance, and healthy relationships. In our highly technical society, one in which children are taught from a young age to "get ahead," it is worth pointing out that empathic individuals have healthy relationships, are less aggressive, and more prosocial. What's more, the ability to pick up on positive emotions displayed by others is a particularly valuable asset in cultivating healthy social relationships.

Dubos's claim takes on greater meaning when we evaluate the changing psyche in westernized nations by looking deeper than stress *per se* and examine select characteristics in our ways of feeling, behaving, and thinking. Empathy is one of them. The research of Dr. Jean Twenge and others shows that critical aspects of humanness—most notably empathic concern and the ability to take the perspective of another—has fallen off the cliff among American college students. For example, the average level of empathic concern declined by 48 percent and perspective taking by 34 percent between 1979–2009. The biggest plunge was noticeable around the year 2000 after holding steady for many years.

As when an invasive species takes over a once diverse ecosystem, the loss of one great personality trait—empathy—seems to be replaced by another. In its stead, there has been an upward creep of narcissism. A scary thing if even marginal increases were to occur on a global scale. Scoring high on empathy is strongly and positively associated with altruistic values, and negatively to values that prioritize the pursuit of personal status. Narcissism is essentially the opposite of empathy. It involves promotion of the self, desire for personal fame and success, forceful exploitation of others, lack of empathy, and inability to see the humanness in others.

Dr. Twenge and international colleagues who have reported these seismic shifts in dispositional empathy and narcissism have defended their research well. Additional research has shown that the narcissism in North American university students was still trending upwards between 2006–2012. Moreover, they have also accumulated research indicating that cultural products—song lyrics over recent decades, for example—reflect the rising tides of narcissism. Interestingly, a new study in *Perspectives in Psychological Science* also shows that shifting cultural products are reflective of our modern disconnect from nature. Starting in the latter half of the 20th century, references to nature have been decreasing steadily in fiction, song lyrics, and film storylines. In contrast, cultural references to the human-built environment have not declined.

In a clever piece of research, psychologists led by William K. Campbell set up a hypothetical business where subjects assumed the role of owners of competing timber companies harvesting trees in the same large forest. Only with careful tree harvesting would the forest remain renewable. Unsurprisingly, those scoring high on narcissism put the self first. They sought profit, obtained it early on, and harvested the heck out of the forest, until there were no more trees. Separate research has found that in both European and American business schools, student narcissism was linked to higher levels of materialism, and, in turn, with lower environmental values.

The implications of this research are enormous. Even if the narcissism is only creeping up at small percentage points in the global population, this would translate into massively compounded threats to the environment. That's before even discussing the ways in which narcissism facilitates incivility and offensive behavior, and worse still, allows for their acceptance.

All boats rise with the tide—narcissists find their way into positions of power and influence. They can often be the loudest voices on social media. As psychologists have shown umpteen times, repeated exposures to media messages leads to many downstream effects, such as dietary choices. But they may also shape the personality traits that in turn shape our lifestyles. Researchers have connected political talk shows (distinct from traditional news programming), reality TV, sports, and horror programming with potential cultivation of narcissism.

We can certainly celebrate the fact that serious violence in most developed countries is close to record lows, but it should be unsettling that generalized incivility has increased. Workplace incivility has doubled in the last two decades. Consider the definition of workplace incivility by researchers Lynne M. Andersson and Christine M. Pearson: "Low intensity deviant behavior with ambiguous intent to harm the target, in violation of workplace norms for mutual respect." We might consider that an erosion of humanness. Research shows that incivility breeds incivility. As we will explain shortly, this is especially true when we are dealing with the high levels of cognitive demands so commonly experienced in modernity.

Social inequities compound the situation. Research shows that socioeconomically advantaged groups are much less likely to consider the economics of the disadvantaged unless empathy is provoked. Thus, a systemic slide in empathy represents yet one more barrier in the quest to narrowing social and health inequalities. The 2016 US presidential election epitomized the empathy deficit and the incivility surplus. It would be tempting to say it was just a one-off outlier. But it may be a barometer of worse to come. It might be the way we've become: unplugged from natural environments, plugged into divisive media, and increasingly unable to take the perspective of another.

Day Is Night, Night Is Day

When Dubos warned the readers of both *Life* magazine and *Scientist Citizen* journal about what it means to lose sight of the Milky Way, he was referring to it both literally and figuratively. He was underscoring that looking up at the Milky Way and a sky filled with several thousand visible stars was a joyful act. Obscuring the stars registered a loss in the ledger book of positive psychology. But it was more than that. Concealing the star-strewn sky cuts the moorings to our ancestral past and the idea of awe, vastness, and something greater than just surviving. Greater than a video game controller in one hand and an energy drink in the other.

> **Urban dwellers never have the chance to see the Milky Way, or a night radiant with stars, or even a truly blue sky. They never experience the subtle fragrances peculiar to each season; they lose the exhilaration of early spring and the delightful melancholy of autumn. The loss of these experiences is more than an aesthetic affliction; it corresponds to a deprivation of needs which are essential to physical and mental sanity, because they were indelibly woven in man's fabric during his evolutionary past.**
>
> **— René Dubos, *Scientist Citizen*, 1968**

In 2016, scientists announced that one-third of humanity, and 80 percent of North Americans can no longer see the Milky Way. We've obscured it with the way we live. The sociosphere has reached its tentacles into the cosmosphere and then back into the biosphere. Our addiction to light creates a form of pollution; it alters life on this planet, including our own. The amount of artificial light we throw into the sky is actually changing the seasons. Buds are now bursting earlier in areas with heavy artificial night light, and this in turn impacts animal life.

Whether we think of anatomically modern humans dating back 200,000 years, or our genus Homo for nearly three million years, our entire experience has been cyclical and circadian. We have been conditioned to the natural cycles of day and night—light and dark—and, with some distance from the equator, familiarity with seasons. The daily cycle of human functions—metabolic, psychological, and behavioral—are regulated by internal clocks found throughout the body.

The primary, or master, clock is located in the brain's hypothalamic suprachiasmatic nucleus and is itself regulated by the timing of light hitting the retina. When natural light makes its way through the pupil in the daytime and its energy is transferred to the "clock," everything is generally as it should be. Natural daylight is especially rich in the blue wavelength and seems to be a critical factor in keeping the 24-hour cycle running properly. Exposure to adequate blue light during the day is associated with blood glucose-insulin balance, better moods, cognitive focus, and better sleep quality later on at night. Blue light keeps the sleep hormone melatonin in check during the day, but experiments show that getting blue light in the day can magnify melatonin levels seven-fold at night, precisely when we need it.

What's more, melatonin isn't just a sleep chemical. It plays critical roles in the immune system, protects cells against stress-induced damage, and may even be involved in the health of the gastrointestinal tract and its microbes. In animal research, appropriately timed blue light increases melatonin and, subsequently, reduces cancer cell growth, especially hormonally-induced cancers. This research matches a growing body of public health studies demonstrating that elevated artificial light at night is associated with higher rates of breast and prostate cancer.

Ancestrally, we've had lots of blue light during the day and would have been naturally removed from it as evening became night. Although modest levels of light—even a few candles—can throw off melatonin levels, that can't be compared to LED lights galore, 60-inch televisions, e-readers and tablets that bathe our faces in blue light. There is little question now that such devices can squash melatonin.

In a recent Harvard study, participants reading an eBook (compared to reading a printed book) for a few hours before bedtime had reduced evening sleepiness and took longer to fall asleep. They also had reduced melatonin secretion, later timing of their circadian clock, and reduced alertness the next morning. The latter bit is critical. Reduced next-morning alertness, as the scientists call it, is code for being part of the walking wounded—being tired and needing an energy drink. Additional research has shown that the effects of evening blue light exposure via devices may be worsened by lack of natural blue light during the day.

This is the type of research that forces bigger questions upon us as a society—the interconnectedness of life and lifestyle. The interconnected "clover story" of sleep. The quality of our sleep can influence subsequent empathic responses, dietary habits, cognitive focus, impulsivity, energy for physical activity, civility, mood, and susceptibility to stress. On the other hand, sleep quality itself can be influenced by dietary habits, time spent in front of screens, mood, physical activity, and stress. Moreover, exposure to nature and natural light, as we will see, can encourage the growth of Darwin's red clover—positive emotions—in our mind. Which, in turn, influences sleep.

Concern for Future Consequences

Researchers who study human decision making often use words like "rewards" and "discounting." Rewards are anything that make us feel good, things that are capable of pulling the levers of dopamine and other neurotransmitters in the emotional centers of the brain. Discounting refers to our capacity for delayed gratification. In other words, our propensity to discount (wave off; minimize; underestimate) the value of a future reward that could be more valuable in favor of something smaller but more immediate. Psychologists often examine this susceptibility by offering little rewards, let's say a small amount of cash, to see if we will wave off a potentially larger prize that could be obtained later. Will we discount the value of future rewards or will we delay our gratification to obtain more later?

Collectively, these studies (under the very academic-sounding names of delay discounting and temporal discounting) are attempts to understand our varying degrees of concern for future consequences. If you have a strong willingness to delay acceptance of cash or prizes in these type of studies, you generally have greater concern for future consequences. Obviously, that's a good thing. Individuals who demonstrate higher concerns for future consequences are more likely to eat healthy, exercise, go for health screening and prevention visits, maintain a healthy body mass, and follow vaccine recommendations for self and children. They would understand that the reward for *not* choosing an additional hour of the deliciously sweet nectar of screen time, and bypassing the chaser of highly palatable but calorie-dense, nutrient-poor snacks and fast food—perhaps opting instead for physical activity, time in nature, and healthy food—would be the larger payoff of future health and vitality.

Modern living, however, can throw off our concerns for the future. For example, researchers have conducted dozens of studies on the aftereffects of cognitive fatigue. When the brain has been subjected to a host of stimulation and lots of complex activities—think a crowded mall on December 24th and doing a bunch of math equations in the middle of the noise and lights, while dodging people in crowds and checking off your list—things can go awry. Results from studies involving experimentally induced cognitive fatigue demonstrate that our decision making, dietary choices, and even our civility wouldn't be ideal. We are less likely to want to exercise, and if we do, it doesn't feel as good and we do less of it. Our attraction to much smaller pleasurable awards and inclination to wave off future gains is magnified.

Unsurprisingly, sleep quality matters to emotional empathy and stronger concern for the future. Diminished sleep quality seems to desensitize us to the emotions of others, the subtle signals of distress that we would normally pick up on. It makes us reach for the smaller short-term prize at the expense of future possibilities.

While it may be nearly impossible to truly determine how long (never mind how well) our ancestors slept, in their remarkable piece entitled "Ancestral Sleep" (*Current Biology*, 2016), Harvard sleep expert Dr. Jean Duffy and colleagues make a strong case for diminished sleep in association with modernity. Examining isolated communities living hunter-gatherer or very traditional lives, they have found sleep diminishes with the encroachment of electric light. Moreover, if only looking at developed nations, even here we may be sleeping three hours less per week than we did before the year 2000.

All of which allows us to wonder: what if our entire modern existence comes with the subliminal undertow of being in the crowded mall on December 24th? While doing some difficult math sums. Compounded by sleep debt. A Groundhog Day of cognitive fatigue. A never-ending low-grade inflammatory soup of highly-scheduled responsibilities, invasive social media, pervasive political rhetoric, and reality TV, the work-home interloping of our devices, confusion of accumulating choices and demands, and the noise and blue-light glow of incessant marketing. Which, in turn, corrodes our humanness.

Screen Test, Time Sink

Among the many things that have changed in our environment in the last several decades, one particular manufactured item has sprung up like dandelions on a lawn—screens. We are surrounded by them. The number of television sets per household in the United States and Australia has doubled since the 1970s, and although these TV numbers are leveling off, they have an alien spawn. Personal computers, tablets, laptops, smartphones, and monitors used for gaming bring the number of screens close to ten per household.

Since there are only 24 hours in a given day, it brings up the issue of time use. Recall Thomson's forewarning: when we are doing one thing, we aren't doing something else. Screens and the amount of time we devote to them represent a massive and very rapid environmental change. Outside of work, our information consumption—largely from screens—has increased by 350 percent since 1980. Our time devoted to information consumption has increased by 60 percent.

In and of itself that might not be a bad thing. Information can be life-saving and uplifting. It can encourage understanding and promote intelligence and growth. But the bulk of information cast off our screens is to intellect what cheese puffs are to the Mediterranean diet. We are gaining weight from it. Literally and figuratively. Bloated, we can't turn away. Quality information gets lost in a pixilated fast-food feast. The ultra-processed info typically serves only to reinforce our beliefs or make us angry in disbelief. It's the additional mouthful of sweet potato pie and whipped cream at the end of the holiday feast, that bite we know we shouldn't have. But we fill ourselves to the brim of the esophagus and loosen the belt a notch.

Studies that have followed children and teens over time have shown that screen time is a predictor of subsequent depressive symptoms, anxiety, and weight gain. The preponderance of evidence for both adults and children is that excess screen time can become an unhealthy habit. Excess can be difficult to quantify but it is typically gauged as more than two hours daily outside of work and school.

Thomson forewarned that when we are doing one thing, we aren't doing another...and that other might be something in the interest of health promotion. The something else we aren't doing, it appears, is spending time outdoors in nature. Time use trends that have specifically evaluated indoor vs. outdoor activities among adults and youth in recent decades, such as that by Canadian scholar Carlyn Matz, indicate more time huddled indoors and less time outside. Other researchers have found similar downward trends when examining, more specifically, nature-based outdoor recreation.

Record numbers of people are visiting the more glamorous nature destinations, including some of the most iconic national parks. But glam tourism is distinct from communing with nature in daily life. In fact, a 2017 study published in the journal *Landscape and Urban Planning* shows that less than one third of urban adults in the United Kingdom are responsible for 75 percent of the direct, regular contact with nature in gardens and parks. In other words, if you live in a city and make frequent, intentional visits to nature, chances are two of your neighbors aren't. Moreover, an individual's emotional connection to nature—something we will spend an entire chapter on—is predicated on frequent contact. Sadly, frequent contact was found to be the exception rather than the norm. This is highly relevant to our journey in *Secret Life*, as nature is a critically important source of microbial exposure, and influences the patterns and health of our own microbiome, as we will see later.

Conservation biologist Oliver Pergams and conservation ecologist Patricia Zaradic have noted international dips in nature-based recreations, as much as 25 percent less since the 1980s. North Americans often visualize the countless shades of green and breathtaking beauty of New Zealand. A relatively high number of its homes have a private garden and the climate is temperate. But new research shows that New Zealanders, like others, are tucked safely inside, spending very little of their time outside their homes.

This was Thomson's fear. As a biologist, he didn't need to enter the semantic argument concerning the meaning of "nature" when he underscored the importance keeping up acquaintance with nature. Researchers and philosophers can get caught up in how nature is actually defined. Thomson's "acquaintance" is what we refer to today as being connected to nature. Of course even while playing World of Warcraft in a basement we are still connected to nature in the sense that there are microbes around us, insects, and who knows, maybe some larger creatures as well. Since we ourselves are nature (even in our sanitized world we are teeming with living microbes on and within us), mingling with other humans in daily life, we are always connected to nature so long as we live and breathe. What we mean by disconnect from nature is withdrawal from expansive contact. Loss of experience. And loss of contact with beneficial microbes we might find in the expanse of green space.

Adaptation: Mere Survival Is Not Enough

Dubos was acutely aware that energy-consuming lifestyles maintained in developed nations were presenting serious and potentially irreparable dangers to life on Earth. His optimistic side suggested that humans may unite in the face of that very obvious threat. We would survive.

We would also survive the expansion of global cities. He constantly reminded his reader about the remarkable adaptive ability of humans, spreading out and surviving extremes of conditions all over the globe. But then he turned from the narrow view of biological survival and reproduction, and focused instead on what it might *mean* to adapt to aspects of modernity:

> Because man is so adaptable he can learn to tolerate murky skies, chemically treated waters, and lifeless land. In fact, he may soon forget that some of his most exhilarating experiences have come from direct contact with freshness, brilliance, and rich variety of unspoiled natural phenomenon. Unfortunately, perhaps, starless skies and joyless sceneries are not incompatible with the maintenance of life, or even with physical health. The only measure of their loss may be a progressive decadence in the quality and sanity of the human condition.
>
> — Man and His Environment, 1966

Here, Dubos was describing the extinction of experience. The loss of what once was. Yes, we would still have a pulse, we would still breathe and reproduce. We would adapt. But adapt to what? Surviving could not be equated to thriving. And as Dubos said many times, "mere survival is not enough." In his conversations Dubos worried that adaptations to synthetic environments and "alienation from nature will eventually rob him [us] of some of his [our] important biological attributes and most desirable ethical and aesthetic values."

The only way to measure the losses incurred by adaptation to starless skies, according to Dubos, would be to examine more subtle health statistics over time: quality of life, low-grade psychological distress, and psychological assets that might take a hit, such as empathy and optimism. Those markers are now making themselves known. Every sign indicates that Dubos was right.

The polar bear examining a melting icecap is a statuesque image of the threat imposed by climate change. But far more than this highly identifiable species is under threat. More than half-a-century after Rachel Carson's book *Silent Spring* resulted in a closer look at the chemicals we dump onto plant life, and by extension all life, pesticides of all sorts are still causing irreparable harm to the environment. Butterflies, fireflies, honeybees are beautiful to look at, and we tend to take more notice when they disappear. We have, in fact, added them to the endangered species list. But throngs of insects and amphibians, the sorts that don't necessarily make the cut for a line of plush toys in the Hallmark store, are also disappearing off the face of the Earth. Creatures without champions, but no less important.

So much of our personal, public, and planetary ill-health—the dysbiosphere—is induced by a collective act of delay discounting on the part of our politicians and policymakers. (Not forgetting our medical-scientific professions that, notwithstanding excellence in treatment and palliative technological developments, provide only lip-service advocacy for prevention.) Collectively, obvious steps could be taken to enact policies to support lifestyles that would ensure the future health of society and the next generations, not to mention the health of the planet itself. But highly compensated lobbyists convince politicians and policymakers to operate in the short term. To take the small reward now. Doctors are bandaging the collateral wounds of diminished health in their clinics, but need to do far more to help heal a diseased system.

It is unfortunate that the meaning of the word "allergy" has been restricted to a limited range of pathological conditions. The etymological meaning of the word allergy is, of course, "altered reactivity." In this sense all aspects of life can be considered as manifestations of allergy, since most anatomical and functional characteristics of man can be modified by the stimuli—physical, chemical, social—that impinge on him throughout life.... High priority should be given to the study and control of the forces that affect the quality of human life and its environment, and are rapidly making the Spaceship Earth a place unfit for human life.

— René Dubos, 1969

Allergy and Spaceship Earth

On St. Patrick's Day, 1969, René Dubos made his way to the podium to provide the keynote address—*Spaceship Earth*—at the 25th annual meeting of the American Academy of Allergy, Asthma and Immunology. He had just completed a decade's worth of study involving the effects of early life nutrition, stress, lack of parental grooming, environmental contaminants, unusual microbial exposures (and much more), and the ways in which these variables influence immune dysfunction in later life. Although his studies were of animals, the results left no doubt that the thoroughly modern experiences being imprinted on the immune system were at odds with its evolutionary experiences. This, he warned, was going to start manifesting itself as higher rates of allergies (altered reactivity) over the following generations. He was trying to help his audience of allergists and immunologists lift their heads up from black-and-white diagnostic lab tests. Allergies are more than the presence or absence of an antibody response measured via rudimentary lab tests.

Half a century on, it is plain to see Dubos was right; life on Spaceship Earth today is one where altered reactivity is an unacceptable new normal. In Europe, childhood hospital admissions for severe reactions to food have risen sevenfold over the past decade. But the allergy epidemic is much more than the sheer numbers of children and adults who are living with severe reactions and the allergies that are more visible through obvious clinical signs, symptoms, and the limited blood tests that are available to clinicians.

Blood tests are used to validate the suffering. But there are countless more people who experience allergies and sensitivities in the ways in which Dubos inferred. People with food sensitivities or reactions to environmental chemicals are suffering no less. Bounced around from clinic to psychotherapist, for too long these patients have been reviled. In the next chapter we will take a closer look at the immune system and see how new scientific discoveries are demonstrating that altered reactivity to the contemporary environment has many physiological underpinnings. Although research developments including new laboratory tests haven't yet entered mainstream clinics, studies are providing plenty of blood work validation for people who have been strong in their convictions that they have reactions to gluten, wheat, and/or dairy.

When we met many moons ago, there wasn't much in the way of solid research tackling the walking wounded with dietary or environmental "sensitivities". Mostly, this was a pejorative term. Code for weakness. Authoritarian medicine dismissed food sensitivities, claiming in a tweet that food intolerance was a shaky diagnosis "created by and for practitioners of alternative medicine in order to make US doctors non-competitive. Sad!" We jest about the latter. But the actual pain and discomfort—not to mention the emotional distress of not being validated—was indeed glossed over.

But in 2000 one solid hint was reported in Europe's finest medical journal, *The Lancet*, that food intolerances were indeed wreaking havoc with the immune system. Patients with self-perceived food intolerance were challenged with foods they had eliminated from their diet, in this case milk and wheat. After the challenge, researchers found significant elevations in inflammatory immune chemicals called cytokines. We will discuss cytokines in more detail in the next chapter. As you will see, and as the researchers of this study concluded, cytokine elevations can account for increased abdominal discomfort, headaches, and joint and muscle pain. The folks in this study were normal on all the run-of-the-mill laboratory tests for allergies. None of them showed them suffering from lactose intolerance. Thus, what the study showed was that subtle food intolerances might only be uncovered by more sophisticated examinations of the immune system.

In the last few years several high-profile studies have shown that gluten may indeed be causing a wide variety of symptoms, even lowered mood, in patients without celiac disease. It also seems to worsen the already bad effects of a westernized diet on the gut bacteria, at least in animal studies (more on that, later). Moreover, immune cells that aren't routinely tested in clinics are pushed into overdrive by gluten. Even if routine lab tests looked normal, what is going on within the inner workings of the immune system is far from normal. Allergy is probably a far bigger concern now than it ever was. It includes everyone from the child showing up in the emergency room with a very serious reaction to a specific food, to the person with chronic fatigue, depressive symptoms, and diminished quality of life due to altered reactivity. These are all linked with varying kinds of immune activation. Bigger still, is the question: what on Spaceship Earth is going on? The answer seems to lie in altered reactivity to our microbial environment.

Ecological Injustice and Grey Space

As we get ready to dive deeper into the immune system in the next chapter, and shortly thereafter take a closer look at the microbe-laden world in which we live, it might be helpful to ponder the word *ecol-*

ogy. It derives from the Greek *οἶκος* or *oikos*—house/dwelling place and *logia* which means discourse on a certain subject. Hence it is the study of the dwelling place. Our entire frame of reference is just that. The biosphere is the dwelling place. It is the largest house of them all.

But there are many spaces within the house, all the way down to the smallest level, at least from the human perspective, which is a tiny encampment of microbes within various anatomical locations. For example, it might be the microbes that make their home using the raw materials and physical structures within the microscopic undulations of minute portions of your gut lining. Much of the intestines are covered in finger-like projections which add to the surface area. If your entire gut mucosa was laid out flat it would cover one-half of a badminton court, so these microbes have lots of room for their little civilizations.

At all points in between, from that tiny internal ecosystem to the single Spaceship Earth there are houses within houses, habitats within habitats. Beyond the tiny communities of microbes on your skin and other anatomical locations, there is the *oikos* that makes your whole being, then there is the residence you occupy (alone or with others), outward to the *oikos* of community neighborhoods, radiating to surrounding municipal, corporate, and national dwellings. These houses envelop and control so many aspects of the state of being of the largest *oikos*, the universally shared residence of Earth. Yes, we only have one of those. Not another sphere to spare. No planet B.

We bring this up now because it's too easy to forget that discussions of the role of microbes in health and disease cannot be detached from these houses. In other words, the microbes sitting on a single intestinal villus, and the ways they might contribute to health, are not unrelated to the larger ecosystems that impact their human host. The policies and practices in the *oikos* of corporate boardrooms (e.g., the need for profits), or the houses of government (e.g., policy or absence of policy concerning corporate practice), insinuate themselves right onto that villus. An advertisement for a kid's sugary cereal cajoles purchases. Acquisition and consumption of said sugary "food" then determines the goings on in the homes of microbes, including the microbial mega-metropolis that's half the size of a badminton court.

The notion of ecosystems takes on greater meaning when we look at the way things are slanted toward dysbiosis. For many communities, it is not a level playing field. Not for badminton or any sport. Socioeconomic inequities run amok. These are major contributors to chronic disease. Although depression and mental disorder can strike anyone, they do not occur randomly in developed countries. Not at all. They are far more common along a socioeconomic gradient which points directly at the disadvantaged. Dr. Michael Marmot has written extensively on this topic. He refers to "evidence-based optimism" because research shows that the social gradient in health can be reduced: "The graded relation between deprivation and health runs all the way from top to bottom of society. The gradient implies that we should not only focus on the most deprived areas, but be striving to improve the whole of society. We coined the phrase proportional universalism to convey that universalist policies are those most likely to create inclusive societies." (*American Journal of Public Health*, 2016.)

We can make changes so everyone wins. Reducing health inequalities is a realistic goal for all societies. It matters in *Secret Life* because some of the health-related inequalities manifest as microbial dysbiosis and barriers to experience nature in all its forms.

As we move through our discussions of nature, biodiversity, and natural environments, it is worth describing the opposite side of green space. We, your authors, have defined the absence of green space and the presence of unhealthy features of the human-built environment as grey space in our scientific papers. Grey space is dysbiotic—distressful to life—on all fronts. Not just because in it green vegetation and clean, natural bodies of water are sparse, but because it also contains disproportionately higher dysbiotic forces that drive stress and ill-health. Grey spaces include residential areas heavy in industrial operations, commercial activity, and major transportation routes, with resultant noise stress and excess light at night (LAN). The commercial activity itself doesn't have to be industrial. It can include bars, liquor stores, convenience stores, fast-food outlets, and tobacco vendors. All of these are more prevalent in disadvantaged communities.

However, it is more than just the mere structural presence of these operations that creates dysbiosis. The toxic ecosystem pushes on the disadvantaged at every level—their intestinal villi, their homes, and communities—by profit-driven marketing, billboards, sidewalk signage, in-store magnification of unhealthy products, and targeted screen media delivery. When we enter a supermarket in a disadvantaged community and the shelf space is oriented to promote unhealthy foods, we are looking at a system rigged for dysbiosis. We can talk about the importance of microbes and green space, but it is important to keep the contextual framework in mind. Who is pushing the buttons of dysbiosis? Who is standing up to speak of the science of inequalities and ecological injustices? Grey space insidiously reinforces dysbiosis by encouraging unhealthy behavior, compromising positive psychological outlook and, ultimately, transgenerational health. We need to break this cycle!

If we are going to discuss the many values of making a connection with nature, including with its ability to feed a healthy allotment of microbes on our skin and in our bodies, we cannot ignore the obstacles to equitable opportunity. Many studies show a dual burden of more grey space and less access to green space and biodiversity in disadvantaged communities. The greatest burden of chronic, non-communicable diseases is shouldered by the disadvantaged, while opportunity to experience urban green space and local biodiversity is often slanted in the opposite direction. That is, favoring access by the affluent and less vulnerable.

As we will discuss later, these ecological injustices manifest themselves in dysbiosis of the gut microbiota. Communities with more green space access generally have healthier dietary habits (e.g., more fruits, vegetables, whole grains, nuts and beans, and less sodium-rich food, sugar-rich beverages, and fast food in general) and lower risk of blood glucose and insulin imbalances. They are also more physically active. All of these things are interconnected! At the other end of the spectrum, fewer parks and open spaces for physical activity operate in tandem with greater density of fast-food outlets (and fewer fresh food stores) to promote non-communicable diseases, especially in the disadvantaged neighborhoods.

Take Back Our Humanness

"Ecological justice" is the term we coined in order to frame the microbiome revolution. We use it to illuminate the policies and practices that either promote or erode the health of *oikos* at all levels, personal, communal and planetary. Healthy ecosystems involve living creatures, seen and unseen, operating within their abiotic (nonliving, physical/chemical) environment. An ideal ecosystem should allow each human to reach their potential, while at the same time ensuring planetary health.

The study of the microbiome has united researchers from virtually every branch of science and medicine. In the process, it is making it abundantly clear that healthy ecosystems at all scales, microbial and otherwise, can be manipulated by individual human behavior and lifestyle. However, the lifestyles of the microbial host (e.g., us) can be manipulated by forces within the larger dwelling place (e.g., neighborhood) with more grey space and less green space, within which high concentrations of fast-food outlets may be considered invasive species that limit growth and potential.

Moving the lens even further out, collective lifestyles promote environmental degradation, climate change, and biodiversity loss. These too, are driven by policies and practices. Cue the jingle for fast foods, sugary cereals, and so much more. These are threats to ecological health. Since the effects of degradation and losses are now—and will continue to be in the future—shouldered by the disadvantaged, this entire discourse is a matter of ecological justice.

Our mantra through the chapters that follow is simple: "No Health Without Ecological Health."

We close on J. Arthur Thomson's thoughts on progress and a better world: "Besides the fulfillment of the preconditions of health and wealth, man's ideal of progress means a balanced all-round movement towards a fuller realization of the True, the Beautiful, and the Good. It is surely not far-fetched to point out that in the ascent that man has behind him there are great evolutionary trends towards these highest values.... Nature crowns the raw material of morality."

3

Meet Your Maestro: The Immune System

The cases which have come under our notice have been from amongst people living in towns and not in rural parts. I frequently asked persons engaged in haymaking if they often suffered from hay fever or hay asthma...I made inquires of some hay salesmen and at others engaged largely in horse hiring, whether or not they knew of frequent distress of any kind referable to the presence of new hay or other grass, but I could not hear of anything lending countenance to the idea of hay fever being a noticeable complaint among these classes.

— William Pirrie, MD, *On Hay Asthma and the Affection Known as Hay Fever*, 1867

As John Arthur Thomson was inhaling his first breaths of life-giving Scottish air, a professor of medicine at Aberdeen University, William Pirrie, was investigating the increasing problem of hay fever. Pirrie grew up on a farm about 40 miles away from Aberdeen, studied in Paris, and at the young age of 23 was back as a professor at the big city university. Although a renowned surgeon, Pirrie took an interest in the differences in frequency of illnesses encountered in cities, towns, and rural areas. He knew that hay fever (or what we now call allergic rhinitis) was said to be a condition of affluence. The physician who coined the term, John Bostock, noted its occurrence in association with wealth. But Pirrie took it a step further and reported in his research that allergic rhinitis seemed to be almost nonexistent among those who grew up and worked with hay!

Today, the observations of Pirrie and Bostock—lifestyle factors meet chronic non-communicable disease—have taken on immeasurable relevance when examined in the light of major scientific developments concerning the immune system. Currently, allergic rhinitis, asthma, and allergies in general, are at epidemic proportions. Their observations fit neatly together with the *vis medicatrix naturae* argument. They are also in line with the changing lifestyles and plague of psychological distress associated with modernity. Thomson focused on the psychopathology of everyday life, Pirrie focused on allergic rhinitis. As we will show, in many ways the functional and physiological focal point of their interests is one and the same. Discussions of health, happiness, and resiliency and the continuum between disease and realizing one's full potential—between emotional wellness and physical health—all involve the immune system. The maestro.

The Final Frontier

Although it is often said that the brain is the final frontier in medicine, we contend that it is actually the immune system. Only now are we beginning to understand the ways in which the immune system is at the interface of mind-to-body, and body-to-mind communications. Even anatomy textbooks are being rewritten. Using sophisticated brain imaging techniques, a remarkable 2015 discovery showed that lymphatic vessels belonging to the immune system are found deep within the brain. Importantly, these vessels connect to immune centers *outside* the brain.

Besides being a boon for publishing houses specializing in medical anatomy textbooks, as well as folks that like to update Wikipedia in their spare time, the new anatomical findings hint at more than just a way for the brain to combat threatening invaders. They point to something more than the brain simply informing the immune system what its needs are. Rather, the discovery of direct conduits between the brain and body-wide immune structures add to emerging research indicating that the immune system may "talk" to the brain, telling the brain what it needs to know. Such "conversations" take place through a number of different mechanisms, and we will describe the implications of that dialogue for mood, cognition, and links to nature and lifestyle.

Although it is not our focus here, from a strictly biomedical viewpoint there are legitimate hopes for the effective treatment of virtually all non-communicable diseases, including cancer, by way of innovative immune therapies. Remarkably, even vaccines derived from innocuous soil-based microbes—something we will focus on—may soon be applied as preventatives in mental disorders via the immune system. The immune system touches every bodily system. It can sense everything in the environment around us, and within us, at microscopic levels. It can very much be considered a sensory organ.

As an independent scientific discipline, immunology, the study of the immune system, is one of the youngest branches of biological science and medicine. In fact, the British Society for Immunology wasn't established until 1956, decades after the formal organization of societies such as psychiatry, cardiology, neurology, and others. Until then, research discoveries involving the immune system were thinly peppered around various medical disciplines. Rather than viewing the immune system as the North Star that actually *connects* all branches of medicine, this only marginalized and isolated its relevancy.

The fact that medicine has been so slow to recognize the relevance of immunology is a salient point in the context of *Secret Life* for several reasons. First, the immune system was historically pigeonholed into its role of defending from infectious disease and thus considered to be of little relevance to non-communicable diseases. It was many decades before these were recognized as inflammatory diseases. Second, the continued rise of subspecialty medicine has been increasingly predisposed to silo thinking about systems and organs, rather than seeing ways in which they are all connected. Third, the immune system is so complex that even basic understandings of its operations have been an ever-changing landscape. Hence, physicians in the early 20th century considered immunology and its application to everyday medicine to be a science of "instability in theory and hypothesis." Fourth, science and medicine are magnetically attracted to neat, easily treated pathological conditions or reproducible findings that apply to all within a species. However, the fact that *potentially* disease-causing microbes are so often found in healthy humans makes the story a bit messy; it forces more difficult questions about how the environment interacts with host (that would be you!) resistance via immune mechanisms.

There is no question that the establishment of immunology as a specialized field of study has led to unprecedented findings. However, there has been a downside, too. When researchers only speak to one another *inside* a discipline, cross-pollination and seeding into other medical branches becomes limited. In the realm of noncommunicable diseases, immunology has been historically lumped in with allergic conditions, or with extremely rare genetic defects that cause immune deficiency and catastrophic infection.

Allergy and the immune system? Makes sense. Contagious disease and the immune system? Sure, it's easy to see the connections. But until recently the immune system in relation to the environment and lifestyle seemed a million miles from cardiology's long-standing focus on cholesterol in blocking arteries, psychiatry's focus on neurotransmitter deficiency or complexes of Oedipus, endocrinology's attention to drugs that palliate the symptoms of type 2 diabetes, or primary care medicine's focus on calcium for osteoporosis, and so forth. For years, immunology textbooks leaned toward the extreme ends of defense against germs or the detrimental friendly fire of immune chemicals directed at pollen and hay.

> **The immune system is, like the brain, an organ of adaptation to the external environment, recording biochemical instead of sensory experiences. Like the brain, the immune system is said to have a memory, but it is a more mobile organ with cells searching out the environmental agents and responding to them. In this sense immunologic memory is like adaptation of the species which records the experience of the ages in diverse characters.**
>
> **— David W. Talmage, "The Nature of Immunological Response," 1969**

Immune System or Unified Sensory Organ?

This quote from one of the foremost immunologists of the 20th century, David W. Talmage (1919–2014), highlights the intelligence of the immune system. In referring to environmental agents, Talmage meant the recording of information from microbes and molecules like pollen and pollutants. But we can take it a step further. It is possible to argue that the immune system is recording its own *sensory* experiences of the total environment we encounter—the stacked fast-food burger, soft drink, and fries, the TV jingle and the overpaid celebrity telling you to eat it, the excess blue glow of your devices as you work into the night to quell the demands of an angry boss. The whole shebang is destined to be sensed by your immune system.

In the scientific annals there have been many older signposts suggesting that the immune system was more than an efficient, self-contained Department of Defense. More than a structural arrangement of functional thorns to protect us from threat. There were hints that the immune system was actually a sensory organ, a massive bodily part that trafficked information to the brain. A sixth sense.

For example, it was already known by the first half of the 20th century that organs and vessels within the immune system—e.g., the thymus, spleen, bone marrow, and the lymphatic vessels/nodes—were innervated. Hence, the immune system and central nervous system were connected. In animal studies, when researchers cut off communication from the brain to these immune centers, the result was compromised immune function. In the 1920s, it was demonstrated by a student of Ivan Pavlov that, at least in guinea pigs, the immune system could be trained by conditioning. While the experiment didn't involve any ringing of Pavlov's bells, it did involve the daily pairing of scratching (or warming) the skin with an injection of a microbe over the course of 20 days. Thus, the guinea pigs were conditioned by scratching or warmth at the same time they were being injected with something that could clearly provoke an immune response. Two weeks later the researchers revisited the guinea pigs and determined the immune system would respond to the scratching or warmth alone. It was a landmark study that essentially proved that behavior could influence the immune system. Mind-body medicine (or mind-immune medicine) in action.

By the mid-20th century European researchers had shown that the hypothalamus, which controls the autonomic nervous system (the part on "autopilot"), as well as hunger, mood, libido, etc., can exert control on immune function. It was determined that the brain was leaning on the immune system function directly via nervous system messaging and also indirectly by circulating hormones produced by signals from the brain. But most of this research was obscured by headline-grabbing discoveries in genetics and antibody production. These studies were highly relevant to allergies and asthma, amazing pieces of knowledge in immunology, but they still tended to maintain the view that the immune system is a mostly self-contained and self-regulated operation.

The relevance of the nervous system's control of the immune system just wasn't ready for prime time. Still, in 1966, Andor Szentivanyi, the author of many groundbreaking immune-nervous system discoveries, proposed that the immune system may actually be more intelligent than the brain. The active immune players—cells and chemicals—are constantly on the move, kicking the backsides of harmful (pathogenic) intruders and taking names; the entire system is dynamic, maintaining continuous plasticity, and since the stakes are so high in striking a balance between defense and surveillance in protecting the brain and all organs, the sophistication of its operations must surely be well beyond many scientific imaginings.

But what about the other direction, immune to mind? It wasn't until the 1970s that researchers began to show that antigens (potential toxins) could dramatically increase the electrical firing of the brain, especially in the hypothalamus. Electrical firing in the ventromedial portion of the hypothalamus indicated the immune system was letting the brain know that environmental toxins (or allergens) are a potential threat. Further, it was revealed that immune responses to antigens influenced mood-regulating neurotransmitters such as serotonin and norepinephrine in the brain. In turn, the brain sent out signals to ramp up production of the stress hormone cortisol and immune chemicals known as cytokines. The enormous consequences of this to daily mood, behavior, and motivation were still far from realized. But researchers were knocking on the door.

The evidence continued to accumulate and by the 1980s it started to become clear that immune chemicals could actually act like hormones and even neurotransmitters. In particular, the cytokines took on a starring role in the new theater production known as psychoneuroimmunology. The psyche, nervous system, and immune system were all intertwined and functionally pushing each other's buttons. From the first observation in 1980 that cytokines can influence the frequency of the beat of heart muscle cells, to the 2001 blockbuster study demonstrating just how influential these special chemicals are in our daily mood, and countless observations since, the cytokines have become Hollywood A-Listers. More on the blockbuster studies shortly; but first a short but essential "biography" on some very basic aspects of these celebs and other actors within the immune studio.

Key Players in the Immune System

As subsequent chapters unfold we will make specific references to various immune chemicals. We will also use "buzz words" like *inflammation, oxidative stress*, and *stress hormones* to highlight the ways in which the immune system is deeply entrenched with these health-associated terms. Later on, when we discuss the new science of the microbiome—the collection of microbes, their genetic material, and little ecosystems they make home—the immune system will become an essential part of the discussion. Some clarity about these terms may be helpful, especially as we start sewing together the threads between biodiversity and *bios* (way of life). Nothing too heavy; you won't need waders to get through a swamp of immune system jargon.

Researchers refer to "innate" vs. "acquired" immunity. Innate, as its name suggests, refers to the functional parts of the immune system that we are all born with, that is, the inherent parts not dependent upon experience. Acquired (aka "adaptive" or "specific") immunity is learned; massive amounts of energy and intelligence are poured into the acquired portion of immunity. The innate immune response is often stated to be the ancient portion of our immune system, and the acquired a more evolved and refined portion that knows that of course the oyster fork sits at the right side of the soup spoon in formal place settings.

Within the sophisticated, acquired branch of immunity, B cells (made in the bone marrow, hence B) manufacture antibodies, which are custom made to fit the harmful invader's shape with precision.

They lock onto the invader like a spacecraft landing on a dock. Once the antibody is locked in place, the pathogen cannot infect your cells. B cells can do the same locking technique with the toxins produced by harmful microbes. What's more, the B cells share intel with and fire up other cells, including T cells which smartly scan the surface of invaders—all of their little undulating nooks and crannies—knowing exactly what molecules to look for in order to stop the threat. It's a two-way street—T cells also help B cells make more antibodies. Remarkably, specialized T and B cells are formed as "mobile minds"; everything is recorded and tracked in memory, so that if the same threat is recognized again, the future response will be swift.

No doubt these evolved cells are really smart and operate with precision. But they would be nothing without their innate ancestors. Cells of the innate system patrol our tissues constantly for threats. Those at the surfaces of the internal pipes that separate us from the outside world, such as our gut or respiratory tract, have long finger-like projections they extend to constantly sample what is going on in the world beyond. These cells send signals to the hot-shot pilots in the T and B ranks. In fact, it's the ancient part of the immune system that is in many ways the instructor. Leukocytes (from Greek *leukos*: white; *cytes*: cells) respond early to threat in the form of neutrophils, monocytes, dendritic cells, and macrophages. The latter two signal for the cytokines which we will mention throughout the book. These are tiny proteins that do mammoth work. Some of the cytokines promote inflammation, while others can damp it down. A true drama in action, balance in conflict.

In the midst of immune system generals and foot soldiers, there are natural killer cells. As the name might imply, they are special agents with license to use lethal force to protect you from viruses like influenza and the common cold. Importantly, they can also recognize tumor cells and play an important role in preventing cancer progression. It seems that natural killer cells also have a memory and are far more sophisticated than once thought. New research is demonstrating their deficiency in depression, and later on in *Secret Life*, we will talk more about these cells and the ways in which spending time in forests and natural environments may benefit their activity.

Through millennia of experience, all sorts of checks and balances have been set down so that the entire immune system works in superbly elegant fashion. It really is quite incredible. Even beautiful. As mentioned, there are anti-inflammatory cytokines to match the inflammatory ones, and there are T regulatory cells (Tregs) that keep other aspects of acquired immunity from running out of control. But it seems that modernity is placing a ton of pressure on these checks and balances. When the scale is tipped too far toward inflammatory cytokines, it gets ugly. When Tregs no longer do their regulatory work, we can start shooting antibodies at phantom threats like our own cells (autoimmunity) and benign elements of nature (allergy).

Last, but by no means least among our notable immune system players, are the microglia. These immune cells do their specialized work within the borders of the brain and central nervous system. They play a vital housekeeping role in the brain, cleaning up, pruning, and scaling nerve cells. They are constantly touching and feeling their environment. Like a kid's Transformers toy, they might move around and do this day-to-day business like a normal looking creature. But at the slightest hint of danger, they transform into a completely different shape—a more aggressive state to deal with the threat. Microglia also produce cytokines in the brain, and although this is effective for short-term protection, the unbridled activation of microglia can contribute to damaging consequences (neuroinflammation).

The Thorn of Inflammation

To illustrate a basic immune response, imagine yourself out for a hike in the wilderness, absorbing biodiversity to its fullest. You've got your well-read copy of *The Secret Life* in your backpack, figuring you'll read it for a third time while nestled on a rocky ledge with a superb vista. On your hike you venture too close to some thorny brush or the spines of a cactus. Just like that, you've connected a little too closely with nature; in this case the teeth of the plant kingdom, a couple of thorns, find their way deep into your leg.

Immediately your innate immune system springs into action; it's on the scene with lights and sirens. Some of these first responders send out signals to ramp up the production of inflammatory cytokines.

The priority is to get the area of encroachment and potential pathogens (e.g., potentially harmful bacteria or viruses carried by the thorn) under control.

The collective action results in inflammation which you can see as swelling and redness in the period following the thorny "bite." Later, the aforementioned B and T cells work their magic. T cells themselves (we are doubling down on our view that they are little living creatures!) can produce cytokines, and cytokines can, in turn, have a tremendous impact on both innate and acquired immune functions.

We can consider inflammation by looking at how it was first described in Latin by the Roman medical authority Aulus Cornelius Celsus some 2,000 years ago: *rubor*, *calor*, *dolor*, and *tumor*, translate as redness, heat, pain, and swelling. In the case of the thorn, it's pretty easy to see and feel the cytokines and other inflammatory mediators at work. But a few years after Celsus, urban legend has it that a fifth aspect of inflammation was tacked on by Galen—*functio laesa*, or loss of function. Galen was a good guy—one of the first proponents of unified mind-body medicine—but there is no historical evidence that he ever actually uttered or penned *functio laesa*. Can someone update Wikipedia, please? But loss of function *is* a massive part of the inflammation story, a very legitimate one. Again, it's easy to see how cytokines, by causing swelling and pain, can influence function. With significant levels of inflammation, simple acts like using your fingers or moving your feet can be functionally difficult. Arthritis is the most obvious example of inflammation impairing function.

But what about more subtle aspects of inflammation, levels at which you might not be able to see redness or swelling? Levels at which its impact on your functioning may not be so recognizable? Could it be that immune-mediated loss of function is happening to us modern urbanites in ways that we are aren't even aware of, putting a glass ceiling over us fulfilling our potential, regardless of gender? In our opinion, the fifth cardinal sign of low-grade inflammation in the modern environment is loss of function.

We can explain this by answering our own question. What if many children and adults living in westernized nations live with small, omnipresent environmental thorns "biting" their bodies? These thorns take the shape of ultra-processed foods, artificial chemicals in foods, environmental contaminants, antimicrobials, psychological stress, sleep disruptions, noise stress, excess light at night, tobacco smoke, stimulants, excess alcohol, and other "exposures" as scientists like to call them. Worse still, the response to the thorn remains unabated because of missing exposures that would otherwise prevent or help resolve the inflammation associated with the low-grade thorny attacks. Lack of physical activity, lack of exposure to friendly bacteria (much more on that later), lack of time spent in biodiverse natural environments, lack of positive emotions and assets (optimism, awe, gratitude, empathy, etc.), and lack of important nutrients (omega-3 fats, vitamin D, magnesium, zinc—again, more later) all compound the directly harmful exposures.

It's Heating Up

Slowly, researchers have started to connect the disconnected research. A 1999 editorial review in the *New England Journal of Medicine* finally spelled it out to all medical colleagues: "Atherosclerosis is an inflammatory disease." But it took a pioneering Harvard physician. Iris R. Bell MD, PhD—a woman who needed no bowtie to be an expert in allergy and neurobehavioral sciences—to suggest that it was necessary to start thinking differently about the immune system's ability to lean on the mind: "Psychoimmunologic research has often had a unidirectional focus from the mind to the immune system—e.g., immune dysfunction in individuals with and without psychosocial stress.... The field should also consider the other direction for central nervous-immune system interactions, i.e., possible affective dysfunction in individuals from additive or synergistic immunological events."

This quote from Dr. Bell is beyond salient. Yes, there were already embers of understanding that immunity and the nervous system are intertwined. But initially, clinicians and scientists seemed to be primarily focused on one direction only, that the immune system was being conditioned by the mind. Mind-to-body. This limited perspective carried unhelpful implications of blame: that some symptoms were all in the head. "Plastic roses cause allergic responses in those with allergies!" screamed the headline. "Patient responsible for symptoms!" was the loudly echoing inference.

There were so many "experts" trying to pull apart the personalities of people with chronic disease to show psyche as cause. These are actual diagnostic discussions we found: "Asthma patients are egocentric." Patients with ulcerative colitis are "fussy." Spastic colon (now called irritable bowel syndrome) was considered to be commonly caused by "repressed hostility, resentment or aggression," especially in women. So said male experts in 1950s medical journals. These practitioners were utterly convinced their diagnoses were correct, yet with less scientific validity than pixie dust.

Obviously, there was much nonsense written. Hysteria, Oedipus complexes, and other fantastical Freudian fables mixed in (see R. M. Lattey MD, "Dr. Sigmund Freud, pseudoscientist," *Canadian Family Physician*, February, 1969). Mountains of these fossilized fabrications are still reverberating in the collective psyche of modern authoritarian medicine and institutional policy makers. The petrified wood of dogma facilitates an all-too-quick orientation toward blame-the-individual thinking. It facilitates the dismissal of societal responsibility for non-communicable diseases. Can't navigate modernity without succumbing to emotional disorders or other non-communicable diseases? Must be "repressed" in some way. Can't figure out a complex disease with one visit to a family doctor and a specialist? Must be "psychosomatic." Obesity? Easy-peasy, it's just a lack of "willpower."

Meanwhile, the disturbed environment that leans on the immune system to drive inflammation is left out of the discourse. This includes everything from the microbes that surround us (or the ones that should, but aren't there!), to a billboard hawking fast food, to the ways in which junk foods are taste engineered and positioned in marketplaces where they are most sure to grab your attention. These and many other environmental forces—industrial pollution, nutritional deficiencies, lack of optimism—are actually influencing the brain via the immune system.

Many of these influences may be operating outside conscious awareness (not to be confused with *the* unconscious of Freudian pseudoscience, the place of dastardly "repressed" desires) and yet are deeply influencing our physiology. All of those little immune cells may be better described as mobile organs, each one a small brain, determining mood and health. The spastic colon of those stereotypically "angry, resentful, repressed" women—as actually described and diagnosed by the fantastical X-ray vision of physicians confident in their classifications—turns out to be a product of the immune system meets environment as well.

The historical roots of psychosomatic medicine remain awkwardly visible. At inception, its primary purpose was to demonstrate *psychic* influences on *soma* (body/organ). And demonstrate they did! As described in *Life* (1945), psychosomatics describes "the treatment of disorders caused primarily by emotional disturbances." Key word: caused. Peptic ulcers (now proven to be mediated by microbes) and mucus colitis (IBS) were said to be ironclad examples of psychosomatics. Cardiovascular diseases were "another major psychosomatic group." To quote further, "Emotions are regarded as the primary cause—or as a precipitating or aggravating factor—in many cases of bronchial asthma, hay fever, hypertension (high blood pressure), arthritis, heart disease, rheumatic disease, diabetes mellitus, the common cold, and various skin conditions such as hives, warts, and allergic reactions.... A percentage of patients [with tuberculosis] impede or prevent recovery...to fill an emotional need. Obesity [is an] act whereby the gourmand compensates for some inner deprivation or frustration."

Finally, biology and physiology have caught up with such conjecture. "Primary cause"? Not even close. Psychosomatics began as some good science mixed in with a juggernaut of assumptions. The *Life* piece wasn't written on the fly. It is a reputable news magazine for the general public, a filter of available science and medical news, and plenty of doctors propped up what the *Life* author was saying. In the process, an even more plentiful supply of patients with deep suffering and legitimate physiological ailments were dismissed for counseling.

If this seems like a digression to a bygone era where doctors appeared in ads promoting their favorite brand of cigarettes, we underscore that such simple notions concerning the ability of the brain/mind to *cause* illness all on its own remains a popularized supposition.

From *Life* in 1945 to 2017 there have been many articles of similar ilk. The "Special Report on Psycho-Immunity" in *Science Digest* (1981) describes claims from a famous Canadian doctor that psychosomatics (as patients with arthritis were described!) have "characteristic unconscious recurring patterns in the dreams: acts of extreme cruelty to others, incidents in which they are the victims of brutality, acts of incest and situations in which others express their own repressed feelings." Without any supporting evidence, articles in psychosomatic-oriented journals and lay press continue to be written on the "neurotic" personality being the *cause* of IBS.

Psyche Privileged

The origins of psychoneuroimmunology are the same as psychosomatic medicine—privileging of the psyche—and this permeates contemporary discussions. Thus, in the latest edition of the textbook *Introduction to Psychoneuroimmunology* the discipline is described as one that "seeks to shed light on how mental events and processes modulate the functioning of the immune system, and how in turn, immunologic activity is capable of influencing the functioning of the mind." No mention of shedding any light on the opposite concept: that immune system events and processes modulate the functioning of the brain (mind), and how, in turn, mental activity is capable of influencing the functioning of the immune system (health/aging etc.).

Of course the psyche and emotions matter to the aforementioned "psychosomatic" conditions. They matter in all aspects of health. That is, they matter in our quest to live life to our fullest potential. However, as we will see, in the cases formerly claimed to be "psychosomatic"—the list above could include many others such as fibromyalgia and chronic fatigue syndrome—the soma-to-mind perspective is critically important. Translation: the immune system is the mind *and* body, thus there is *no* mind-body dichotomy. It's all one. This is what Dr. Iris R. Bell was trying to shout about.

The "outside" environment of the so-called obese gourmand described in *Life* is not separate from him/her. It's not really outside at all. The total environment, the theater in which we reside, is an ecosystem which is continuously imprinted onto the immune system. Thus, it is imprinted onto the sufferer's total person. Yes, the allure of fast-food logos and enticing ads for unhealthy comfort foods make their mark on the mental processes. Reaching into the vending machine can therefore reverberate into the immune system via those conditioned mental processes. On the other hand, there is now plenty of evidence that countless aspects of the environment touch the immune system and subsequently signal the psyche.

All of this is essential background for our later description of how certain environments influence immune function in ways that escape awareness. Inflammatory cytokines can be fired up by aspects of the external environment that do not run through the psyche. Beyond the toxic chemicals we hear so much about, we will refer to beneficial exposures that represent *vis medicatrix naturae* in chemical form. Walking in a forest, for example, can expose you to chemicals that escape visual and olfactory sensory detection. Humans are exquisitely sensitive to oxygen, carbon dioxide, and ions (not necessarily toxic) in the air. Slight atmospheric shifts toward carbon dioxide can promote inflammatory cytokine production completely independent of psychological stress. Areas rich in biodiversity can influence the diversity of microbes on your skin. As we will discuss later, this can influence the immune system throughout the body. As well as mood. And motivation. So, too, moisture in the air and other aspects of the external environment can charge up the skin cells to manufacture chemicals that can potentially influence emotion from the outside-in.

Privileging the psyche in the psychoneuroimmunology equation, without qualifying that the psyche can be shaped and dominated by nerve signals via the immune system, is a slippery slope. It unwittingly allows for callous descriptions such as "glutton" and for presumptions that feeling low motivation and depressive symptoms are either a matter of intentional malingering or attention seeking. If the hypnosis sessions, cognitive-behavioral therapy, and/or psychopharmacological drugs don't work, then surely—so say some of the next-gen psychosomatic experts—"the patient doesn't *want* to be well." Cherry-picked anecdotes of patients who laughed and willed themselves into cancer remission are gorgeous. But it doesn't mean that the psyche of those who have died from cancer were any less strong, or somehow broken. And it doesn't mean individuals struggling with chronic depression and anxiety don't want to be well.

Top Down: Distress to Inflammation

With appropriate background and cautions in place, we can examine some of the ways in which the immune system might be influenced by stress and emotional outlook. Let's imagine a scenario where you are up to your eyeballs with work, deadlines loom, files are piled high, and the boss's favorite golf buddy, your same-ranked peer who should be helping share the load, is playing Tetris on his phone all day. The thorn here isn't a pointy spine of an acacia tree, but it may as well be. It is a *cerebral* threat, but still well capable of turning on the immune fires.

A cognitive assessment is made, and this particular situation is interpreted to be beyond the pale. It has pushed the outer limits of what you can accommodate for the day, and the fight-or-flight system ramps up. And what a beautiful system it is! Signals from the brain rapidly make their way to the adrenal glands perched atop your kidneys. Like protective gargoyles, the adrenal glands shoot out a spray of stimulating adrenaline, noradrenaline, and a larger than normal amount of the stress hormone cortisol.

Adrenaline and noradrenaline (aka epinephrine and norepinephrine) increase your heart rate, elevate your blood pressure, and mobilize energy. Cortisol shifts all priorities to make way for the acute threat: it pulls glucose into the bloodstream, shunts blood away from low-priority digestive processes, and doesn't give a hoot about your reproductive system in that moment, only your future ability to reproduce. It wouldn't matter how many milligrams of Viagra or Cialis were on hand—if you throw a piranha into the tepid waters of the exquisite white claw-foot bathtub, the last thing a dude will think about is copulating. The "moment would be right"—but for flight.

An important aspect of this immediate cascade is that it enhances several key aspects of immune preparedness. This system has been developed with extreme proficiency to promote the survival of the human species throughout our ancestral past. For nearly three million years, our genus has faced acute threats. Serious ones. It would have made a ton of sense to gear up the immune system in preparation for potential damage while fighting or fleeing. Indeed, with a good degree of consistency, research shows that acute stress prepares us for the elimination of potential pathogens and the healing of wounds.

If the stress is relatively short lived, say a few hours, all should be good—as long as you don't keep poking the system. When our previously described stressful office scenario becomes a rinse-and-repeat cycle, the immune story changes. Your co-worker keeps playing Tetris, your boss gets more ornery, and your files keep gaining height on all days of the week that don't begin with an S. Throw a traffic ticket into the mixer, then a chaser as a home appliance just decides to stop chilling your perishables. Now acute has become chronic, and the immune system is being stretched in ways that are not in your best interest. The system wasn't developed to accommodate what psychologists call "daily hassles." Cumulatively, the constant, steady stream of these minor-to-modest stressors can easily be interpreted as a collection of teeth within the jaws of a predator known as modernity.

The immune fallout of chronic stress can be generalized to translate as suppression of some aspects of the immune system and uncontrolled elevation of others. The suppression includes compromised ability to fight off viruses, diminished natural killer cell activity, and loss of effective antibody production. In the short term, cortisol can help regulate inflammation. However, when the adrenal glands keep pumping out cortisol, eventually the immune system stops listening and inflammatory cytokines aren't shut down. The cytokines promote higher levels of oxidative stress—a state where the production of so-called free radicals (a normal process of living and breathing) cannot be adequately compensated by the body's antioxidant defense mechanisms. Oxidative stress begets more inflammatory cytokine production. Over time, cellular damage can result.

Low-Grade Inflammation, Complex Causes

We have provided a relatively simple depiction of the path from psychological stress to cellular damage. The fiery combo of chronic oxidative stress and inflammatory cytokine production is referred to as low-grade inflammation. From acne to osteoporosis, low-grade inflammation is an important player in virtually every chronic medical condition. This is mostly a subtle process, one that doesn't pose an immediate threat to life. The cellular damage here is more like waves eroding a shoreline. The damage may be slower, but the erosion still cuts into fertile ground of your lifespan. This kind of chronic erosion equates to early mortality.

The difficulty with simple top-down brain-to-immune scenarios is that it is very difficult to take psychological stress out of the context of the daily living environment. The theater of life includes fast-food outlets, vending machines, high-calorie/low-nutrient foods, energy drinks, the TV remote control, and a bunch of other factors that look more appealing when psychological stress bears down. It's as if stress causes us to put on a virtual reality headset that makes us see unhealthy lifestyle choices as shiny, glittery objects. Must. Have. Now. So when we talk about psychological stress, we are often talking about the unhealthy, sedentary lifestyles that result from it. Stress can directly impact the nervous system, stoke up inflammation, and it can indirectly maintain the inflammatory flames by adding kerosene via lifestyle.

When it comes to inflammation, emerging studies show that psychological stress and unhealthy dietary patterns are worse than the sum of their individual parts. They work in detrimental synergy. For example, a westernized diet (high sugar and refined fat, low in fiber and colorful fruits and vegetables, as discussed in detail later) has been shown to interfere with some of the physiological mechanisms that would otherwise shield our cells from the damage done by psychological stress alone.

These are also practical considerations as to how our lifestyle might impact on our assessment of stress in the first place. How deeply daily hassles get under your skin—and your assessment of situations in terms of demands vs. ability to navigate them—may be determined by the amount of sleep you had the night before, the type of breakfast you had (if you even had one), the amount of caffeine in your system, the microbes in your gut and on your skin, the exposure to biodiversity you've just had (or didn't have), your level of physical activity, your level of mindfulness, and a host of other factors that feature into what we call lifestyle. Not to mention social support and whether or not you have a living wage, a level of income that would allow you not to have to work two jobs, and so forth. And to what degree are your daily hassles offset by daily harmonies?

Harmonies Not Hassles

Within the volumes of medical and scientific literature involving the study of the psyche, so much has been devoted to pathology and evaluation of the disturbed mind. The slant toward trying to fix the source of distress and mental anguish is completely understandable. We have spent a good deal of this chapter examining links between distress and the immune system. But what about states of positive emotion? If subjective distress, tension, and low mood push down on the immune function over time, could the feelings we often equate with happiness actually help maintain immune system balance?

Positive emotions are completely distinct from the absence of negative emotions. Feelings associated with the expression of positive emotion are also distinguishable from the flavorless mid-point where emotions are numbed. Positive emotions or states that are linked with positive wellbeing include, but are not limited to: awe, love, hope, joy, gratitude, trust, contentment, amusement, interest, inspiration, optimism, forgiveness, and positive empathy (sharing in the positive emotions of others).

It turns out that various positive emotions—measured distinctly from depression/anxiety—are associated with lower inflammation. Optimism, the tendency to look upon the favorable side of how conditions and events might shake out, is a powerful asset that is associated with lower blood markers of inflammation. Awe is yet another emotion that seems to be particularly important in keeping the flames of inflammation in check. The wonderful aspect of positive emotions is that they can be primed through experience and learning. Interventions that focus on gratitude have been linked to lower inflammation. Empathy has been, too. Educational efforts that nourish an empathic perspective have been shown to reduce blood levels of inflammatory immune chemicals.

Lifestyle: Immune-to-Mood

The immune system pathway leading from biodiversity and *biosis* (way of living) to your mental health can be discussed from the low-grade inflammation perspective. Over the last two decades it has become increasingly clear that mental disorders are associated with the duo of oxidative stress and inflammation. The argument in favor of inflammation playing at least some sort of causative role is strengthened by a variety of different findings. We can use the example of depression:

1. Higher levels of inflammation predict subsequent depression over time.
2. Induction of inflammation in otherwise healthy individuals promotes depressive symptoms.
3. Drugs known to have anti-inflammatory properties lift depressive symptoms.
4. Plants rich in antioxidant/anti-inflammatory chemicals have antidepressant properties. Turmeric with its curcumin chemicals is an example.
5. Dietary patterns that include anti-inflammatory/antioxidant foods are linked to fewer depressive symptoms over time.

We underscore that it is quite possible to have depression in the absence of inflammation and we repeat that the psyche can certainly rev up inflammation. However, an important study from McGill University in Montreal has shown that the evidence of causation points more heavily in the direction of low-grade inflammation messing with the mind, rather than it merely being a downstream consequence of psychological distress. The study involved over 2,200 adults assessed at baseline and then again, five years later. Baseline inflammation (measured through blood testing, including a primary marker of inflammation known as C-reactive protein) proved to be a better predictor of subsequent depressive symptoms and stress in comparison to psychological distress predicting subsequent inflammation.

The mechanisms linking inflammation to depression have been the subject of intense scrutiny over the last several years. At this point, it is clear that inflammation can disturb normal communication between nerve cells by throwing off the availability of mood-regulating neurotransmitters such as serotonin and dopamine. It can also diminish the resiliency of brain cells by reducing the amounts of chemicals that normally tend to them, nurture their survival, and encourage new growth. Brain-derived neurotrophic factor (BDNF) is one such enriching chemical.

BDNF is a key player in maintaining brain plasticity, the ability of the brain to constantly grow and prune itself throughout life. Although once thought to be a fixed entity beyond adolescence, the brain undergoes constant change; these adjustments, small as they may be compared to the rapid growth of early life, are functionally important to the maintenance of memory and aspects of executive functioning, areas of the brain that govern planning, regulating, and carrying out "jobs" to completion. Higher levels of BDNF are associated with brain health and cognitive functioning through old age, while BDNF levels are known to be low in depression.

Later on, we will have much to talk about concerning microbes and their influence on health. For now, we will focus on a single study that was a stunner when it was published in 2001. Up to that point scientists were well aware of the serious, potentially fatal consequences of bacterial sepsis—bacteria in the bloodstream leading to massively damaging inflammatory responses. What they didn't know, was that tiny amounts of broken bacterial bits (the membranes or coatings around microbes) could also lead to a more subtle form of sepsis, one that might present itself as lowered mental outlook, tension, anxiety, problems with cognitive focus, generalized fatigue, and increased sensitivity to pain. The reason for all the "walking wounded" symptoms wasn't the bacterial bits making their way to the brain; the symptoms were shown to be a product of the immune system response to the bits.

The bacterial membranes in question are referred to as endotoxins (the most commonly discussed is lipopolysaccharide endotoxin or simply LPS). In the blockbuster study, soon after small amounts of endotoxins were injected into the blood of otherwise carefree adults they had an altered cognitive and emotional take on their environment. Endotoxins get the inflammatory cytokines fired up and the cascade works its way to the brain. Symptoms ensue.

Since this first observation in 2001, the package of walking-wounded symptoms in response to low-level endotoxins has been demonstrated in many replication studies. In 2015, another blockbuster study showed just how bad low-level endotoxins can be for the brain. Remember the microglia? Using sophisticated PET scans, researchers showed how the microglia become hot (activated) in conjunction with small amounts of endotoxins in the blood and the aforementioned emotional symptoms typical of low-grade inflammation. Since continuously firing up microglia—a side effect of too much inflammation in the brain over time—is damaging to nerve cells, the implications are massive. Repetitive microglia activation and neuroinflammation have been linked to depression and many brain-related conditions.

You might be wondering how small levels of endotoxins might find their way into the bloodstream when researchers aren't injecting it into the veins of their study volunteers. One primary pathway is an intestinal lining that is just ever so slightly more porous than it should be. With multitudes of bacteria in the gut, there are plenty of microbial bits and membranes that could squeeze through, and as we will see, most lifestyle factors can interact with the intestinal lining in one way or another!

In addition to the cytokines, the acquired immune cells may also play a role in mood and mental focus. Narrowly defined, allergic diseases are characterized by altered acquired immunity. Antibodies run amok. The antibodies produced in allergic diseases have been shown to activate areas of the brain implicated in emotional disorders. Recently a study showed that optimally functioning lymphocytes are essential to shutting down inflammation *after* endotoxin (LPS) challenge. Dysfunctional lymphocytes—as might be the case in allergic diseases and emotional disorders—were shown to be a very specific factor in low motivation. Again, this means more walking wounded.

Finally, exciting research published in 2016 adds yet more support to the notion that the immune system is actually who we are. The study showed that animals deficient in lymphocytes were less social than their compatriots with healthy immune functioning. When the deficiency was corrected, normal socialization returned. When researchers looked more closely they found it was the way in which T cells interact with the cytokine interferon-gamma that leads to social behavior. In fact, this cascade actually fires up the GABA pathways in the brain—the parts responsible for keeping you relaxed and calm. These GABA thoroughfares in the nervous system are associated with both relaxation and cognitive focus. GABA communication is dysfunctional in many anxiety and behavioral disorders.

Needle in the Haystacks

These new findings on immune disturbance—the broken immune system *is* the broken brain—bring us back to William Pirrie's hundred-year-old haystacks in the northern British Isles. In his quest to understand the origins of asthma and hay fever, he couldn't reconcile its absence among those who were living among the haystacks, and its presence among town- and city-folk. This conundrum looms large now, especially since asthma, allergies, and autoimmune conditions have risen to epidemic proportions.

When Pirrie pondered this question, the percentage of the world's population in urban areas was barely in the double digits. Today, in the same developed nations where these non-communicable diseases are at unprecedented levels, about 70 percent of us live in urban centers. There are probably more people engaged in updating the wiki pages devoted to "hay" and "hayrides," than actual people working with hay.

As we have been pointing out, these epidemics are not neatly confined into diagnostic check-off boxes. Asthma isn't *just* recurrent airway inflammation. Atopic dermatitis isn't *just* a compromised epidermal barrier. There are volumes of studies linking asthma, rhinitis, dermatitis, and other allergic diseases to depression and anxiety. Even when assessed very early in life—between one and five years old—allergic disease is linked to subsequent behavioral and mental health problems in the teen years. Conversely, when mentally healthy youth are evaluated for allergic diseases and followed over time, those with hay fever and atopic dermatitis are more likely to be using antidepressants by the time they reach middle age.

Understandably, this has often been assumed to be a consequence of dealing with a difficult chronic illness. But there may be more to it. The same immune abnormalities that may be driving airway inflammation, epidermal barrier disruptions, and various detailed manifestations (take your pick) of autoimmune diseases, may be driving the development of emotional difficulties. Our argument, and the seriousness with which we should be taking the allergy epidemic as a canary-in-the-coalmine marker for a disconnect from biodiversity, will become more cohesive as the following chapters unfold.

Environmental Imprint on Immune Cells

Researchers often look for ways to measure the resonance of the external environment, stress, and lifestyle as they manifest in human tissue. The stress hormone cortisol is often present in saliva and in blood. These levels wax and wane throughout the day and can't tell us much about long-term cortisol levels. But hair cortisol levels, measured close to the roots, is emerging as a nice marker of chronic stress in a range of months, not days and hours. Researchers can take samples reflecting approximately three months' worth of cortisol production.

Higher hair cortisol levels have been linked to depression, obesity, anxiety, circadian rhythm disruptions, blood glucose-insulin imbalance, and cardiovascular disease. When measured during pregnancy, the hair cortisol levels ended up being an accurate measure of high levels of perceived stress reported by women during the second trimester. The external environment matters, too. Although we will talk about natural environments and green space at length in the next chapter, for now we will point out that living in areas with higher levels of vegetation-rich green space has been associated with lower hair cortisol.

Repetitive bursts of excess cortisol can do some serious damage, especially in the brain. Sometimes the carnage is a directly corrosive action on the nerve cells. Nerve cells start "over-breathing" when cortisol levels are high; that is, they burn through more energy and this sets off a cascade of oxidative stress and inflammation that requires an emergency response. As previously mentioned, cortisol is deeply intertwined with the immune system. Although modest levels of stress are generally not harmful and indeed may build resilience, our modern world is not one in which we are deficient in stress exposures. Thus, when we refer to elevated cortisol levels in the context of the following chapters, it generally signifies a threat to cellular health.

In addition to hair cortisol, there is an even more reliable way to see the resonance of life and lifestyle. Remarkably, your immune cells can provide a fossil record of the adversity you've endured, as well as your resilience and the extent to which daily hassles have been offset by daily harmonies. White blood cell telomeres have been discovered to be the documentarians of life and lifestyle. Telomeres are like plastic caps on the end of shoelaces, except in this case their function is not to prevent cotton laces from fraying. The precious cargo under telomere protection is your ancestral past and your future self in the form of DNA strands. Cells divide and new ones are formed throughout life. With each replication, as we age, the telomeres get a bit shorter until they finally tap out. The length of telomeres varies with different animal species and correlates with life expectancy between species. Although there are individual differences in telomere length in humans, we have not yet got to a point where we can use these to predict lifespan. But there are many studies showing how lifestyle stresses can accelerate shortening.

The white blood cells in our blood stream are an ideal place to look at the length of telomeres because, unlike more static tissue in the heart, liver, and kidneys, they divide more frequently throughout life. The shortening of DNA telomere material known as base pairs happens to all of us. However, the combination of oxidative stress and inflammation can accelerate the normal rates of telomere shortening in white blood cells over the course of a life. Low-grade inflammation interferes with an enzyme that usually helps to offset telomere shortening—an enzymatic seawall that would otherwise protect an eroding shoreline. The shortening is easily visualized by researchers, and they have linked accelerated shortening to many medical conditions. Shorter telomeres have been noted in most mental and neurological disorders. In keeping with the haystacks, they have also been noted to be shorter in asthma and allergic and autoimmune diseases.

Research shows that the immune consequences of long-term stress gets placed upon our precious telomeres. On the other hand, serious forms of trauma and adversity in early life have been linked to shorter telomeres much later on, even at mid-life and beyond. The shortening can be kicked into higher gear by the total environment in which we live, and the way we live. Exposure to synthetic chemicals, excess alcohol, smoking, processed meats (key word: processed), sugar-rich foods and beverages, refined or highly-processed foods, poor sleep, traffic noise, and other environmental factors may accelerate telomere erosion. On the other hand, exercise and healthy dietary habits, especially lots of deeply-colored plant foods, are associated with longer immune cell telomeres.

In keeping with our theme of positive aspects of the psyche, it has also been shown that emotional intelligence and optimism are associated with longer immune cell telomere length. Here again, the immune cell fossil record reflects contemporary mortality records. Emotional intelligence and optimism are associated with well-being and longevity. Mindfulness—appreciation for the present moment and viewing the environment with a child-like sense of wonder—is also linked to longer telomeres. Thus, the entire environment, and how we navigate our way through its beauty and blemishes, can imprint its collective force on your immune cells via cortisol, oxidative stress, and inflammation.

Ancestral Mismatch with Modernity

In the context of the three-million-year journey of our genus, the time we have spent in the modern industrial, highly technological environment is but a thimble of experience poured into the Great Lakes of evolutionary wisdom. The Greek root of atopy (*atopia*) means "out of place." The roots of the word allergy stem from *allos* (strange; different) and *ergo* (work; activity). Which is quite appropriate regarding the immune system in modernity. The evidence suggests that we are all "atopic" or "allergic" to some degree—out of place and reacting strangely because our immune system currently operates as a function of our ancestral past.

Through millennia of experience, the immune system has come to expect certain things, especially early in life. It has come to expect to meet up with a kaleidoscope of biodiversity. Microbes of untold variety. Early in life these microbes challenge the system and set the tone for an educated and emotionally intelligent immune system for the rest of our days. Disciplined and strong. Lack of exposure to diverse microbes—our old friends, as some scientists call them—is a form of impoverishment.

Waiting in anticipation of microbial diversity, the immune system instead gets an ultra-hygienic environment and, worse still, foods and beverages that further discourage diversity. Overuse of antibiotics, antimicrobials in our personal products, heavily-chlorinated water, environmental pollutants, medications, artificial sweeteners and emulsifiers all work against the diversity of microbes that live on, within, and around us and limit our opportunities for that essential contact. That's before we even discuss our indoor, sedentary lifestyle that disconnects us from soil and our microbial educators.

Our immune system has no historical precedent for what gets thrown at it in the modern environment; it is missing what it expects, and doesn't know what to make of what it gets. We have an innate longing for contact with microbial diversity, being outdoors in nature and eating relatively unprocessed foods like our ancestors. Instead, our immune system gets indoor screen time and cheese puffs. How could three million years of experience prepare the immune system and its microbial partners for cheese puffs? It didn't. How could our ancestral experiences prepare us for a squeaky-clean environment? It hasn't.

This mismatch manifests in the ever-expanding rolls of non-communicable disease epidemics. In addition to Thomson's psychological perspective on the healing powers of nature, *vis medicatrix naturae* runs its channels through the immune system. But only if the conditions are right. Elie Metchnikoff, the 1908 Nobel Prize-winning biologist who literally wrote the book on inflammation, recognized that the immune system was a product of its environment. Inflammation, he said, was a double-edged sword. It helped us when we needed it, and was destructive when run amok. Right now, in the human experience, all indicators suggest it's running amok.

Telomeres and the Biosphere

Now that we have the context of the immune system in place, we can conclude this chapter by returning our thoughts to the biosphere, urbanization, and our fellow creatures. In a rapidly changing world, the white blood cell telomeres reveal that we are all in this together. Consider the birds that increasingly try to eke out a living in urban environments. European scientists have shown that compared with their same-species, countryside-dwelling counterparts, the urban-dwelling birds have shorter telomere lengths.

This matches separate research showing that urban birds have higher levels of stress hormone production and elevated oxidative stress. What's more, their urban food sources may not be as nutritious. Urban caterpillars, for example, have lower levels of antioxidant carotenoids compared to their rural caterpillar counterparts. Which is actually quite similar to our own human situation as our dietary practices in modern urban settings are often devoid of antioxidant sources! More on that later. The point is that the resonance of modern urban lifestyles can manifest in ways that sit below the surface, unseen to the naked eye, in biological manifestations representing another form of unity with the creatures that surround us.

Another recent study on urban birds provides more evidence of urban adaptation being good—until it isn't. Urban birds use whatever they can get their beaks on to build nests. In many ways nests are like a womb for the delicate eggs, their offspring. Green plants contain chemicals that keep insect predators away. But in certain urban environments green plants may not be available so the birds grab cigarette butts and use them instead. The tobacco leaf chemicals and nicotine from the butts perform the same protective function. All seems good. In fact, researchers find better hatching success in the butt-lined nests. They seem to serve their purpose. However, later on the chicks born in nests lined with more cigarette butts show signs of cellular stress.

As well, the "lifestyle" of urban birds may compromise gut microbial diversity and lead to less exposure to microbial diversity and parasites. For example, same-species seagulls hanging around urban environments have gut microbes that are seven times more likely to be resistant to one or more antibiotics vs. their rural coastal counterparts. Meaning, the urban gulls are likely exposed to a steady supply of antimicrobials in the environment and potentially harmful gut microbes are learning how to outsmart antibiotics. The ability of environmental chemicals to compromise the friendly, health-supporting gut microbes is part of this story. One that we will discuss in detail later.

Compared to forest-dwelling birds of the same species, birds dwelling in the more sanitized world of urban asphalt, brick, mortar, steel, and glass appear to have diminished exposure to ectoparasites (parasites that live on or in the skin). Which might *sound* like a good thing. But it's very similar to our children losing contact with the soil, leaves, sticks, and mud. Lack of exposure to microbes in early life has immune consequences later on. The costs of *not* having early-life exposure to a variety of environmental microbes (and even parasites) are a central part of the scientific story connecting haystacks to modern allergy epidemics—and neurobehavioral epidemics. Haystacks are merely a symbol of contact with a multitude of different environmental microorganisms. The sort of microorganisms and parasites that have been properly training our immune systems for the last three million years.

Just like us, birds that are under increased oxidative stress in urban areas may have altered longevity. Their jet-set lifestyle—no longer hanging out with their millennia-old microbe and parasite friends, all cozied-up with the soft cellulose of cigarette butts, and the newfangled, less nutritious caterpillar fast food—works its way towards their telomeres. Works its way toward shortening their lifespan. Urban life is something we and our fellow creatures can adapt to—but at what cost? Are these costs inevitable? "Urban" isn't a bad word. Viewed optimistically, the hidden costs of urban adaptation and modernity can be eliminated if we place more focus on the type of urban environment we want to call home.

The human attraction to urban centers is obvious. We are a social species. Global cities can be marvelous stages for the theatrical production of human social life. At their best, cities are hubs of education, arts, intellect, commerce, and efficient healthcare delivery. Ideally, cities can help undo poverty and reduce the consumption of resources through efficiency. The practical benefits are obvious. On the other hand, there can be pollution, crowding, excess light, noise, crime, the deterioration and infrastructure neglect that defines urban blight, and lack of open, natural environments. Health-promoting cities are far from fantastical utopian ideals. But we do need to understand how the hassles and harmonies of urban life can get under the skin.

Immunology and the immune system are relevant to the entire epidemic of non-communicable diseases. They are also relevant to how we see the world, and how we see each other. If there is any doubting this fact, even after all that we have discussed in this chapter, consider the effects of over-the-counter painkillers: these acetaminophen/paracetamol medications, consumed by tens of millions of adults every day, work by altering cytokine production. Even in the short term, they can diminish levels of empathic concern and joy. Clearly, this massive mobile cranium we call the immune system is connected to the most pressing issues of our time.

We will close on the words of the famous microbiologist and environmentalist René Dubos. He pleaded with fellow scientists, government policy makers, and the public to think beyond health as the absence of disease, and to truly imagine designing the types of urban places that support health. To imagine urban nests for ourselves, our fellow creatures, and even the larger societal nests that keep us linked to the unity of all life:

> Man needs other human beings and must maintain harmonious relationships with the land; he is also indirectly dependent on other creatures—animals, plants, and microbes—with which he evolved and that form part of the integrated patterns of Nature.... The ultimate goal [of conservation] should be to help man retain contact with the natural forces under which he evolved and to which he remains linked physiologically and emotionally.
>
> An immense amount of money and effort will certainly be expended in the years to come on programs of environmen-

tal control. It is therefore essential that we try, collectively, to imagine the world in which we want to live.... Improving the environment should not mean only correcting pollution or other evils of technological and urban growth. It should be a creative process through which man and nature continue to evolve in harmony. At its highest level, civilized life is a form of exploration which helps man rediscover his unity with nature.

— Man and His Environment: Scope, Impact and Nature, 1966

4

Biodiversity: Getting It Under Your Skin

Man is of the Earth, earthy. The quality of his life is inextricably interwoven with the quality of the Earth and the life it harbors.

— René J. Dubos, "Civilizing Technology," 1971

Biodiversity, if we are talking about it on the grand scale, is the variety of species on our planet. It also includes the multitude of genes carried by those species. Locally, it is the variety of species and ecosystems formed within a select environment. There is little question that greater biodiversity and the word "flourishing" are well suited to one another. There is little doubt that the sustainability of life on Earth is maintained by diversity. Hence biodiversity is health.

Until recently, scientists have conveyed the message that biodiversity is an essential prerequisite for human health from a very unemotional perspective. They have told us how important biodiversity is for the development of future pharmaceutical drugs. They have told us how important biodiversity is for biomedical research. They have told us how important biodiversity is for global food production and clean water. They have told us that it is important to keep infectious disease in check. These are facts, of course. But in the westernized world where water flows freely from taps, where the food supply is abundant and conveniently available, where there are 1,500 different FDA-approved drugs, and where, for the moment, serious infectious disease is much less frequent than the non-communicable sorts, it's tough to appreciate the saliency of the message.

Our ancestral past is a long history of trying to survive in the here and now. Hunting and gathering in the natural environment, with no stainless steel refrigerators in which to place the rewards of that effort. No meteorological reports and five-day forecasts to determine whether we should bring a golf umbrella to the forage. Our ancestors developed jaw-dropping knowledge of the daily, seasonal, and annual movement of sun and stars but their day-to-day survival was linked, in their minds, to the more immediate environment. This experience has privileged an immediate "seeing is believing" environmental voice that still runs through our daily lives. So long as water runs out of the tap, and the corner store is filled with "food," climate change and biodiversity loss seem like a million miles away from relevancy to health.

Consider a recent study that shows how the subtle placement of two living vs. dead plants in a room can influence beliefs and confidence in global climate change. Research subjects answered a variety of different questions on current affairs in a room with two floor plants in a corner. The difference was that for one group of participants the floor plants had no foliage—they were dead. The other half of the group answered their questions with two living plants in the corner of the room. Regardless of the political ideology, economic viewpoint, or social outlook participants entered the room with, dead plants being in the room increased their confidence in the legitimacy of global climate change. A second portion of the study added a couple more smaller table-top trees (again, living vs. dead) and showed that participants with dead plants on the table became even more firm in their beliefs that climate change is legitimate.

The point is that humans often need to see threats to the environment, and, by extension to their health and survival. Progressive losses over time can dilute urgency and awareness. So can progressive gains. One supersized burger with fries and a soft drink doesn't lead to immediate obesity and type 2 diabetes. From our perspective, the missing part of scientific communications surrounding biodiversity loss is that which taps into cognitive awareness. Even while clean water still runs from the tap, biodiversity is actually a massive part of our everyday wellness, vitality, and mood. Recognition of this often overlooked fact is essential.

Humans don't like threats to our real and perceived sources of daily *joie de vivre.* Imagine the one billion currently active Apple devices as species. If the company phased out its products and shuttered its doors, it would be immediately perceived as a global threat not just to employment and efficiency, but to vitality and wellbeing. While this horrifically unimaginable scenario is fantasy—never mind shuttering doors, there are global outcries when Apple moves a headphone jack—bees, fireflies, butterflies, and amphibians are all disappearing from our midst. We are told, factually and correctly, that since bees play critical roles in pollination, their losses are a threat to the future of global food supplies. Isn't there more to it, though? Even if scientists can find a way to safely and effectively chemically pollinate us into a synthetic utopia—though feeding a human population set to reach nine billion by 2050 is a tall order—bees and other forms of disappearing life are important to your life in other ways, today.

The View

The scientific exploration of the ways in which nature, including all its (biotic) life-forms and its (abiotic) geological theatre, influence your daily health, mood, and wellbeing is relatively new. Since the mid-20th century, most studies examining the environment have been confined to the effects of obvious toxins, or, if positive influences, the effects of pure aesthetics. Studies in the 1970s provided some important hints that humans prefer certain aspects of natural environments that would be in line with our deep ancestral experiences. Savannah-like scenes with some trees and a vista. And water. We have a strong preference for scenes with water. These preferences appeared to transcend culture or birthplace, suggesting an innate appeal. Remarkably, nature scenes were preferred over urban brick-mortar-and-glass environments even when such images were presented only for a split second. There were also hints within those early studies that even within nature scenes biodiversity mattered the most.

Although nature scenes were all preferred over urban scenes, the nature scenes were not equally rated. When the number of independently perceived elements in a scene were moderate to high, the nature scene was more preferable. Scientists call this complexity. An image of a forest full of same-species evergreens, for example, is a low complexity scene because they are perceptually grouped into one. Looking back, it seems clear that the researchers were actually finding that total eco-diversity—the sum of species and geological diversity—was highly preferred. In addition, the older studies showed that human responses to nature scenes were sensitive to signs of life and death. Healthy forest undergrowth increased preferences, while downed wood and tree damage had negative effects on scene preference.

Mostly, these research findings had very little relevance to day-to-day health and wellbeing. That is, until a geography professor at the University of Delaware entered the fray in the 1970s. Roger Ulrich had found that, in general, motorists take longer routes to their local mall, bypassing a quick route on an open highway and instead choosing more scenic routes rich in vegetation. He was fascinated by people's fascination with nature. He formulated a "nature restoration hypothesis" which argued that the attraction to nature wasn't simply a matter of aesthetics, it was reinforced because it reduced stress. In a series of studies he found that nature scenes (as opposed to urban scenes) can lead to a quicker return to normalcy after mild stress. Electroencephalogram (EEG) readings of brain activity backed up subjective reports; nature scenes were linked to higher alpha-wave activity indicating relaxed wakefulness.

Inspired, Ulrich took advantage of a novel research opportunity. The design of a suburban hospital in Pennsylvania happened to provide two very distinct views out of the recovery room windows. Dependent upon which side of the wing a patient was placed, they had a view of a miniforest or of a brown brick wall. The rooms' sizes, window dimensions, and furniture were identical. Ulrich complied a decade worth of data on patients who had undergone identical surgeries, in this case to remove their gallbladder. Chance determined the post-op recovery room, but chance did not determine recovery success. Shorter hospital stays, fewer post-surgical complaints, and less use of potent painkillers were noted in those who had the view of nature. These patients also seemed less ornery during recovery: they had significantly fewer negative comments placed in their chart by nurses.

Published in 1984 in one of the most prestigious journals, *Science*, this study was a fountainhead of subsequent research into the nature and health connection. It provided more questions than answers, and that is always good for stirring scientists' imaginations. Although its primary application was for the planning and design of healthcare facilities, the study stoked up many lines of research. In the following years other studies have supported the use of various aspects of nature in hospital recovery and/or the reduction of pain and distress during surgical procedures. These studies have included potted plants in rooms, nature scenes on display monitors, and the audible sounds of nature. With good reason, scientists break up parts of nature—sights, sounds, aromas—with the goal of further understanding, and for practical application when patients are largely immobilized. In this chapter we will examine some of these parts, too; but we intend to sew them back together.

Earthrise

> **The vast loneliness is awe-inspiring and it makes you realize just what you have back there on Earth.**
>
> **— William Anders, Apollo 8 Astronaut, Dec. 24, 1968**

Across cultures and ethnicities, no matter how many times we see it, humans find awe in the sunrise. On Christmas Eve, 1968, as Apollo 8 moved through lunar orbit, its three astronauts suddenly clamored for their cameras. They were supposed to be taking pictures of the moon, and instead went off script and used their film for something they weren't expecting—the sight of the Earth rising over the moon. By chance William Anders found himself with the one camera containing color film. If there was ever a time when color film was a necessity, this was it.

Although Neil Armstrong is best known for stepping on the moon first, taking that giant leap for humankind, from the perspective of biodiversity it was Anders who gave us the greatest visible gift. Earth's colors exploding out of the black atmosphereless background and the grey, lifeless foreground powder of the moon surface. Whether Armstrong walked on the moon the following year or not, we already knew what the moon wasn't. It wasn't Earth. It was a dustbowl. His holiday snapshot is now known as *Earthrise*, easily rated one of the most important photographs ever taken. Now in his 80s, Anders recently said,

"Here we came all the way to the moon to discover Earth." But his words on *that day* are equally important. Those in the quote above aren't retrospectively added to the Earthrise photograph. They were expressed in the moment. They encompass awe. Appreciation. Life contrasted with Lifelessness. Loneliness contrasted with Vitality.

Although it would take a few years, scientists started figuring out that they could take advantage of satellite images of regions and neighborhoods and start matching them to health data. It was a transition from a single Earthrise photograph, and everything it represented about life and lifestyle, to high-precision space photographs that could help decipher whether or not vegetation and trees in neighborhoods matter to mental health. And matter they do. The first greenness studies started to emerge about ten years ago; using satellite images similar to Google Earth, as well as locally filed land-use records, researchers began to see fairly consistent patterns—more greenness in neighborhoods is associated with health in general, and mental health in particular. For example, residing in areas with 10 percent or less green space within one kilometer of the home was associated with a 25 percent greater risk of depression and a 30 percent greater risk of anxiety disorders vs. those residing at the upper end of green space concentration near their homes. Even increased mortality—the risk of dying early—was significantly higher in areas where green vegetation was more sparse.

More recently, scientists have employed even greater levels of sophistication in their studies. High-resolution satellite imagery has been combined with individual tree data in cities. For example, in Toronto a greater number of trees in a city block is associated with decreases in cardiometabolic conditions (that is, high blood glucose, diabetes, hypertension, high cholesterol, heart attacks, stroke, cardiovascular disease, and obesity). Of critical importance to our future generations, studies in various cities have linked proximity to green space to normal birth weight and/or decreased risk of preterm birth. Similar findings have been reported for residential proximity to green space and a decreased risk of depression during pregnancy. And not by a little bit. Researchers found a 20 percent reduction in the risk of depression. If there was a drug that could safely lower the risk of depression in pregnancy by 20 percent it would be an international bestseller. Then we have research reminiscent of Dr. Pirrie's haystacks. In a study examining a decade of birth records in Vancouver, Canada, the surrounding greenness of the parental home is associated with decreased risk of asthma at preschool age.

The volumes of research connecting urban green space and various aspects of health are growing on a monthly basis. Shortly after the latest state-of-the-science review is written, it becomes outdated. Green space has been linked to sleep quality, less need for psychiatric drugs, lower rates of type 2 diabetes, prostate cancer, and obesity. Green space has been associated with decreased risk of children needing eyeglasses, decreased risk of depression in a study of adult twins (meaning environment matters!), better cognitive development in school children, better academic grades on standardized examinations, and decreased risk of mortality in older adults. Thus, green space research now spans the entire life course, from a healthy birth to education, to quality of life, and onwards to length of life.

Narrowing the Gap

Ah, but wait. You might be thinking that greenness is just a marker of socioeconomic privilege—well-heeled people do indeed tend to live on the leafy side of town and/or with views of water. Up to a certain point, financial assets bring health-protective assets. Maybe higher greenness simply measures homes where individuals *aren't* holding down two different minimum wage jobs. Maybe greenness is a marker of less consumption of ultra-processed foods, less sedentary screen time, and greater access to quality health care. Maybe greenness is also a marker of what *isn't* in the neighborhood. Clusters of fast-food outlets, factories, pollution, billboards cajoling an unhealthy lifestyle—all the elements of what we have defined academically as grey space. If you are thinking these things you would be right. Satellite images can help to identify green and quantify trees, but they have a hard time revealing the lifestyle intersections of green space and grey space. That's why we can't always pick apart the oneness of life and lifestyle.

However, something remarkable emerged from the deep green when Scottish researcher, and our friend, Rich Mitchell and colleagues looked a little more closely at neighborhood greenness across most United Kingdom cities. In areas of low income, if residents had higher levels of neighborhood greenery, then the usual mortality differences (i.e., dying early from cardiovascular disease) compared with the affluent were minimized. However, the combination of low income plus little surrounding green space was literally a deadly combination—it was associated with much higher mortality.

In 2015, Rich and his collaborators in public health examined data from over 21,000 urbanites from 34 different European cities. These subjects were part of the large-scale European Quality of Life Survey and had completed standardized mental health surveys. Just like cardiovascular disease mortality, risk of mental distress can be determined by socioeconomic inequality. As mentioned earlier, although depression can strike anyone, it does not occur randomly in westernized populations. There is a wide gap in risk between the poor and the affluent. Mitchell found that the usual divide in socioeconomically-determined mental wellbeing was narrowed by 40 percent if an individual had good access to green areas. In addition, many of the new green space studies, including those that have found connections to better academic performance, pregnancy outcomes, and mental health have factored in socioeconomic status. In fact, the green benefits may be most pronounced among disadvantaged groups.

Since social, economic, environmental, and ecological inequalities are among the most pressing issues of our time—and are no doubt politically complex—this research is nothing short of amazing. The need to turn off the valves of inequalities and injustices at all their points of origin remains no less urgent. However, as Rich and his colleagues state: "If societies cannot, or will not, narrow socioeconomic inequality, research should explore the so-called equigenic environments—those that can disrupt the usual conversion of socioeconomic inequality to health inequality." In an urban environment, a vegetation-rich shared commons with some water and diverse wildlife is a living, breathing, omnipresent disruptor of inequality. It doesn't undo other obvious societal problems, but it represents hope and joy.

Getting Into the Green

As the annals of medical science have been expanding with research on the quantity of satellite-measured green space around the home or school, other researchers have been examining the effects of actually spending time both in forests and patches of urban green. Our colleagues from Japan were the first to start examining the physiological effects of spending time in nature in comparison to the same activities in a laboratory or urban setting.

When we met each other for the first time in Tokyo in 2003, we chatted about a relatively obscure Japanese study that had been published in the *International Journal of Biometeorology* several years earlier. The title of the article was "Shinrin-yoku (forest-air bathing and walking) effectively decreases blood glucose levels in diabetic patients." The term *shinrin-yoku* was fascinating to us. What the researchers meant by it was the absorption of the forest into the body and mind. Shinrin-yoku was described as a sensory experience, taking in the "components emitted from the forest." Bathing in the forest atmosphere. A shower without rain or a showerhead.

The first formal study of shinrin-yoku began in 1990 on Yakushima, a subtropical island and internationally recognized hotbed of biodiversity. Pioneering researcher Yoshifumi Miyazaki found that walking in a forest environment, compared to walking in a lab setting, could indeed improve mental outlook and reduce objective markers of physiological stress. Much like Ulrich's hospital room view study, it provoked far more questions than answers. If walking in a forest improves mood, why? With over 25 years of accumulated shinrin-yoku research, it has become clear that walking through forests isn't just bathing in forest air. Rather, it is an opportunity to visualize, touch, listen to, and inhale nature. It means, as our colleague Jeffrey M. Craig from the University of Melbourne says, "bathing in biodiversity."

Over the years, Japanese researchers have studied the way nature influences our different senses. They have wondered whether or not the benefits of nature are simply visual. Remarkable studies have demonstrated that, individually, the sounds of nature, the sights of nature, the invisible chemicals secreted from trees (phytoncides, or phytochemicals), and the touch of natural products like wood (compared to synthetic resin), can positively influence stress physiology and our parasympathetic nervous system, the "rest and digest" branch of the nervous system that cools the jets of over-stimulation. The sum of research shows that our sensory system understands nature like an old friend.

Researchers have also looked at invisible negative air ions which are found in relatively higher amounts within natural environments, especially vegetation-rich areas, forests, and near "blue space" where water is present. These ions are air molecules broken apart or sheared due to the movement of air and water, as well as sunlight and radiation. Several studies have found that negative ions have beneficial effects on mood; indeed a 2013 meta-analysis of all published controlled studies concluded that negative air ionization was associated with lower depression scores. It also lowered the level of inflammatory cytokines. Conversely, when negative air ions are experimentally lowered by researchers, anxiety, suspiciousness and even mania have been noted.

Areas around waterfalls, in particular, are a fountain of negative ions. In a recent study in the *Journal of Physiological Anthropology*, researchers from Austria found that the benefits of spending time in nature could be enhanced by taking routes that included a modestly sized waterfall. These benefits included enhanced psychological well-being and, critically, immune function. The air within the area around the waterfall contained up to 70 times more total air ions compared to other green hiking areas. For perspective on the differences in the type of air ions even within an urban environment, consider the kind of air characteristic of office supply stores. Inside photocopy centers positive ions are almost five times higher than outside air!

Writing in *Science Reports*, Likun Gao and colleagues from China's Northeast Forestry University suggest that the negative ion concentration in many contemporary environments has shifted enough over the last century to make a meaningful difference to health and wellbeing. What's more, since total air ions can influence microbial growth, this may be yet another untold way in which the microbiome of the urban built environment is being manipulated by modernity. On page 124 of his remarkable book *Man Adapting* (Yale University Press, 1965),

René Dubos makes reference to animal research showing that air ion concentration can influence the growth of beneficial microbes in the intestinal tract. Remarkably, negative air ions also appear to increase phytochemical release from plants. It would seem certain that this involves an interaction between plants, microbes, and ions.

Although the shinrin-yoku studies (also referred to as "forest medicine" or "forest therapy") have tried to examine the "bits" of nature in various lab studies (the subject, for example, sitting in a lab listening to sounds of nature, or masked and touching actual wood vs. synthetic wood-like material) the truly exceptional part of the collaborative work by Japanese, Korean, and Chinese experts is the field work. The real world environment. Getting out into green spaces and comparing the same activity carried out in urban settings.

They have completed dozens of studies over the last two decades. The results are clear—spending time in a forest environment vs. performing the same activities in an urban built setting (most studies have examined sitting or walking) has been linked with higher parasympathetic activity, changes in brain blood flow indicative of relaxation (they have ways to measure that in the field), decreased cortisol, lowered blood pressure, and improved heart rate variability. The latter is a marker of good heart health and physiological resilience in responding to stress. Forest experiences are also associated with lower urinary markers of norepinephrine (adrenaline) and dopamine breakdown, further suggesting that the natural environment takes the foot off the accelerator pedal and allows for relaxation.

Other groups of researchers, including those in North America, have found that walks in urban green space are associated with lower depressive symptoms and better cognitive focus compared to the same physical activity conducted in an urban setting. Symptoms of attention deficit hyperactivity disorder (ADHD) have been reduced after a green urban walk. In fact, when researchers conduct post-walk cognitive testing, they find that the nature walk influences cognitive focus on a par with the leading ADHD drug. Most of these studies involve relatively short walks in nature, 20 to 50 minutes. Other studies have examined extended periods of time, a few days or so. Brief excursions into wilderness leave participants in a more creative state, less anxious, and with higher levels of positive emotions.

In addition, new initiatives have been designed to challenge nature-disconnected children and adults to get outside into nature for 30 minutes every day for a month. Examples include Canada's David Suzuki Foundation and their 30×30 Nature Challenge, the Wildlife Trust's 30 Days Wild program in the United Kingdom, and ReWild Your Life, also in the United Kingdom. Researchers have tracked participants in these programs and found that engagement is associated with sustained increases in happiness, overall health, mindfulness, reduced stress, and improvements in mood and vitality. Vitality alone is worth the price of admission. Psychologists describe vitality as approaching life with energy, enthusiasm, and zest for life. You're probably not going to find that in an 8.4 fluid ounce tinned energy drink. There are also enhanced personal connections to nature and pro-nature behaviors which we will discuss later on. For now, we can conclude that there are multiple returns for those who invest relatively small amounts of time in nature.

Remarkably, these field studies have shown that forest experiences are associated with decreased oxidative stress, lower inflammatory cytokine production, and elevations in natural killer cell activity. Based on our discussions in the preceding chapter, the implications of this should leap off the page. Spending time in nature penetrates your skin and influences your mobile cranium—your immune system. When Japanese researchers examined the effects of a weekend trip to the forest, the benefits on natural killer cell activity were still noted a month later. A simple day trip brought about a week's worth of subsequent natural killer cell benefit.

What is responsible? Is it the ions? The sounds? The smells? Or the vegetation-secreted chemicals we can't overtly smell? The sights? Surely it is more than mere aesthetics. The beauty of the reductionist research demonstrating that sounds and invisible airborne chemicals and air ions influence physiology is that it proves we are dealing with far more than mere visual appeal. The more researchers peel shinrin-yoku apart, the more it seems the mechanistic answers are laying in the collective. In biodiversity.

Signs of Life

Interestingly, more than a few of the top-down satellite studies have compared green space with the actual use of these natural environments; results suggest that the benefits of having green space around the home or school are not predicated upon physical activity or specifically using natural areas. Of course, there are plenty of studies connecting use of green space with higher levels of physical activity and social connectedness. Thus green space's benefits might be mediated through obvious lifestyle influences. Green space *quality*—not only in design, but also in the absence of litter, graffiti, animal waste, etc.—certainly has aesthetic appeal as well. All green space areas are not alike, some are more readily accessible and better maintained than others, and some may afford greater health benefits than others. Why?

Clean, safe, and well-maintained parks and green areas are obviously important, but the health benefits may be predicated on life within the green. Upon closer inspection, it seems that biodiversity does matter in the human preference for nature images. When photographic images are presented, preferences increase with plant species variety. What's more, a clever study used the sounds of biodiversity to see if they could influence visual assessments of environmental images. While research subjects were engaged in rating the likeability of scenes of urban environments, researchers piped in recorded birdsongs in the background. Subjects had more favorable impressions of the urban landscapes as the diversity of audible birdsongs increased.

Some researchers have gone a step further than merely examining window views of nature vs. views of brick walls. Even though they aren't in the original study, we've seen the color images of the hospital room views in the 1984 Ulrich study. The nature view, especially from spring to autumn, was certainly more attractive than the brown bricks, but far from rich in diversity of plant life. German psychologist Jasmin Honold and her colleagues wondered if visible biodiversity might be a factor in the physiological responses to the window views. In other words, are all outdoor window views equal, so long as they have some green? Her results suggest that they aren't. She found that in the same neighborhood, adults who had a greater variety of trees and vegetation in the view outside their primary living room window, also had lower hair cortisol (now you know why we spent so much time last chapter explaining the immune system and hair cortisol!).

Other lines of evidence point to the importance of biodiversity in the view. Researchers in the United Kingdom took advantage of a unique opportunity to follow the refurbishment and restocking of a large exhibit in the one of the world's best-known aquariums, the National Marine Aquarium in Plymouth. The restocking (adding new life) was conducted in three major steps. Researchers saw their opportunity to gather data on visitors' mood, heart rate, and interest as measured by length of stay in the vicinity of the exhibit. Put simply, it was a manipulation of the window view by incremental increases of life within the tank. The results indicated that higher levels of biodiversity stimulate positive mental outlook, greater interest and fascination, and signs of lowered stress physiology. It would seem we are more aware of signs of life than we may think.

When researchers in the United Kingdom looked closely at 15 different urban green space settings, the extent to which each one was capable of improving the emotional state of visitors was predicated upon its biodiversity. Specifically, psychological benefits were associated with greater richness of various plant and bird species. Similar results were recently reported in Sweden where the health-promoting value of six urban green space settings was reported to be higher as biodiversity increased. Personal connection to nature mattered, too. We will talk about that later on.

The variety of bird species in a neighborhood has been found to be an excellent marker of good health in the United Kingdom. In 2011 Australian researchers reported that wellbeing within urban neighborhoods was associated with species variety, abundance of local birds, and totality of vegetation cover. A massive study in Mexico looked at national statistics on adult mental health and compared them with a nationwide Biodiversity Index; depressive symptoms were much higher in areas where biodiversity was low. Adding together the research involving images of nature, sounds of nature, urban communities and their proximity to nature, as well as in-the-field studies, a jigsaw puzzle starts to take shape. Humans are searching for life, and life is reflecting itself right back inward to the immune system. Which is to say, vitality and meaning are the essence of life and our immune system has been sensing this for as long as we have walked on Earth.

Outside-In; Inside-Out

One of the themes we have been exploring is that everything on Earth is connected. Which means connected to you. You can enter green space and receive benefits, but might the forest and vegetation also reach out and touch you from beyond its fixed roots? The answer is yes. In the next chapter we will discuss microbes in detail, but for now, we can set the stage by illustrating how the diversity of outside plant life surrounding your home acts as a launch pad for microbial friends destined to make their way home to you.

Since the 1970s, Tari Haahtela from Helsinki University Hospital in Finland has devoted his career to the study of allergy and asthma. Like Pirrie in the late 1800s, he has questioned why these conditions are not evenly distributed throughout the world. A lover of butterflies, he began to note an ominous trend—as butterfly species declined in certain areas, rates of allergies increased. In 2009 he warned his fellow physicians and scientists in the pages of the journal *Allergy*; the title of his piece, "Allergy is rare where butterflies flourish in a biodiverse environment," was obviously not shrouded in scientific jargon. Here is how he concluded: "When the last tiger is killed in India, it may hit the headlines, however, shrinkage of biodiversity also takes place in the microworld close to us, but without notice. Preserving biodiverse life might have a preventive effect on allergy and other diseases of modern civilization."

But just like the study where the subjects had to see dead leaves on a houseplant in order to really consider climate change to be legitimate, very few seemed to comprehend the seriousness of what Dr. Haahtela had to say. His article went essentially unreferenced by the scientific community. But that's not where the story ends.

Undeterred, he joined with other Finnish colleagues and had a closer look at the types of plant life surrounding the homes of a random sample of 14- to 18-year-old school children living in apartments, row homes, and individual houses ranging from isolated rural homes to villages and an urban center of 73,000 people. They wondered if the types of land around the home and the diversity of vegetation might influence the types of bacteria on the skin and the risk of atopy. It turned out that living in areas closest to forested and agricultural land, as well as within the proximity of diverse flowering plants in the yard, was associated with a decreased risk of atopy.

The researchers discovered an additional nugget of information. The microbial family Gammaproteobacteria, found within soils and above-ground vegetation, was identified as particularly important for skin health. Higher diversity of Gammaproteobacteria on the skin of the residents was associated with decreased risk of atopy. Although more research is necessary, the inference is that the diversity of microbes, and their ability to protect the skin, is a product of greater exposure to diversity in the great outdoors. Sort of like Darwin's bees, mice, cats, animal-loving women, and the growth of clover. We are just starting to make the health connections. One thing is for sure, international researchers are now taking Dr. Haahtela very seriously. And butterflies continue to disappear.

You Don't Know What You've Got Until...

Yet another way for researchers to evaluate the effects of bathing in biodiversity is to observe what happens when it is taken away. The reverse of an experiment examining the progressive restocking of an aquarium would be to examine the effects of environmental degradation and a faster than usual loss of life and ecodiversity as they happen. One of the reasons it's hard to appreciate the global loss of various butterfly species is that butterflies are still flapping their wings around us. But there are situations where the losses occur at a faster pace and in more obvious ways.

When visible environmental degradation takes place, especially at a fairly rapid pace, the effects on physical and mental health make themselves known. Examples include the already apparent effects of climate change on indigenous groups living in circumpolar regions. In northern Canada the effects of changing climate are a direct threat to the vegetation and wildlife that are the basis of Inuit livelihood. However, there is an even deeper level of loss beyond food on the table. Researcher Dr. Ashlee Cunsolo Willox and colleagues have studied the deeply rooted connections and comfort provided by the land. As one of the Inuit research participants said in 2013: "I think for the Inuit, going out on the land is just as much a part of our life as breathing. Really, we are so close to the land. We are land people. So if we don't get out then, for our mental well-being, it's like...you are not fulfilled. There is just really something missing. I think we take great pride in being able to go on the land and just feel that energy when you get out on the land. For some people, it's just like taking medicine."

The loss of millions of trees to the handiwork of the invasive emerald ash borer beetle provides another example of the effects of degradation. In September 2002 the Michigan Entomological Society released its thin newsletter. By the look of it, you wouldn't know it from the local Pennysaver. The headline read "The Emerald Ash Borer: A New Exotic Pest in North America." Butterflies may be disappearing, but other species can show up in their stead. Exotic pests. That's the thing about biodiversity loss, it doesn't always mean total loss of life. Invasive species can walk in when the door is cracked open. Or maybe they just hitch a ride on a cargo ship. However they got there, the eagle-eyed scientists from Michigan, USA, and bordering Ontario, Canada, spotted some small metallic-green buprestid (wood-boring) beetles within ash trees in the summer of 2002. The find was of monumental significance. Fifteen years on and these beetles have sawed their way through untold amounts of bark. Net result: 100 million less ash trees.

The emerald ash borer, *Agrilus planipennis* if you prefer, may live in ash trees, but this creature has also made its way into the health statistics of the people living where ash trees once stood tall. Ash trees are concentrated in the upper Midwest to eastern American states, as well as Ontario and Quebec in Canada. Researchers have made clear links between the death of these trees and rates of cardiovascular and lower-respiratory-tract illness. Not just disease, which is bad enough, but risk of mortality. In areas with the highest tree mortality rates, humans are experiencing increased risk of dying from cardiovascular disease or lower respiratory tract diseases.

To further underscore our argument that life and lifestyle cannot be separated, we point to research showing that loss of the ash trees has been linked to diminished time spent outdoors in nature. Trees die, humans stay inside. Which is precisely what the Inuit population in northern Canada described to Dr. Willox. Changes to their environment encouraged an indoor lifestyle. Thus, environmental degradation may have negative impacts on lifestyle behaviors such as leisure time spent outdoors in nature.

There are other examples, such as human-generated industrial activity, where environmental degradation from mining leaves visible scars on the land, and potentially on the less visible minds of those living in the affected areas. This may be especially so when the degradation occurs in longstanding communities with a strong sense of place such as Upper Hunter Valley in Australia and Appalachia in the United States. Bearing witness to altered lands may magnify risk of depression. The effects of this visible intrusion into a healthy sense of place has been described as "solastalgia" by Australian philosopher Glen Albrecht. Combined from the Latin *solacium* (comfort) and the Greek *algia* (pain), it is distress resultant from the loss of place that provided comfort or solace. It is similar to the melancholia of nostalgia, except those suffering from it have not actually left their homeland. They are pining for what once was, even though they remain in the same geographic location.

We have been making the case that so much of our day-to-day quality of life and vitality is driven by our connections to the world around us. Much of the connectivity occurs outside of our primary thoughts, and parts of it are driven in unseen ways, including the effects of microbes and airborne chemicals that make us feel good in direct and indirect ways. The presence of life—biodiversity—fuels vitality, communicates with our immune system, and provides a layer of resiliency, a coat of flexible armor. Since all of life is connected, a stepped-up removal of biodiversity should not be without consequence to human health. The evidence mounts. Loss of biodiversity is taking its toll.

Remediation, Restoration

We have no intention of ending this chapter on a sour note of what we've lost, and the sobering realization that these biodiversity losses and invasive species gains are having profound impacts on our health. The good news is that we are increasingly aware of these connections. When we see degradation, we can fix it. When we see blight, we can add the world's greatest disinfectant—light. We mean light in every sense of the word: physical, emotional, spiritual. Something that is broken is something that has the potential to be transformed.

I believe the time has come to realize that concern with the environment should not be limited to the damage done by pollution, or to the depletion of natural resources. We need a much more creative approach, one that emphasizes not only the maintenance of what we have but also the creative interplay between man and nature. It is this creative interplay that constitutes the spirit of the place, and that makes it come alive.

— René Dubos, "The Despairing Optimist," 1974

René Dubos had much to say concerning the creative interplay between humans and the natural world around us. His primary message was simple. Throughout our evolutionary history we have always worked with nature in creative ways, from moving stones to develop shelters to the representations of the human-nature interface depicted upon the walls of those shelters via "crayons" made from natural materials. Why stop at merely turning off the valve of toxic pollutants? Why not work with nature to illuminate the life within the places we dwell?

Just as surely as there are examples of environmental degradation and biodiversity losses firmly linked to ill health, so too are there examples of human interventions leading to transformed landscapes and health promotion. For instance, the transformation of urban vacant lots has been shown to improve general perceptions of the neighborhood and is associated with increased physical activity. But the transformation also gets under the skin. Incredibly, the simple remediation of vacant urban lots with trees and vegetation has been associated with less stress and decreased levels of violence. Researchers from the University of Pennsylvania School of Medicine and their colleagues have conducted many of these transformative studies. The financial return to society on violence reduction alone, never mind healthcare in general, is estimated at hundreds of dollars for every single dollar invested in remediation.

We can also look to the remarkable transformation of the Glenridge Quarry Naturalization Site and to a newly published study in the journal *Ecopsychology*. Located a stone's throw from the so-called eighth wonder of the world, Niagara Falls, Canada, Glenridge has quite a history. First it was marvellous untouched land. Then it was a limestone quarry. Once the colossal machine-generated hole coughed up all that was deemed valuable to mining, a nearby city saw opportunity: dump some 1.5 million tons of waste into the void. Nasty stuff, including commercial waste. The dump was closed in 2001, and a naturalization project began. Native species were seeded and given room to grow and thrive.

The area was restored to reflect "Carolinian Canada," a term describing the rich biodiversity of southwestern Ontario. Although only a fraction of Canada's total land area, it contains the highest diversity of flora and fauna species within any ecosystem in Canada. By bringing back nature as it most likely appeared before the quarry, species are returning on their own. For example, 125 species of bees have returned to the site. The planning also made room for humans to experience Carolinian Canada. Paths work their way around the over 100 acre site.

Which brings us to the *Ecopsychology* study. Researchers from Brock University followed a group of adult subjects as they walked through this one-time dump or a walk of a similar length through a nearby urban center. Within a week, all those walking in the urban area switched to the Glenridge Quarry, and vice versa. Walking in either the urban or remediated land lowered stress, but only the nature walk improved mood to a significant degree. In addition, the walk through this biodiversity-rich land increased scores on individual connection to nature. We will discuss the value of nature connectivity later on. Interestingly, those reporting low levels of nature-connection prior to the start of the study actually had distinctly poorer mood after the urban walk, and great benefits from the nature walk. This hints that those who are really disconnected from nature and biodiversity might have the most to gain.

Onwards now to a closer look at the tail that might be wagging the dog—microbes!

5

Your Microbial Orchestra

In terms of the total economy of nature, the creative associations in which microbes are involved are probably far more important than are the diseases that they cause.... Indeed, the entirely new creative processes that these associations represent, give to the phenomenon of symbiosis a significance which transcends analytical biology and reaches into the very philosophy of life.

— René Dubos, *The Unseen World*, 1962

Microbes are the ensemble of life. They are unseen and quiet, and yet this belies the true extent of their influence on day-to-day existence. As Dubos said in 1962, they call out vibrantly like a Greek chorus, commenting on the plays that unfold in the theater of life. The latest discovery by Australian scientists suggests that they have been "singing" a life song for at least 3.7 billion years. Somehow, even in the chaos of Earth's early molten existence, in the days before Bruce Willis could save it from the Armageddon-inducing impact of asteroids, microbial life seems to have eked out an existence.

The history of the genus *Homo* is far shorter, but microbes have been with us for every step of that evolutionary journey. *Symbiosis* describes living in close proximity with mutual advantage—an apt description of our relationship with microbes. The critical importance of this symbiotic relationship is reflected by the fact that our gut microbes have always been passed on to our offspring. We can trace several of the key microbial families in our gut today to those inhabiting our common ancestors living a minimum of 15 million years ago.

Interestingly, one bacterial family, *Bifidobacteriaceae*, has been with us all the while. Which tells us it must be important. More on that later.

However, it is also important to note that there are vast numbers of species and strains of bacteria within the major microbial families. Some of these may be of critical importance to modern human health. When humans split from our hominid ancestors, spreading out over the globe, the species and strains found within the gut coevolved with us, interacting with the maestro, the immune system, and furthering our development. Collectively, the friendly microbes that reside in the gut, skin, and genitourinary tract—the microbes that actually protect us from the bad (pathogenic) microbes—are referred to as *commensal*.

The remarkable companionship between human and microbe provides interesting trivia. Not only are there about the same number of microbes residing in and on the human body as there are actual human cells, the microbes bring about a hundred times more genetic material to our symbiotic relationship. From a genetic perspective you are microbially-dominant, but this is not mere trivia—it is where microbial form meets function. The human microbiome is a term used to describe microorganisms, and their genetic material, that reside in various locations on and in the human body.

The genetic dominance translates into functional benefits such as nutrient extraction, protection against potentially harmful predatory microbes, regulation of metabolism, and the production of important biochemicals that support *your* life. So, we let the health-supportive microbes live, and they—the vast majority, anyway—let us live. We cannot underscore this functional relationship enough. Because when the commensal apple cart gets upset, the consequences are manifest in illness.

Although each of us maintains a unique microbial profile, we share a core composition including members of four primary phyla, or grand clans—Actinobacteria, Proteobacteria, Firmicutes, and Bacteroidetes. Below this level there are other families (genera) that dominate and several hundred species below the genus. Gross overestimates of species have consistently been repeated by well-meaning researchers. But there are many strains within each species, and this lingo matters because these strains are functional. That is, they do lots of things for us, and some of these strains might be keeping you healthier than your neighbor.

All of this helps to define the adult intestinal ecosystem. Scientists don't yet know what an ideal microbiome looks like, and those test kits you order over the internet certainly aren't equipped to let you know. They are good for cocktail party talk, but tell us precious little about the species and strains that are pulling the strings of health. Still, researchers are starting to see that a relative expansion of the Proteobacteria is a rough marker of chronic disease. On the other hand, at the genus level, *Bacteroides* and *Prevotella* are breaking through as possible indicators of healthy lifestyle and reduced risk of chronic disease. But nothing is etched in stone at this point.

The anthropocentric view sees humans as the most important entity in the world, thus we see a human host and its microbiome (trillions of microbes and their collective genomes) as an ecological community. But maybe the microbes are the host and we are in their world! Put simply, as you walk down the street, keeping up with all the latest fashion trends, all decked out in the coolest designer gear, you aren't really one species, but rather an assemblage of different species. These assemblages are referred to as holobionts (Greek *holos*: entire, complete; *bios*: life), and this view of what might be called the humicrobe underscores that we humans and our microbes make up a multi-species ecological unit.

As far-fetched as it may sound, those microbial assemblages may influence your mood and even cravings; thus, they could influence lifestyle choices. Meaning, if microbes influence mood, and given that lowered mood increases the likelihood of unhealthy lifestyle choices (e.g., eating more comfort foods), then the tail might be wagging the dog. In fact, if microbes have a detrimental influence on mood, then it's quite possible that the fallout could manifest as a desire for retail therapy in the form of purchasing over-priced clothes—a futile attempt to mitigate microbe-mediated stress and lowered mood!

Microbes may thus be picking out your ties, scarves, and other accessories at Bloomingdales. We will explain. But we must put the spoiler out there first. Despite the many internet stories suggesting that we are merely marionettes on the strings of our microbiome, our human side can decide for itself what type of microbial friends we hang out with. These are important lifestyle decisions that can help decide the circle of friends that are truly in your corner, and separate them from those that weigh you down, or worse still, land you in trouble.

Ancestral Differences

Realizing the importance of microbes to human health, researchers have begun to examine the increasingly fewer groups that live very traditional lifestyles. Before it's too late, and westernization and urbanization engulfs all in its path like Pac-Man, there is still an opportunity to see what microbial ecology was once like. Researchers have worked with groups that resemble hunter-gatherer (Paleolithic) or early farming (Neolithic) lifestyles, and, remarkably, even a group in South America who have been largely isolated and uncontacted by any western influences. Isolated for a minimum of 11,000 years, that is, until a bunch of scientists emerged from the Venezuelan jungle and asked for their fecal samples.

Whether the groups studied were completely isolated from westernization or clinging to tradition, the long and the short of these unique studies is the consistency of the results: traditional groups, in relation to European or North American populations, are almost always differentiated by far greater diversity of intestinal/stool microbiota. There are exceptions, but that is the general rule. Which is important, because the general theme in nature is that diverse ecosystems are always much healthier than those where just a few species start crushing everything around them.

Adding support to these findings, researchers have also analyzed the microbiome within ancient human coprolite samples. Coprolites are fossilized poop or paleostool that can be reconstituted to reveal dietary forensics and, thanks to advanced techniques, even microbial DNA. The reconstitution of paleostool can sometimes include its characteristic reek, reminiscent of a truck stop toilet on I-70!

When our North American ancestors went to the back-of-the-cave latrine to take care of some private business 1,400 years ago, in what is now northwest Mexico, they surely would have laughed if they knew their stool would one day be collected, packaged, and transported over many miles. Destined for a place where men and women in lab coats would be gleefully poking through the products of their bowel movement. The microbial DNA results from the ancient cave coprolites show a remarkable microbial kinship with the modern hunter-gatherers in various parts of the world—and lack of alignment with that of the contemporary North American.

When researchers look closely at populations that are transitioning from traditional to western lifestyles and from rural to urban settings, they find a decline in microbial diversity. For example, in research from the Central African Republic, BaAka hunter-gatherers have distinct microbiomes compared to their neighbouring Bantu relatives that maintain more of a early agricultural subsistence lifestyle. However, the Bantus' microbiomes are still distinct from the industrialized microbiomes that characterize North Americans. In research from the Nicobar Islands, migration from the remote to rural and urban areas highlights microbial shifts, with the remote dwellers maintaining a diverse and stable microbial profile. Compared to westernized groups, the previously uncontacted Amerindians in Venezuela have magnificent differences in microbes living on the skin, in the oral cavity, and gut. Here, researchers find high levels of microbial "dark matter," a science term for unidentified microbes. In simple terms, we are likely witnessing a disappearance of microbes, an extinction of sorts.

We underscore that researchers do not have a complete understanding of what constitutes an easily defined, healthy intestinal microbiome. There probably isn't one. Again, a generalized finding is that diversity may prevent metabolic dysregulation and work toward health promotion. There are exceptions. For example, urbanization doesn't always lower the overall biodiversity within the environment—but that doesn't mean it is healthy. Non-native and invasive species can bring diversity to an environment, but not health. Same in the gut or skin. In some situations higher levels of diversity in unhealthy populations might reflect the presence of microbes that simply don't belong. Diversity in this case represents instability. Overall though, the isolated, traditional communities are united by microbial diversity *and* their extremely low levels of non-communicable diseases.

It is clear that dysbiosis is a relative term. We could argue that everyone is dysbiotic in relation to the Amerindians. The urban dwellers of the Nicobar islands are dysbiotic compared to Nicobarese tribal communities, and the subsistence farming Bantu to the hunter-gatherer BaAka, while North Americans are dysbiotic in comparison to all of the above. But as we proposed several years ago in the *Journal of Physiological Anthropology*, even in North America dysbiosis exists on a sliding scale, and that burden is shouldered by the disadvantaged. Lower levels of microbial diversity are found among socioeconomically deprived groups. Lifestyle is a large factor. Therefore, dysbiosis is a matter of ecological justice and equity. For the sake of health, equity of the microbiome is an important discussion.

Out of Touch

In 2016, biologists from the University of Indiana estimated that the Earth is home to one trillion microbial species. Can you imagine if there were a trillion different kinds of zebras? There are three. Of course we don't make contact with each of these trillion different microbial species; however, we have always been in touch with many of them. And they in touch with us. Even though environmental microbes may not make their home in the gut, it doesn't mean that making contact with the vast sea of microbes that surrounds us is without relevance to health. Neither does it mean that avoiding contact is irrelevant.

> **With the simple exercise of a little prudent oversight, the soil never did a child any harm... far richer is what it gives than take away. Far better are dollars spent on children's clothes than pennies given to doctors... every breath of leaf and soil makes finest fiber, every moment gives pure and healthful delight. The soil is the child's best friend.**
>
> — Louisa Knapp, editor, *Ladies Home Journal*, 1898

Staying in touch with the grand diversity of environmental microbes wasn't something our ancestors had to do with intent. It was simply a product of how they lived their lives. Embedded in nature, hunting, gathering, foraging, and farming, they were forced to make contact with microbial variety. With soil—that would become known as dirt, grime, and filth. As cities expanded during the primary phases of the Industrial Revolution, hygiene became an urgent matter. It was a period of time when microbes were considered to be villains. Somehow *soil* became synonymous with *dirt*. Despite Louisa Knapp's plea to keep children connected to the soil, we didn't.

The progressive drift from nature, especially over the last two decades, includes the unseen. As we spend more time indoors, our cultural relationship with microbes has also changed. Your authors should point out that we both highly appreciate the public health importance of hygienic and sanitary practices. Taking appropriate steps to limit the transmission of communicable diseases is a good thing. Scheduled vaccines are good things. Yes, we want the "Employees Must Wash Hands" poster to remain on the bathroom wall in places that prepare food. But somewhere in a sea of sanitizers, we may be losing the plot. The sales of "Kills 99.9% of All Bacteria" products and the get-your-squirt sanitizer stations popping up like dandelions on a lawn reflect a cultural shift.

We should have seen this coming. In 1998, the *Wall Street Journal* printed a small piece in its advertising column. Underneath the title "Dial Soap Campaign Aims To Soothe Fear of Germs," a few small nuggets of information foretold an impending zeitgeist. The 50-year-old Dial soap company, a one-time industry leader, was watching its profits go down the drain. So, Dial spent tremendous sums of money to gain access to the minds of North American consumers, who seemingly had a shifting relationship with soap. The consumer insight was clear, it wasn't the 1950s and '60s anymore. We had become germ-a-phobes.

Here are some snippets: "We aren't simply primping, or worried about offending others in the outside world. Rather, we want to scrub the outside world away." "It used to be, 'I'm trying to make myself presentable to you.' Now it's more about 'Hey, I've got to wash you off of me.'"

Dial decided on an emotional appeal to 25–49-year-old women with families. "You might feel your very best when you get out of the shower in the morning, and after that, you start to come into contact with a world that has germs," said Mark Whitehouse, Dial's vice president of marketing. "We can restore you to a safe haven."

Today we live in a world where everything can be antimicrobial. Appetizer trays and dessert carts are filled with antimicrobial wipes, gels, lotions, sprays, pastes, creams, and allsorts—anything that can be perceived to help maintain a safe haven. As mentioned, hand hygiene reduces communicable disease transmission, so we aren't making this a thing about hand cleanliness. However, many physicians and scientists share our concerns about a landscape where *all* microbes are marketed as germs. The ecosystem of modernity is increasingly one in which there is an urgent need to wipe everything down and continuously scrub off the outside world.

Thus, at one end we have changed our lifestyles such that opportunity for microbial contact is diminished simply because we aren't in touch with the natural environment. At the other end, we live in fear, checking under our bedcovers for germs. In the middle, as we will discuss below, we have changed our microbial community of friends by the way we eat, drink, and choose merriment. From a health perspective, the "winnings" from avoiding microbes and the "losses" from killing them off isn't a zero sum game. Going out into the natural world, or bringing elements of it indoors, can put us in contact with microbes that can confer health. It's bad enough that we miss out on that by living indoors in recycled indoor air. But then we make it even worse with the fear-based notion that "anti" is the solution to what ails us. It actually might be the problem.

Antibiotics

Each year, thousands of tons of antibiotics are directly consumed by people in westernized nations. Many of them, about half of all prescriptions outside of hospitals, are unnecessary. Thousands more—about 70,000 tons—are consumed by the animals we eat. Not just meat, but farmed fish as well. Questionable chemicals like triclosan have found their way into countless antibacterial personal products. In 2016, the US Food and Drug Administration (FDA) banned triclosan and 18 other antimicrobials from hand soaps and body wash products. Urinary levels of triclosan have been connected to allergic diseases, hormonal disturbances, and obesity. But triclosan still remains in the toothpaste we swallow! Bacterial resistance to antibiotics is ultra-serious. That is, repetitive use of unneeded antimicrobial agents leads to superbug development and antibiotics that are rendered useless when we really need them.

Antibiotic residues find their way into water supplies, onto plants, and into soils and sediment. Even residual levels of antibiotics have the potential to disturb the normal intestinal microbiome in animals. Even though triclosan is being taken out of some personal care products, the sheer scale of its use should be more frightening than the fear of germs. One million tons are produced annually in the United States, and have been for decades. In 2008 it was shown that 75 percent of Americans have triclosan in their urine. Not surprisingly, triclosan has been demonstrated to alter the gut microbiome; indeed, the mass entrance of triclosan into the environment has been implicated in developing antimicrobial-resistant mutant strains of bacteria. In response to triclosan, bacteria can alter the expression of genes involved in fatty acid metabolism, antibiotic resistance, and amino acid metabolism. Thus, triclosan adds another layer of the natural environments story—this time the inner gut ecology. It forces us to try and think about "what once was."

Then we have actual antibiotic prescriptions—266 million courses dispensed in US community pharmacies. These numbers don't even include those dispensed in hospitals! If Ben Franklin were alive today he would more likely say, "In this world nothing can be said to be certain, except death and taxes, and, of course, an antibiotic script." Antibiotics are far from precision killers. They are to a single pathogen what dropping Agent Orange is to killing poison ivy. Antibiotics eliminate hordes of different microbes directly, disturbing other microbes that depended on the lives of the ones killed off. Just like outside in wild nature, many species depend on other species for their own survival in their ecological niche. Antibiotic use, especially early in life and with greater frequency, has been linked to a wide variety of chronic conditions, including asthma, allergies, autoimmune conditions, obesity, and mental health disorders.

As antibiotic prescriptions increased in the second half of the 20th century, so too did allergies, asthma, and stress-related disorders. The epidemics we are dealing with now have occurred in tandem with antibiotic prescriptions. Although this doesn't prove that antibiotics have a causative role, experimental (animal) and cohort (following large groups of people over time) research supports the idea of at least some degree of causation. Teasing out the causes of complex non-communicable diseases is difficult. We can't pin everything on antibiotics because even before they were doled out like candy, there were still noticeable differences in rates of NCDs per region. But the studies on antibiotics do confirm that microbes matter.

The value of microbes to human health has been discovered in many ways. Researchers have observed the effects of losses of microbes via antibiotics and other factors. They have also seen what happens when they compare germ-free mice to animals with healthy commensal bacteria. In 1986, the outstanding scientists Linda Hegstrand and R. Jean Hine reported that animals with the usual gut microbes have higher levels of histamine in the hypothalamus than animals that are germ-free. Thus microbes were proved, for the first time, to influence brain chemistry in the area of the brain governing many aspects of physiology and emotion. Histamine provides essential functions in the brain, including regulation of the sleep-wake cycles, anxiety, and cognitive focus. Too little, as seemed to be the case in animals without microbes, isn't good. Histamine-blockers, the sort used in allergic conditions, have been connected to fatigue, sleepiness, lowered mood, anxiety, and poor memory.

A century ago, Pirrie couldn't understand why people living in rural areas, and especially those working around hay, didn't have hay fever. Among other reasons, it may be because hay was a surrogate marker for a way of life. A lifestyle that placed an individual around microbes that trained the immune system. Hay meant microbial diversity. It meant breathing in microbes. Touching them. But it also meant different dietary habits. Different vitamin D levels from being outdoors. Different stressors. Different levels of physical activity. Different sleep habits. And more.

Diet

The intestinal microbial diversity found among those maintaining traditional lifestyles, and apparently that of our distant ancestors, is likely to be explained in large part by dietary practices. It is not a product of some sort of defined, universal Paleolithic cuisine. The one where our ancestors dined almost exclusively on meat, blubber, and some tubers. That's fantasy. Our distant ancestors ate all sorts of food, including grains. Evidence from stone-grinding tools found in Italian caves shows that if Federico Flintstoni lived 32,000 years ago, he would have started the day with some oatmeal.

Evaluation of evidence found at Paleolithic archaeological sites, some dating back almost 800,000 years, shows that our ancestors ate a wide variety of plant foods. No fewer than 55 different kinds of plants were on the menu. Talk about diversity. Despite regional differences, our ancestors and those living traditional lifestyles are united by much higher intakes of fiber and phytochemical-rich foods. They are also united by what they did not (and do not) consume—the refined sugar, added fats, and high sodium content common to ultra-processed foods.

Nutrition will be explored in detail in the next chapter, but for now we will simply point out that changes to our dietary patterns appear to have decimated the diversity of our commensals. The Latin roots of the term "commensal" (*com*: together; *mensa*: table) reflect an incredibly apt image of eating at the same table, as a family. You and your microbes—the humicrobe clan. Or, as defined in a 19th century dictionary: "Commensal—term used in regard to an animal living like the messmate of another, that is, sharing the food of his host without being parasitic upon him." Sharing the food. But haven't we drastically altered the foods we consume, the ones we share with our microbial messmates? Haven't we introduced a bunch of artificial chemicals and synthetic emulsifiers into our food supply? The answer to our rhetorical question is a resounding yes.

Westernization has changed what we and our commensal microbe messmates actually eat. The foods, the beverages, and all of the associated chemicals that go down the gullet have been changed in the blink of an eye. Our ancestors wouldn't know what the heck some of our pseudo-foods are. For example, the twin-pack "cream"-filled golden cakes known as Twinkies, described as an American icon. An icon of what? Of a nutritional world gone mad? Of separation from nature? With almost 40 ingredients, many of which defy pronunciation without elocution lessons, they boast a remarkable 65-day shelf-life. That truly is a chemical feat for a pair of sponge cakes sitting at room temperature, separated from the environment by a thin, transparent plastic film. If our intestinal microbes could be heard as the Greek chorus they are, they would probably be shouting in unison "What *the*...?" upon encountering this churned-up bolus of sponge.

There is no shortage of research showing that dietary choices impact our intestinal microbial mates. We will discuss this in more detail in the nutrition chapter but for now, suffice it to say that high levels of refined fats, sugar and sodium, artificial sweeteners, and the emulsifiers (for example, polysorbate 80 and carboxymethylcellulose) found in mass-market baked goods have all been shown *individually* to dramatically alter the microbiome in mammals. So do even very low doses of the environmental contaminants known as phthalates, found in fast foods. Biologist Dr. Claire Joly Condette of the Academy of Amiens, France, has demonstrated that pesticide residues on foods may be influencing gut microbes and intestinal permeability in detrimental ways. Moreover, in 2017, research published in *Genome Biology* showed that organophosphates, the most frequently and widely applied pesticides in the world, may be contributing to diabetes by the way in which the chemicals are degraded by the gut microbes.

Risks with these dietary factors showed up when they were studied alone. But, of course, the westernized diet represents the synergy of *all* these forces—a little dab of emulsifier, scoop of synthetic sugar substitute, and a shaker of pesticide residue on top of our sodium, sugar, and refined fat. Highly processed foods, those furthest away from nature, act, in a sense, as an Agent Orange against microbial diversity. The key word is diversity. It's not that you lose all your microbes by adhering to the highly processed diet; you lose microbial diversity. Once again, we return to the story of biodiversity; as mentioned, with a few exceptions, the preponderance of evidence shows that diversity of microbes found in and on the human body equates to a healthy state.

Thus, the twin-pack golden cakes are a semi-food surrogate marker for the loss of nature. They are a painted portrait of the nutritional forces that work against a healthy, diverse ecosystem. We cannot talk about how stunning the microbiome revolution is, with all of its optimism and the hope of resetting and undoing loss of microbial diversity, without thinking of the all the ways in which life and lifestyle interact. We cannot separate this conversation from discussing natural environments because the landscape of the gut and skin is being altered by the greater environment in which we live.

As we will explain in more detail later, a variety of antioxidant-rich dietary items are associated with fatigue reduction, positive mood, and lowered risk of depression. This is the basis for the rapidly expanding field known as nutritional psychiatry. Yet, these same dietary items can also influence the growth of beneficial bacteria and prevent dysbiosis. What's more, once inside the gut, or if fermented beforehand by microbes used in fermented foods and beverages, many dietary phytochemicals undergo transformation by intestinal microbes. This transformation appears to determine the absorption and delivery of important phytochemicals to the target tissue—your brain.

Overall Lifestyle

Diet is joined by a number of other environmental forces that push toward dysbiosis. Psychological and physical stress can disturb the microbes on and within the human body. Environmental pollutants such as airborne chemicals (particulate matter) found in urban environments can also cause gut microbial dysbiosis. So can noise, crowding, excess heat, tobacco smoke, excess alcohol, sedentary behavior, circadian disruption (that is, deregulated sleep-wake cycles), or chronic constipation.

The influence of physical and psychological stress on microbiota isn't new. For at least half a century we have known that stress hormones can influence microbes, including the bad guys that cause infections. Interestingly, it took a series of studies in the 1960–70s involving astronauts to garner some much needed attention to this area of research.

Astronauts (and Russian cosmonauts) were found to have marked alterations in their intestinal microbiota while under extreme forms of stress.

Stress hormones can increase the growth of certain bacteria. For example, experiments have shown that *Escherichia coli* (*E. coli*) can multiply 10,000-fold in the presence of stress hormones. Stress also influences the ability of select gut microbiota to adhere to mucosal surfaces. Today we know that it doesn't take the extreme stress of hurtling around space to upset our microbial family. For example, high reported stress and high cortisol (that is, subjective and objectively measured) during pregnancy was associated with gut microbial dysbiosis in infants. The dysbiosis was also associated with infant health complaints.

Another example can be seen in the experiences of individuals with addictive disorders. There is wonderful research on the benefits of very modest levels of alcohol intake, and this includes favorable effects on gut microbes. However, excess consumption—typically defined as more than two to three drinks daily—is the world's third largest risk factor for disease and disability. It accounts for about six percent of all deaths worldwide. Anxiety and depressive symptoms fuel alcohol cravings. In a 2014 study published in *Proceedings of the National Academy of Sciences* (USA), when individuals with alcohol addiction entered a detoxification program, the first three weeks of abstinence seemed particularly difficult for some participants and their gut microbiome. They experienced increased gut permeability and altered composition and activity of the gut microbiota.

The potent stress of the detoxification experience appears to manifest itself in the gut. But it hints at the possibility that gut microbial dysbiosis is itself a contributing factor, compounding the mood symptoms during the most difficult period of recovery from addiction. It hints that microbes may be pulling one or two levers in numerous aspects of cravings in the modern environment, whether for alcohol, food, or other substances. Several studies in humans show clear evidence of gut dysbiosis among users of cocaine. However, a new animal study shows that disturbing the gut microbiome with antibiotics subsequently enhances sensitivity to cocaine reward. In other words, dysbiosis might prime addiction. This opens the door to possibilities for treatment or even prevention. We will expand on this in Chapter Eight.

This whole life and lifestyle thing gets complicated. Stress can alter the way our gastrointestinal tract moves in its wave-like patterns. It can change the production of normal gastric secretions. But, importantly, it can also shift our dietary patterns. And mostly, that isn't a good thing. As we will discuss in the next chapter, we tend not to make kale our primary choice when stressed and in need of comfort. That's when the golden cake and arches seem more attractive. Highly processed foods can be highly palatable and, for many, they help put a temporary bandage on stress. That temporary fix can become a chronic habit. "Hello, Mr. Dysbiosis? I've got urbanization on line one, and westernization on line two."

We also have to consider all the medications we use. Over-the-counter and prescription drugs are often used to prop up or palliate the lifestyle of westernization. It was a pretty big shock to the medical community a couple of years ago when, as naturopathic doctors have been saying for years, acid-blocking medications were shown to be causing microbial dysbiosis. Acid-blocking drugs (proton-pump inhibitors or PPIs) are among the most commonly prescribed drugs and are widely available over the counter as well. In developed countries stomach acid has been treated much like microbes—something to be eliminated and eradicated. But researchers are starting to uncover the downsides of PPIs and the vitally important role played by stomach acid. It has become quite clear that taking stomach acid away has caused a host of issues further down the digestive pipeline, including diminished absorption of nutrients, additional gastrointestinal discomfort, and overgrowth of bacteria in the upper part of the small intestine. The latter condition—small intestinal bacteria overgrowth (SIBO)—has been linked to irritable bowel syndrome and chronic abdominal bloating.

But acid-blocking medications are apparently just the tip of the iceberg. For example, animal studies are showing that commonly used pain-relieving drugs can cause dysbiosis. The animal studies are translating to humans. In 2017, a new study in the journal *Gut Microbes* showed that a laundry list of drugs, including antidepressants and drugs routinely prescribed for cardiovascular disease, appear to alter the human gut microbiome, Scientists will need to figure out if some of these drugs are actually providing their *benefits* by modulating gut bacteria, but the safer bet is that the common gastrointestinal side effects of many drugs are driven by the way they tip the microbial scales.

Sleep, too, is a massive factor. Research in animals shows that sleep disturbance—the equivalent of jet lag—can cause dysbiosis. On the other hand, inducing dysbiosis itself appears to disturb our circadian cycles. Thus, the gastrointestinal microbiome may be an internal clock, one that goes awry when it isn't taken care of properly.

Interconnectivity of Life, Microbes in Nature

In the late 1800s, one of Maine's better-known physicians was doing his own talking about hay and its wondrous aroma. An accomplished musician and orchestra member, Dr. Searle was underscoring the essentiality of microbial life to all of life. The idea of dung on a cart may be repulsive, even frightening, especially in the sanitizing world in which Dr. Searle lived. But, as he said, "the bounty of the soil" was dependent upon it. Humans, animals, and plants are all "indissolubly linked," in his words, with the lowest forms of life. No grasses and no distinctive aroma of hay without the work of microbes.

> **Do you know what those piles of ordure are, collected at the corners of the street, those carts of mud carried off at night from the streets, the frightful barrels of the night-man? All this is a flowering field; it is green grass, it is mint, and thyme and sage; it is game, it is cattle; it is the satisfied lowing of heavy kine at night; it is perfumed hay; it is gilded wheat; it is bread on your tables; it is warm blood in your veins; it is health, it is joy, it is life.... The one great fact that everywhere permeates nature is the truth that all life is interdependent.**
>
> — Frank W. Searle MD, *Journal of Medicine and Science*, 1897

Returning again to the isolated and traditionally living groups, it is noteworthy that they have been found to have higher levels of certain bacterial families on their skin, like Proteobacteria. Recall from last chapter the European research demonstrating that compared with healthy individuals, atopic individuals have lower environmental biodiversity in the surroundings of their homes and significantly lower genetic diversity of Proteobacteria members (specifically, Gammaproteobacteria) on their skin. On the other hand, the isolated and traditionally living populations have lower levels of the *Propionibacterium*, certain species of which have been linked to acne. As you might expect, chronic skin conditions, including acne and atopy, are virtually nonexistent in the traditional groups compared to North Americans.

Scientists are defining the functional connection between microbial diversity on the skin and many aspects of health. The health-promoting effects of microbes residing on the skin and inhaled through the nasal passages may be far reaching. For example, the presence of Gammaproteobacteria on the skin appears to keep inflammation in check by its effects on anti-inflammatory cytokines in the blood. Long gone are the days when the dermis was considered a complete barrier to the outside world. In some ways it is like the intestinal lining—a filter. Thus, microbial access to deeper dermal tissue can touch upon the mobile brain, the immune system.

One particular species within the Gammaproteobacteria family is *Acinetobacter lwoffii*. It has been shown to induce anti-inflammatory gene expression. When injected just below the skin surface it reduced inflammation as far away as the lungs. Researchers are now working closely with a particular strain isolated from a German farm (*A. lwoffii* F78). Several studies have shown this strain to have anti-allergic properties. Intriguingly, when miniscule amounts of *A. lwoffii* F78 were placed into the nasal passages of pregnant mice, asthma protection in offspring was noted.

Then we have *Mycobacterium vaccae*, a generally nonpathogenic microbe commonly encountered in the soil of natural environments. It's easy to imagine that our ancestors would make contact with all sorts of microbes while foraging, getting soil underneath their fingernails along with incidental contact. As you might expect, researchers have found that traditional groups have a higher frequency of soil microbes on their hands. Much like *A. lwoffii* F78, *M. vaccae* has been shown to lower inflammation in the blood. Remarkably, *M. vaccae* can activate the serotonin pathways in the brain of animals. It has been shown to reduce depression and anxiety-like behavior, improve stress coping, and prevent stress-induced dysbiosis in animals. If you need more, animal research also shows it can reduce asthma when inhaled. First isolated from Ugandan soil 50 years ago, *M. vaccae* is taking us back to nature, back to our origins.

The timing of exposure to beneficial microbes may be critical. One recent Canadian study involving over 300 infants showed that in the realm of asthma, the window of opportunity might be rather small. Scientists first showed that the relative abundance of the bacterial genera *Lachnospira*, *Veillonella*, *Faecalibacterium*, and *Rothia* during the first three months of life was significantly decreased in children at risk of asthma. Then using an animal model, they showed that a cocktail of these same microbes could reduce airway inflammation. Importantly, the microbiome differences they found—that is, the genera protective of asthma—were noted at three months of age but had mostly disappeared in the fecal samples taken at one year. Put simply, if they didn't look early, they might have missed identifying the microbes that seem to afford lifelong immune benefits. Knowing what we know now about the immune system and mental health, it is highly unlikely that this story begins and ends with asthma risk.

Big Picture

The emerging research on the asthma-protective microbes, the farm microbes like *A. lwoffii* F78 and soil microbes such as *M. vaccae*, is so exciting because it represents a big picture framework. It would be naive to think that in the entire five-mile stretch from the bottom of the ocean depths to the upper atmosphere in which microbes live, these two are the only environmental microbes that promote human health. No. These microbes represent at once the optimistic potential of what we have to gain by contact, and despair at the thought of what we might be losing. We have to make contact. As Henry David Thoreau said in his journeys through the Maine woods: "Talk of mysteries!—Think of our life in nature,—daily to be shown matter, to come in contact with it,—rocks, trees, wind on our cheeks! The solid earth! the actual world! the common sense! Contact! Contact! Who are we? where are we?"

Thankfully your authors aren't standing alone on some island making claims that conversations about biodiversity and microbial diversity are connected to one another. And that contact matters to health. Our line of thinking, the ideas we have shared and written about over the last 14 years since we met, are now becoming more widespread. Dr. Tari Haahtela, the pioneering researcher who tried to warn global scientists that butterfly losses are a portent to non-communicable disease, has gathered experts and formalized a consensus statement for the World Allergy Organization: "Biodiversity loss leads to reduced interaction between environmental and human microbiotas. This in turn may lead to immune dysfunction and impaired tolerance mechanisms in humans." Impaired tolerance means allergic diseases and asthma. We take it a step further. Since the immune system is the mobile cranium, we see a place for non-harmful environmental microbes in brain and behavior.

In 2015, the World Health Organization released its massive 364-page Biodiversity Report. They did a little talking about contact as well: "Reduced contact of people with the natural environment and biodiversity, and biodiversity loss in the wider environment, leads to reduced diversity in the human microbiota, which itself can lead to immune dysfunction and disease. Considering microbial diversity as an ecosystem service provider may contribute to bridging the chasm between ecology and medicine/immunology, by considering microbial diversity in public health and conservation strategies aimed at maximizing services obtained from ecosystems."

Isolated communities and those living very traditional lives do not have the comfort of modern medicine and all of its technological advances. That translates into health risks early in life. Developed nations have relatively low rates of infant mortality. Most live births are healthy and infection rates are low. But the early-life microbial exposures, more time spent outdoors, and the absence of highly processed foods among those living traditional lifestyles provides benefit later on. As adults, they have lower levels of inflammation in the face of stress compared to their modern, urban-dwelling relatives.

Air-O-Biology

In the late 1990s researchers started to turn their attention away from a near-exclusive focus on toxic airborne chemicals. They began to consider some of the good things that might be in the air, including the microbes that train the immune system. Several notable studies had shown that growing up on or within close proximity to a farm seemed to protect against allergy and asthma, essentially backing up Pirrie's observations. There were hints that inhaled microbes associated with natural environments, especially in early life, were helping to normalize immune responses. In 1998 the implications of airborne biodiversity—aerobiology—for allergy were described as follows: "In recent years, special attention has been given to the biodiversity of airborne biocomponents. These biocomponents are present in the air in the form of pollen grains, fungal spores, bacteria, mycelium, cysts, algal filaments and spores, lichens, insects and their parts, plants and animal tissues, and several other microorganisms. The presence of such biomatter in the air depends upon the occurrence of diversity of flora and fauna in the surroundings" (A. K. Jain and T. R. Datta, "Biodiversity of angiospermic taxa in the air of central India").

Almost 20 years later we now know with more certainty that there are indeed distinct microbial differences in the air as determined by biodiversity. There are differences between urban and rural environments. The microbes found in the vicinity of vegetation you walk by are not the same as the ones in most urban homes. Green spaces contribute unique and diverse bacterial signatures to the urban environment, and such vegetation makes a significant contribution to the airborne microbial content—up to tenfold higher than nearby non-vegetated built areas.

Researchers have found that contemporary home design is separating humans from the external microbial environment. This is happening along a gradient of urbanization. In other words, rapid global urbanization has been more concerned with "rapid" and not so much with unintended consequences. Thus, urban homes bathe us in our own human-associated microbiota. They are well-insulated but do little to keep us in contact with the outdoor environmental microbes with which the humicrobe evolved. Research indicates, for example, that vacuuming *increases* airborne microbiota. Higher frequency of vacuum cleaning is associated with differing gut microbiota among mothers during pregnancy and also in offspring when evaluated two years after birth. Key word: gut. What this means is that airborne microbes might influence gut microbes. And that our lifestyle practices intersect with life.

Research in animals also shows that airborne pollutants can cause gut microbial dysbiosis, so there are many implications for the food we eat and the air we breathe. Even the pollen in the air appears to be experiencing microbial dysbiosis! Pollen from trees, weeds, and grasses is, of course, the facilitator of allergies in the form of seasonal polinosis and so-called hay fever. But the actual allergen potential of pollen is, among other things, a product of the environmental stresses faced by the plant. For example, birch trees exposed to high ozone levels produce pollen grains with a shape and structure that is more allergenic. In 2016, scientists from the German Research Centre for Environmental Health reported that pollution and urbanization alter the microbiome of pollen itself, which in turn, appears to provoke a more allergenic plant pollen.

Some scientists and urban designers are getting wise to all of this. Brilliant minds are engaged in efforts to understand what constitutes healthy outdoor urban air, and ways in which to bring nature indoors while maintaining efficiency. Vegetation can remove airborne pollutants indoors and out, and the microbes that hang out with diverse vegetation may provide many additional bonuses. Given the massive amount of time we spend indoors in our modern caves, this research cannot happen fast enough.

Gut to Brain, Microbe to Mood

In 1909, Hubert J. Norman, the Assistant Medical Officer of Camberwell House Asylum in London had a novel idea concerning the treatment of depression. As he said: "Many have noted the invariable concomitance between disordered conditions of the alimentary canal and melancholic states. The latter, it is true, are frequently looked upon as causes, the former as effects. But, for the sake of argument, let us assume a contrary hypothesis." In other words, even in 1909, before

Freudian pseudoscience and psychosomatic theories would dominate, it was widely accepted that depression and gut problems often occur in tandem...and that depression causes the gut problems.

Dr. Norman swam against the current; he theorized that gut problems were causing at least some cases of depression, and a possible solution could be found in fermented milk and probiotic tablets containing friendly microbes in the family of Lactobacilli. He wrote up his cases of successful treatment of depression in the *British Medical Journal* and encouraged other mental health experts to study the topic in more detail. The following year, additional cases of mood improvement using live Lactobacilli in a gelatin-whey formula were written up by another London physician, this time in the *British Journal of Psychiatry*.

Sadly, these two reports didn't spur on further developments. Although fermented milk containing live *Lactobacillus acidophilus* was still touted to help adults "Keep a Cheerful Outlook" (Walker-Gordon Acidophilus Milk ad campaign, *New York Times*, 1932) through the 1930s, any notions that such claims were scientifically legitimate were scoffed at. Even the dairy industry warned of making such inferences and advised manufacturers to stick to the nutritional value of yogurt and fermented dairy.

Rising Sun, Dogma Undone

Although strangers to one another, when we, your authors, first met on a bus in Japan in 2003, we had lots to talk about. We had been invited there by the internationally recognized yogurt/fermented milk company Yakult. As we sat together, moving through the megalopolis of Tokyo and outward into its forested countryside, we talked about shinrin-yoku, we talked about early-life exposure to microbes, we talked about nutrition and we talked about biodiversity. All within the first 30 minutes!

But we also talked about mental health, and the remarkably high overlaps between allergic disorders, chronic fatigue, defined chronic fatigue syndrome (CFS), depression, and anxiety. There simply had to be an immune and intestinal microbe connection. We were convinced that dogma was getting in the way of progress. If beneficial microbes could influence the immune system, then it was entirely plausible that they could also play a role in neurocognitive functioning and behavioral disturbances via immune and other pathways.

To fully appreciate how radical our ideas were, they can be considered in juxtaposition to the discussions on probiotics and prebiotics at that time. In 2002, as we were gathering our thoughts and brewing them in the pages of *Medical Hypotheses* and other frontier outposts—the places where ideas go when no other journals will take them—the legendary microbiologist Dr. Gerald W. Tannock had just edited and published a 10-chapter, 333-page textbook entitled *Probiotics and Prebiotics: Where Are We Going?* (Caister Academic Press, Norfolk, United Kingdom). The text contained contributions from 21 global experts in probiotics. This was a text about the future. But there was nary a mention of terms such as "depression," "anxiety," "brain," "mental," "neurological," "mood," "fatigue," and the like.

One thing was clear from *Where Are We Going's* futuristic view of transforming the intestinal tract for health promotion—the good ship probiotic was gathering steam, but it wasn't *Going* to be sailing toward the brain. It just wasn't on the nautical charts of the experts; the topic of microbes *for* mental health was the stuff of Atlantis. Even the mere mention of it—as we both learned in the early 2000s, as we whispered about our ideas at conferences—was the scientific equivalent of saying the Earth was spherical at a time when dogma said it was flat.

Just prior to our meeting in Japan, pioneer research by Australian scientist Hugh Dunstan, Director of the School of Environmental and Life Sciences at Newcastle University, indicated that certain fecal microbial communities were correlated with neurological and cognitive dysfunction. This wasn't some pseudo-objective rodent study. It involved adults with unexplained fatigue, pain, and digestive complaints. Unexplained: code for the sort of people who have been offhandedly and confidently dismissed as having "somatoform" disorders or some other nonsense.

Dr. Dunstan and colleagues painstakingly reviewed the data on almost 1,400 patients with these types of symptoms. The patients didn't have marked gastrointestinal inflammation, as in colitis, but they were certainly suffering. In particular, those with chronic fatigue had significantly lower *Bifidobacterium* species. Their research garnished very little attention. However, when viewed in the context of other research, we were able to leverage it to develop our theory concerning the ability of beneficial microbes to support mood and physical energy.

As described earlier, researchers had also just discovered the profound health implications of small amounts of endotoxins. If these bits of microbial membrane gain access to the bloodstream, they can fire up the immune system and provoke fatigue, depressive symptoms, tension, and brain fog. Studies were also starting to show that probiotics could promote nutritional status in ways that could benefit mental health. For example, probiotics could boost the omega-3 fatty acid levels that are responsible for normal nerve cell communication. Omega-3 fatty acids had been the subject of research in depression, with very good results. Our research suggested that omega-3 could control levels of nerve growth factors like brain-derived neurotrophic factor (BDNF). Some researchers consider BDNF to be the WD-40-like lubricant that keeps brain cells healthy throughout life. But what if beneficial microbes could help support omega-3 uptake into our cells?

At the same time, scientists were learning how important commensal bacteria are for the maintenance of a healthy intestinal lining. Remember, the way for endotoxins to find their way into the bloodstream is through a slightly more porous than usual intestinal lining. As Dubos had shown decades earlier, stress could alter the intestinal microbes and the permeability of the intestinal lining. Animal studies were showing that the intestinal barrier was improved by probiotics while under stress. Researchers began to show that oral probiotics could turn down the dial on body-wide inflammation and oxidative stress. This was at the same time that mental disorders were beginning to be recognized as inflammatory conditions. Fascinating animal studies had been published showing that the most miniscule amounts of orally administered *Campylobacter jejuni* can promote anxiety.

This latter finding was a stunner. *C. jejuni* is a nasty little thing. Not at all like those friendly sea-monkey-like microbes we have been harping on about. *C. jejuni* will find you when your neighbor shows up with an undercooked chicken salad that he whipped up way too quickly. It will stab your gut lining with its wee trident and send you running for the latrine. But the new discovery showed that *C. jejuni* was promoting anxiety at doses too low to be at food poisoning levels, and it was stimulating the emotional centers of the brain. The researchers picked apart the pathways and figured out that the vagus nerve appeared to be the information highway connecting gut microbial messages and carrying them directly to the brain. The vagus nerve innervates the GI tract and sends sensory information to—and receives motor information from—the brain.

As mentioned earlier, researchers had already discovered in 1986 through work with germ-free animals that microbes were required for normal brain neurotransmission. But in 2003, we approached Yakult scientists with a simple idea: to determine if beneficial microbes can influence BDNF. Since BDNF plays a critical role in the plasticity of nerves throughout life, and its levels are low in mood disorders and many brain-related conditions, it would mean that, maybe, just maybe, there is something to the notion that probiotics might be good for the brain. The vagus nerve innervates the GI tract and sends sensory information to—and receives motor information from—the brain.

Subsequent to our visit to Yakult, they did fund a study that showed, once again, that intestinal microbes are essential for normal brain function. They also showed that intestinal microbes are essential for the WD-40 effect. In germ-free animals, the gene expression levels of BDNF were lower in the emotional centers of the brain. The study also showed that mice without microbes had exaggerated physiological response to stress. Thus, commensal microbes were capable of influencing brain structure and function. The study also indicated that the disturbed physiological stress response in germ-free mice could be rectified by the addition of the probiotic *Bifidobacterium infantis*.

Since our visit to Japan we have seen a massive scientific interest in the relationship between gut microbes, allergy, and mental health. No longer is there laughter at the idea that probiotics can influence depressive symptoms. As we look back on our bus ride and reflect on the converging roads of an overly sanitized existence, microbes for mental health and allergic diseases, we can smile knowing that we played our own small parts. Today, with a decade of rodent studies in place, young scientists take for granted the common dialogue about microbes and mood. It's got to the point where there are almost too many animal studies demonstrating a probiotic-behavior relationship. We don't need another two decades of animal studies. Although they have brought about tremendous insight, now we need exceptional human studies. But one recent rodent study is really cool and worth a mention.

Dr. Annadora Bruce-Keller, an expert on the relationships between brain inflammation and cognitive decline, decided to raise two groups of mice, one fed standard lab chow and the other a westernized high-fat diet. Her team removed the fecal microbes from the group fed the high-fat diet and transplanted them into the intestinal tract of the donor mice who had been contently dining on standard lab chow. The result? Altered neurological function. Behavioral changes indicated anxiety, increased stereotypical behavior, and decreased memory in the lean mice after fecal transfer. Similar fecal transfer studies have reported similar results.

Another noteworthy finding is that oral probiotics have been shown to increase blood levels of the social-bonding hormone oxytocin in animals. The pathway by which probiotic bacteria in the gut can increase oxytocin isn't exactly clear but likely happens by communication from the vagus nerve traveling from gut to brain. Oxytocin is manufactured in the emotional centers of the brain. It's important because this chemical also helps regulate our circadian rhythms and keeps the intestinal lining in good shape. Synthetic oxytocin is now being studied for the treatment of anxiety and other psychiatric disorders.

At least four recent studies have shown that the gut microbiota in humans with depression and/or anxiety is distinct from that in other healthy adults. But one study stands out because it set out to determine what might happen if the dysbiotic microbiota derived from patients with depression was transferred to mice. They chose germ-free mice so that the ecosystem playing field would be level. Of course they also gathered fecal material from healthy human donors and transplanted that into a separate group of germ-free mice. The human microbes from adults with depression induced depressive behaviors when transplanted into the germ-free mice. Not so with the transplantation of healthy human donor stool.

Similar research from McMaster University, Ontario, Canada, showed that when human fecal matter obtained from patients with diarrhea-predominant irritable bowel syndrome (IBS-D) was transplanted into germ-free mice, a host of physiological and behavioral changes followed. The study, published in the journal *Science Translational Medicine* (2017), showed that compared to animals receiving stool from healthy humans, recipients of stool from IBS-D patients had faster gastrointestinal transit (that is, food moved through the intestinal tract more rapidly), leaky gut, low-grade inflammation, and anxiety-like behavior. Based on changes that were observed in the blood of these recipient animals, it appears that certain bacterial strains in the IBS-D stool may be producing metabolites that ultimately influence brain and behavior. Yet more evidence that the microbial tail is wagging the emotional dog.

Mental Health Translation to Humans?

The history of science is replete with animal studies where some sort of chemical or even natural agent was looking so good, potentially the next cure for cancer or obesity, but when the human studies began, it all fell apart. There are many cases where fantastical findings in rodents were followed by no benefits, or worse, gross side effects and various previously unseen hazards in human research.

Even when things seem so obviously slanted toward potential benefit, it doesn't always work out that way. A case in point is the use of isolated antioxidants like vitamin C, E, or beta-carotene to ward off disease in smokers. Since smokers are under massively increased oxidative stress, it almost seemed like a no-brainer. Why even bother doing the study, right? Even the usual supplement skeptics said stop wasting money on useless studies. But it turned out that beta-carotene actually made things worse. Then the studies on vitamin E alone started to pour in, also negative. Antioxidants, we now know, work together like an orchestra.

The gut ecosystem is likely to work in the same orchestral fashion. Plus, unlike rodents, humans are confronted with a marketing-rigged mental environment that cajoles adults and their children to make unhealthy, dysbiotic choices. The ad on TV for a sugar-filled cereal is in reality an ad for a bowl, or should we say bowel, full of microbial dysbiosis. So, too, the fast-food jingle and the logos that act as a giant dysbiotic Venus fly trap. The point being, to what end can a biotechnologically-inspired fix (translation: pill) for dysbiosis help with mental health when all the forces of dysbiosis remain in place?

We will discuss the state of the science on probiotics (beneficial microbes) for mental health in the Chapter Eight. For now, we can say that we are cautiously optimistic about the future of this area of research. The volumes of rodent studies are now being supported by an increasing number of human intervention studies showing that our microbial friends can provide benefit to general aspects of emotional health, including mood, stress, anxiety, and quality of life.

Some of these studies have used oral probiotic capsules and others, just like Hubert J. Norman in 1909, fermented beverages that contained live microorganisms. Some studies have looked at subjective mood, and others have included objective markers of stress physiology. One study using brain imaging showed that one month after consuming a fermented dairy beverage there were changes in brain activity suggestive of a reduction in vigilance to negative environmental stimuli. That's important because individuals with either depressive symptoms or anxiety tend to orient toward awareness of the negatives rather than the positives in an environment.

Studies have involved healthy adults, stressed adults, patients with fatigue, and even major depressive disorder. Specifically, patients taking a blend of *Lactobacillus acidophilus*, *Lactobacillus casei*, and *Bifidobacterium bifidum* for eight weeks had significantly reduced Beck Depression Inventory (BDI) scores versus placebo. What's more, patients in the probiotic group also had signs of more balanced blood glucose-insulin levels, as well as lower body-wide inflammation and oxidative stress.

But this is the point at which we should once again point out that major depressive disorder is serious business. As we originally proposed to our friends at Yakult 14 years ago, probiotics might be expected to play an accessory role in depressive disorders, but are never to be considered an alternative to well-accepted mental health evaluation and care. One day there might be prescription drugs targeting the microbiome for mental health, but that day is a ways off. But in the meantime, there is much we can still do with a lifestyle that serves the interests of both a healthy microbiome and mental health.

As the idea of targeting the microbiome with probiotics for mental health matures, it really can't be separated from discussions of diet. Or lifestyle. Probiotic microorganisms are going to find themselves in an ecosystem that might be rough going. It would be a tall order to think that some probiotic pills or a fermented drink are going to undo a diet top-heavy in pesudofoods and twin-pack golden cakes. The gut microbiome *is* the diet. The skin microbiome, which may very well influence mood by way of its influence on the body-wide immune system, *is* the external environment. In the next chapter we will illustrate this in more detail. Maintaining a westernized dietary pattern disturbs our ecosystems, and is as much a story of nature disconnect as kids not playing by the creek or in their local parks.

We began this chapter with René Dubos making the claim that microbes reach into the philosophy of life, and later with Frank Searle arguing that a cart full of dung represents the *joie de vivre*. Mostly they were referring to the essentiality of microbes in the food chain and the web of life. But could microbes influence joy more directly? The preliminary research described in this chapter suggests that the answer is yes. We will explore the topic of probiotics and mental health in more detail later. But first we need to capture the context in which all discussions of intestinal microbes take place—nutrition.

How does the microbiome help nourish us?

The microbiome plays direct roles in human nutrition. For example, gut microbes help to extract nutrients from food. They also help produce vitamins. One of the most exciting discoveries in recent years is the way in which the microbiome can help transform dietary phytonutrients into biologically active forms. In addition, gut microbes are involved in essential fatty acid metabolism.

Researchers have shed new light on the ways in which microbes act upon dietary and supplemental prebiotic fiber to produce short chain fatty acids and a host of other potentially bioactive metabolites.

Is there an ideal microbiome that equates to health?

Scientists haven't discovered the elusive ideal microbiome. It may have existed when we were Paleolithic hunter-gatherers or during our early Neolithic farming days. Researchers have made attempts to study the microbiome of the rapidly shrinking groups who continue to live very traditional lifestyles in a fashion somewhat similar to our ancestors. In general these isolated groups, living in various parts of the world, have remarkably low rates of chronic disease. While their microbiomes are not identical, in general they share one microbial attribute in common: diversity. It appears that westernization is shrinking microbial diversity.

But other dietary factors are rising in the ranks of scientific query with regard to a healthy microbiome. Studies have shown that there may be a two-way relationship between omega-3 fatty acids and the microbiome. For example, the levels of omega-3 fatty acids in your tissue might control the activity of an enzyme, intestinal alkaline phosphatase, that can improve and maintain microbial diversity. On the other hand, omega-3 fatty acids added to the diet appear to promote the growth of friendly bacteria in the class of *Lactobacillus* and *Bifidobacterium*.

Phytochemicals also appear to share this same relationship, and here the research is actually even stronger.

What is the relationship between phytochemicals and the microbiome?

There is a tremendous body of research indicating that phytonutrients can have a beneficial influence on human health and wellbeing. But scientists have been puzzled because the absorption of the phytonutrients found in foods is often very limited. But there may be differences in the extent to which humans extract the benefits of these natural chemicals that give plants their taste, vivid colours, and textures. It turns out that the microbial transformation of phytonutrients plays a significant role in bioavailability. Friendly microbes, including *Lactobacillus* and *Bifidobacterium* have been shown to help convert the phytonutrients found in a variety of foods, by breaking them down into metabolites. These new structures are better absorbed and make their way to where they are needed. And if animal studies are anything to go by, about half of the polyphenols transformed by gut microbes are making their way to the brain.

But it works the other way, too. Foods that are particularly rich in polyphenols have proven to be exceptional at promoting the growth of beneficial bacteria and contributing to microbial diversity. Of course fiber from these foods is likely at play, but researchers have also found that blends of polyphenols derived from foods can have favorable influences on gut bacteria. Even single isolated phytonutrients derived from healthy greens—like quercetin found in many green leafy vegetables, resveratrol from grapes, and tannins from pomegranate—have been shown to stimulate the growth of *Lactobacillus* and *Bifidobacterium.*

Stunningly, it appears that the microbial transformation of dietary phytonutrients into metabolites not only makes for a better absorbed, more bioactive chemical, it seems that the microbially transformed chemicals, that is, the new metabolite made by microbes, can promote the growth of beneficial bacteria.

Is this a story of fermentation?

Yes. This is brand new research and at this point scientists aren't exactly sure how the phytonutrients fermented by gut microbes end up acting, essentially, as prebiotics. Fermentation of polyphenols can stimulate the growth of *Bifidobacterium,* change the ratios of certain classes of microbes in a direction often observed in traditionally living groups, and stimulate the production of short-chain fatty acids that are essential to gut health. Looking more closely, researchers found that it was the polyphenol metabolites—the breakdown products of phytonutrients—that were causing the beneficial modifications to the gut microbial balance. In the study, researchers had a look at the way gut microbes go to work on quercetin. Again, this is a primary phytonutrient in the flavonol class, found in green foods. The quercetin metabolites formed by microbes were acting as prebiotics.

In addition, a new animal study published in late 2016 shows that westernized dietary patterns may impact our ability to metabolize phytonutrients. Specifically, the researchers showed that gut microbiome changes induced by a westernized diet lead to less bioavailability of select polyphenols. Unlike friendly microbes that seem to help us extract the most out of our phytonutrients, microbial dysbiosis in this case was not favorable to bioavailability. In particular, the polyphenols noted to be important in increasing energy expenditure by keeping our body lean and resilient to obesity appeared much less bioavailable. But the researchers were able to undo this cycle and restore normal energy expenditure when they added in a blend of flavonoids. So, the relationship between polyphenols and the microbiome seems much more complex that we might think.

So should we just take the isolated quercetin?

Phytonutrients work through synergy in the promotion of health. It's wonderful that scientists have started to document the effects of phytonutrients on the microbiome, but the research from nutritional sciences points toward their benefits when delivered as whole food, or, if in supplement form, a blend of powdered whole foods with essentially just the water removed.

The lesson from the research on phytonutrients working together in the promotion of health extends to the gut microbiome as well. Phytochemicals have differing and often very selective activity on gut microbes. Many of our herbs contain antimicrobial polyphenols. For the most part these chemicals will not cause dysbiosis; on the contrary, experimental studies show that many different polyphenols can keep less desirable microorganisms like *Candida albicans* in check. On the other hand, they help support the growth of beneficial microorganisms in selective ways. Although polyphenols are not fiber, they may meet the definition of prebiotic in the sense that they are acted upon selectively by certain microbes in the promotion of health.

We can make the same argument that isolated prebiotic polysaccharides are probably not as effective as the whole plant from which they were removed. For example, inulin is an excellent prebiotic found in thousands of plants, including many that are well known to us, such as leeks and asparagus. But when you remove and purify inulin from its food source, phytonutrients are lost. Often you will see chicory in herbal teas; it has a whopping 40 percent inulin in its roots. But there are reams of phytochemicals in chicory as well. Those are lost in purified inulin.

Hence, although isolated inulin can promote the growth of friendly bacteria, it isn't leeks or artichokes or chicory. To underscore this, a new study in the journal *Cell* shows that isolated prebiotic polysaccharides alone cannot undo all of the intestinal problems associated with fiber deprivation. Although the isolated prebiotic was good at restoring gut microbial ecology, it couldn't make up for missing whole food fiber. It was unable to prevent the damage to the mucus layer that protects the intestinal lining. For that, the fiber-rich whole food, inclusive of the plant cell wall, was necessary. The plant cell wall is exactly where the phytonutrients are most abundant! Put simply, purified fiber can undo some of the problems of our westernized dietary patterns, but fiber *per se* isn't the whole story.

In yet more intriguing animal research, in 2017 scientists reported that polyphenol metabolites appear to stimulate the conversion of the parent omega-3 fatty acid alpha-linolenic acid (ALA) into the more biologically active omega-3 fats eicosapentaenoic acid (EPA) and docosahexaenoic acid (DHA). The key word there is metabolites. So, beneficial microbes work on polyphenols in the gut, they make new chemicals that are biologically active (in this case chemicals called hydroxyphenylpropionic acid-O-sulphate and 3-hydroxyphenylpropionic acid) which influence liver conversion of ALA to EPA and DHA. The increases in omega-3 in the blood were coincident with decreases in the more inflammatory n-6 fatty acids. This new study might help explain why moderate red wine intake is associated with higher EPA levels.

6

Traditional Nutrition

Trying to apply any treatment modality whether psychological, pharmacological, or social, to a brain that cannot function normally because of lack of an essential nutrient is like trying to run a 220-volt electrical appliance on a 120-volt system.... No psychiatric patients have fluoxetine or paroxetine deficiencies.... Mental health professionals should be well advised to learn some nutritional biochemistry to keep abreast of this area of scientific development.

— David F. Horrobin, MD, PhD, 2002

Traditional Nutrition and AstroFood

Psychiatry has been so slow to get with the program. It's still trying to catch up. For decades, cardiologists have recognized the importance of nutrition to heart health. Endocrinologists have long known of its importance to hormone release, blood glucose-insulin levels, and an individual's waist circumference. Yet somehow, even in 2002, Dr. Horrobin's words seemed so outlandish. So controversial. Which is so strange given the brain's critical dependence upon quality nutrition—protein, essential fats, complex carbohydrates—for its energy, structure, and functioning. Why did we think the brain could be run on Twinkies, so long as its human carrier had all its deviant Freudian urges attended to through years of psychoanalysis? So long as it had its "vitamin Z" in the form of Zoloft.

When we say "run on Twinkies," we mean that literally. In 1971 the mega-multinational International Telephone & Telegraph Corporation

(ITT) was trying to squeeze more profit out of its Continental Bakery subsidiary, the makers of Twinkies. ITT had made some friends in high places. Their food-engineering scientist Robert Cotton showed up before US Congress and convinced lawmakers that adding in a sprinkle of vitamins and some protein to the company's cream-filled Twinkies (on this project they were named AstroFood) qualifies them as a nutritious breakfast for American kids. Minority kids. When combined with a glass of milk, the idea was to "provide the nutritional equivalent of the following breakfast: a glass of orange juice, two slices of bacon, an egg, one slice of toast and butter... the body doesn't care what form these building blocks are in, it is the availability and the substance that matters—not the preconceived, traditional format." Never mind that the breakfast desserts were loaded up with sugar, emulsifiers, and highly processed omega-6-rich vegetable shortening. As Dr. Cotton explained to a group of government scientists, "We achieve what nature could not do." Indeed.

AstroFood was dressed up as a means to tackle hunger on the cheap. Eight cents for a breakfast cake. Meanwhile, as Cotton was convincing Congress how important eight-cent "nutrition" is for childhood academic success, on the other side of the globe, in Vietnam, every single B-52 sortie was costing $30,000. The total Vietnam experience was $177 billion dollars (not adjusted for inflation). But kids' hunger? Nah, the tolerance for that taps out at eight cents. Said Cotton's buddy, Ed Koenig, then Deputy Director of the Agriculture Department's Food and Nutrition Service:

> We wanted to devise a food that would be cheap and easy to handle. We got a brainstorm. ITT's biggest seller with kids is Hostess' Twinkies.... We figured if we make something as appealing as a Twinkie, yet as nutritious as fruit and cereal... a truck can back up to the school door, dump hundreds of cartons of milk and cakes in the hallway if need be, the kids can grab them and eat them. No forks or spoons, none of this bowls or glasses of juice spilling over.... The gut issue is getting nutrition into kids' bellies.

Yes, the *gut* issue, indeed. It turns out that what was being consumed by the kids was literally a gut issue. A microbiome issue. A mental health issue. At the time, a lone microbiologist and whole-food advocate M. F. Jacobson pointed out that the product was laden with sugar and "furnishes no roughage for the child's digestive system." About 50 percent of the cake was carbohydrate, and since there was essentially no fiber within it, you can figure out that it was a sugar feast. Subsequent research on AstroFood showed that it didn't help one iota with academic performance among the disadvantaged schools where it was dumped.

Evelyn Roehl, an expert on whole-food nutrition, saw the writing on the wall. Roehl knew AstroFood was just the beginning. Writing in the *North Country Anvil* (Jan–Feb, 1978), the literary outpost for holistic thinkers, she had this to say:

> ITT wants children to be healthy...every morning you can be assured that if youngsters skip breakfast at home, they will be fed in the public school system by the private kitchens of space-age technologists. ITT also wants children in other countries to be healthy.... Robert Weiss [ITT subsidiary President] believes "we'll have to depend on artificial foods to feed the world's population. By using artificial ingredients now, we're helping in their development". The question I have is: whose development—the world's population or the artificial ingredients? The poor people in developing countries are being convinced to "depend on" ITT's Twinkies (artificial foods) to feed them. Obviously, the "helping" by ITT must refer to the development of artificial foods.

AstroFood for All

The AstroFood saga isn't over. Yes, these specific cakes were eventually removed from the school breakfast program, but the problem remains. In fact, its the way we've become. Our entire food landscape is one of interlocking spurs of heavily marketed AstroFood-like products. Moreover, these pseudofoods are now being engineered and rejigged to take advantage of human cravings for fat, sugar, and sodium. Then to top it off, for every psychologist that wakes up every morning to help those who suffer, to understand the human condition and connection to nature, a little army of their classmates are going to work to try and help marketing gurus get you to consume stuff you don't need. Like AstroFood.

We've become consumers of everything that ITT's AstroFood represents—nature stripped away with little bits of it added back in as "fortification." Adding a sprinkling of a multivitamin to nutritionally hollow food-like substances doesn't make them food. And we are surrounded by many non-foods. This is especially true in disadvantaged communities where fast-food outlets and convenience stores are a-plenty. In one of the most expansive research reviews of its kind, scientists examined the increasing role of ultraprocessed, additive-laden foods now in global production. The study, published in *Public Health Nutrition* (2017) made reference to the fact that the United Nations has dubbed 2016–2025 the Decade of Nutrition and concluded the following concerning the displacement of minimally processed foods by ultraprocessed foods: "The ever-increasing production and consumption of these products is a world crisis, to be confronted, checked and reversed as part of the work of the UN Sustainable Development Goals and its Decade of Nutrition."

Contrary to what chemical engineering hucksters boldly exclaimed to Congress, the traditional delivery of nutrition does matter. In their arrogance of food alchemy, the scientists dismissed the evolutionary experience between anatomically modern humans and the nutrition of nature. This 200,000 year journey was an epic struggle to obtain nutritional wisdom. The immune system learned to separate friend from foe. The traditional delivery means all the parts of whole foods enter the gastrointestinal tract. The lessons of contemporary nutritional science is that synergy matters. Nutritional context matters. No, you can't devise food with artificial chemicals, add in a nutrient, and expect it to be nourishing. Polysorbate 60 and Red No. 40 are alien to the human experience. The further a whole food is engineered away from nature, the less healthy it is.

Here, as we explore the connections between food and health, especially mental health, AstroFood provides an important jumping off point. The rationale for delivering high quality whole-food-based nutrition to children so that they may have a shot at fulfilling all potential in life is ironclad. ITT, the makers of AstroFood, and a company tied to military coups in various nations and backroom political deals in Washington, exploited this fact under the guise of feeding the needy. Overly confident, they thought that mere calories and a few isolated nutrients would support the brain. They thought nature didn't matter. That the chemicals surrounding the protein and vitamins didn't matter. It was the stuff of the sorcerer's apprentice.

The emerging research in nutritional psychiatry, led by Australian expert Dr. Felice Jacka, underscores the importance of nutrition to vitality, resiliency, and zest for life. It matters in academic performance. It matters in childhood behavior, and even matters in more serious cases of mood disorders.

Back to Basics

It seems that hardly a day goes by when there isn't another sensational media story or a self-appointed nutritional guru telling us that a specific food, beverage, or isolated chemical from a leaf in the Amazonian rainforest is the next fountain of youth. On the other hand, it also seems there are just as many nutritional naysayers who paint a Pollyanna picture of "it all comes down to a balanced diet." Some health experts ignore or pay lip service to the critical role of nutrition in health and wellbeing; others suggest that it's simply a piece of cake for everyone to eat healthy in the modern era.

Perhaps reality sits somewhere in the middle.

Research shows that there are profound ways in which fundamental aspects of nutrition can transform lives. We will explore the ways in which foundational nutrition—the solid nutritional framework upon which health sits—can support resilience, influence mental outlook and motivation, foster vitality, and provide physical energy. The beauty of what constitutes foundational nutrition is that it really isn't complex; it is distinct from the turf wars that go on night and day within the dietary blogosphere, the often legitimate battles on everything from gluten to soy.

Dietary Patterns: The Essence of Foundational Nutrition

One of the easiest ways to demonstrate the overall makeup of foundational nutrition is to examine some of the research on dietary patterns. When researchers examine dietary patterns they are making their best effort to get a big picture snapshot of how people eat over long periods of time (months or years, rather than days). Of course, the ways in which people in different parts of the world fulfill their protein, carbohydrate, fat, fiber, and colorful plant-derived nutrient (i.e., phytonutrients—more on that later!) needs can vary. However, it really doesn't matter where someone resides, there is an uncanny consistency to the constituents of the dietary habits that are clearly healthy.

Dietary patterns deemed healthy include fruits and vegetables (lots of them!), fish and seafood, whole grains, lean meats, some dairy products, fermented foods, and overall dietary fiber. This is not to suggest that foods such as wheat, oatmeal, meat and/or dairy are for everybody. At this point we are talking generalities and it is quite obvious that food sensitivities, changing environments, modern agricultural techniques, and personal preference will allow for an elegant dance around individualized versions of this overall dietary pattern. The main point for now is that whether examining the traditional Mediterranean, Japanese, Nordic, or unique hunter-gatherer diets, these dietary patterns are united in being rich in plants and fiber.

Just as importantly, these healthy dietary patterns are united in what they are not, and this goes to the heart of the matter. These more traditional ways of eating are completely distinct from the modern westernized diet. This diet with all of its highly processed foods, rich in the combination of refined sugars and fats, and devoid of colorful fiber-rich foods, is what is pushed upon us at every turn.

Researchers have shown that the more one adheres to the western diet, the greater the likelihood of health problems. Although there may be no fountain of youth, adherence to traditional dietary patterns has been associated with reduced risk of mortality (that is, dying early), and, as you might expect, quite the opposite has been linked to making a habit out of the western diet.

But when humans visualize a long life they tend to imagine one filled with health and vitality. So it's not simply about living longer as if the goal were to get to some sort of numerical finish line. It's about feeling fully enriched, vital, resilient...and that is where foundational nutrition makes all the difference. Researchers have not only shown that adapting and maintaining healthy dietary patterns is connected to positive mental outlook, when they actually switch people on to a healthy diet there are major improvements in daily mental sharpness, mood, and energy. Remarkably, a new study from Finland shows that a transition to a low-sugar diet rich in fruits, vegetables, and essential fatty acids can lower depression scores and improve sleep.

The Shaky Foundation

Within traditional healthy dietary patterns, scientists have identified the specific omissions that are common to the typical North American diet. It's not really rocket science. When researchers examine the finer details of healthy dietary patterns such as the Mediterranean diet and how they might protect, let's say the brain, they find that the more colorful the better. Color diversity is a good indicator of variety in fruits and vegetables and important naturally occurring food chemicals described below. They also find that greater omega-3 essential fat and fiber consumption by individuals within the overall diet is linked very clearly to better health. Not just major health outcomes like cardiovascular disease and risk of cognitive decline later in life, but also daily mood as well.

If we rely on highly processed foods for a good chunk of our total dietary intake, shunning colorful fruits and vegetables, fish and nuts, fiber-rich foods like oatmeal and ancient grains that are characteristic of traditional patterns, then this would leave obvious deficits. And we are doing plenty of shunning. And the deficits are obvious. Here are some examples:

1. Phytonutrients (*phyto:* plant; also called phytochemicals) are miniscule molecules that give plants their colors, tastes, and textures. When you look at the luscious purples in blueberries and reds of apple skin or cherries, you are looking at phytonutrients. Most phytonutrients are found under the two main categories of polyphenols or carotenoids. Scientists have identified nearly 25,000 different phytonutrients in the various natural foods and beverages that *could* be consumed by humans the world over.

But we are not consuming those foods and beverages, at least not nearly as often as we should. The intake of highly processed foods within westernized nations has steadily increased over the last half-century. In Canada, for example, there has been a 23-fold increase in ultra-processed foods; this means refined flour breads, salty snacks, canned and bottled sauces with added shortening and mayonnaise, sweetened cereals, processed meats, sweetened beverages, pastries, cookies, ice cream, candies, chocolate, jellies, and cake mixes. With increases in these categories, we can assume that other, more nutritious items are being passed over...and they are.

It's just like J. Arthur Thomson stated: if you are doing one thing, you aren't doing another. Hanging in the basement playing Nintendo displaces time outdoors. Consuming ultra-processed foods displaces nutrient-dense foods. This displacement means loss of phytonutrient diversity. It means we get a smidgen of the potential 25,000 phytonutrients. In North America, even when we do consume phytonutrients, the majority are sourced from only four plant foods—tomato, orange, strawberry, and carrot.

Consumption of staples like unprocessed (translation: *not* French-fried) roots and tubers have declined by five-fold over the last few decades, and we are far removed from the minimum of seven daily servings of fruits and vegetables recommended for most adults. In fact, three-quarters of all Canadians do not meet these minimum guidelines. Deeply colored foods are lacking. The daily intake of dark green vegetables provides a good surrogate marker for the void—adults (18–50) in Canada consume only half a serving on an given day. Unsurprisingly, meeting guidelines translates into higher phytonutrient intake.

For several decades scientists spent much of their time looking only at the ability of phytonutrients to act as antioxidants. Simply by living and breathing humans are generating chemicals that can cause oxidative damage to our own cells. Our wonderful antioxidant defense system provides brilliant buffering against this damage. But there is a major caveat. The modern environment can place a more significant burden on the system and as oxidative stress gets ratcheted up we need more defense. But the effective antioxidant defense system operates entirely through nutritional support (vitamins, minerals, and phytonutrients).

The road to Ponce de Leon's fountain is not paved with antioxidants. Yes, antioxidants are important but phytonutrients do far more than act as antioxidants. They interact with the immune system and send out a cascade of information to cells, and the cells respond in ways that promote health in a variety of ways. Later we will discuss the new focus on phytonutrients—their ability to promote the growth of gut microbes that ultimately work toward health promotion.

Phytonutrients can improve blood flow throughout the body and appear to help balance insulin and blood sugar. The influence on metabolism is just now being appreciated. In a recent study a low-calorie meal replacement with necessary amounts of carbohydrates, protein, and fats was compared with a similar meal replacement that also included phytonutrients. After about a month, the meal replacement that included the phytonutrients more effectively reduced the waistline and changed blood fats in a beneficial way.

Remarkably, research shows that even small increases in phytonutrient intake could have profound health impacts. For example, a recent study has found that adding just six to ten milligrams of flavonol (a primary polyphenol phytonutrient), could lead to significant reductions in cardiovascular disease risk. Here is where culinary herbs can help provide a major boost. Edible plants like dandelion have seven times more phytonutrients than spinach. Green tea, turmeric, ginger, olives, cumin, and other dietary supports can make massive contributions to the gap in phytonutrient intake. Although vibrant colors such as those found in blueberries and grapes get well-deserved attention, the ground-up coatings of flaxseeds are also loaded with phytonutrients.

As mentioned previously, gut microbes are altered by the westernized diet. Normally the healthy, diverse microbes act upon dietary phytonutrients and slightly alter their structure so that they are absorbed more efficiently and make it to where they are needed. The westernized diet shifts the microbes and this, in turn, diminishes phytonutrient bioavailability. But dietary phytonutrients are required for healthy, diverse microbes in the first place! The fallout is now making itself known. Phytonutrients act upon enzymes and influence the expression of untold numbers of genes. This means that phytonutrients can potentially influence your very being—the immune system, energy expenditure, cognitive focus, and mood.

2. Fiber recommendations for North Americans vary with age, although for most adults the daily intake should be 30 grams or more. Since whole plant foods are rich in fiber, it shouldn't be surprising that a deficit in fiber is also common. Typically adults are only about halfway to where they need to be—usually around 10 to 15 grams short of recommendations. Fiber is so important because it provides the fuel for the microbes that work for us in maintaining the terrain of health.

3. Omega-3 essential fatty acids are consistently recognized as a nutritional void. There are three primary omega-3 fatty acids—alpha-linolenic acid (ALA), eicosapentaenoic acid (EPA), and docosahexaenoic acid (DHA). These are referred to as essential because we cannot make them on our own. We must obtain them from dietary sources. ALA is found in plant sources such as flaxseed, walnuts, and certain green leafy plants such as purslane. EPA and DHA are found in fish and seafood, and are especially in oily fish such as sardines, anchovies, and mackerel.

ALA is converted into the more biologically active EPA and DHA. However, this conversion can be less than adequate in many adults because it can be compromised by high intakes of vegetable oils common to westernized processed foods and various nutrient deficiencies. EPA and DHA have been linked to good heart and brain health. The research continues to grow showing that EPA is especially important in maintaining a healthy mental outlook. Since most chronic medical conditions are rooted in low-grade inflammation, and given the ability of omega-3 fatty acids to keep low-grade inflammation in check, these special fats can potentially be applied to a broad range of medical conditions.

Expert scientific panels have suggested that North Americans should be striving for a minimum of 250 milligrams of EPA/DHA, to say nothing of being proactive about addressing any specific health issues. Yet, the latest 2014 research showed that North American adults consume only 50 milligrams of DHA per day and 18 milligrams of EPA. This is far from the 1,000 to 2,000 milligrams of EPA and DHA that have been used very effectively in multiple health conditions, including for improved mental outlook.

Why It Isn't a Piece of Cake

As far as providing health and vitality, there is no shortage of research showing that quality protein, essential fats (especially omega-3 fatty acids), fiber, and colorful phytonutrients are the foundations to build upon. But there must be a serious, magnetic lure to the westernized diet if so many people are willing to forgo the potential of a longer, healthier, and more vital life for the taste and convenience of engineered combinations of fatty, salty, and sugary foods and beverages. Put simply, why do so many of us have such a difficult time meeting even minimum requirements for fruits, vegetables, and fiber? Why are we far removed from our essential fatty acid intakes? Why do the gurus with their 25K Twitter followers make it all sound so easy?

Make no mistake, highly processed foods are tinkered with to maximize the odds that all of us will return to them. Throw a bit of stress into the mix, some fatigue, and perhaps a chaser of distraction due to the multitasking of our devices and screen time (television, computers, mobile devices), and the likelihood of taking part in the highly processed food roulette becomes amplified. And that's not all—all sorts of heavy-duty multinational marketing drives the sensory overload. The environment is carefully manipulated so that impulse buys become the default choice. Before you know it, less than healthy foods and beverages are in your hand. It is a rigged system.

It is also easy to lull ourselves into overconfidence that we are indeed eating really healthily on a regular basis. Research shows that we both underestimate our daily intake of calories and overestimate our actual healthy eating habits. In fact, even self-described advocates of healthy lifestyles have been shown to have nutritional voids. And that is before we even begin to talk about the erosion of nutrition within the modern fruit and vegetable aisle!

Diversity, and Changes in the Produce Section

Our theme throughout *Secret Life* is biodiversity. As we have discussed previously, contact with biodiversity begets biodiversity. For example, we highlighted that varied vegetation predicts airborne microbial diversity and greenness surrounding the home, which influences the diversity of skin bacteria. Dietary diversity encourages diversity of microbes in the gut, which, in turn, helps to keep the mobile cranium—the immune system—in a good mood. This is *vis medicatrix naturae.* But since J. Arthur Thomson took the podium in 1914, the diversity of the global food supply has taken a serious nose dive.

Farming isn't what it was when Thomson warned of a disconnection from nature. According to the United Nations Agrobiodiversity Report, some 75 percent of plant genetic diversity has been lost as farmers worldwide have left their multiple local varieties and landraces for genetically uniform, high-yielding varieties. Today, 75 percent of the world's food is generated from only 12 plants and 5 animal species. At best estimate, there are about 400,000 plant species on Earth, of which around 4 percent, or 16,000, are edible. Yet, only 150 to 200 of these are currently used by humans. But even more startling, only three—rice, maize and wheat—contribute nearly 60 percent of the calories and proteins obtained by humans from plants.

The loss of crop and dietary diversity translates into loss of microbial diversity in the gut. Think about uncontacted Amerindians and various other isolated communities and traditional-living cultures. Their lives are different than Westerners' in many ways, but we can be certain that they aren't diversifying their diet in the form of Astro-Food. The available scientific research shows that in humans and animals dietary diversity is healthy, especially when inclusive of a variety of different plant-based chemicals.

Today we may need to eat even more fresh produce to overcompensate for what might be missing *within* modern fruit and vegetables. Several studies have found that a variety of nutrients are lower in modern plant foods compared to those which our grandparents may have eaten, including vitamins and minerals. This may be due to farming practices that deplete soil nutrients. Although there are many scary headlines on this topic of vitamin and mineral declines in fruits and vegetables, it is the loss of phytonutrients that might be most alarming.

Once again, this is a human-generated problem that is driven by the desire to please the delicate westernized palate. Phytonutrients are bitter, so food scientists selectively breed them out of plants. Fruits and vegetables are therefore made more tolerable, but at what cost? The consumption of whole fruits and vegetables at any level is of course a good thing. However, many scientists are concerned that the health-protective properties of fruits and vegetables are being eroded along with the steps taken to diminish the phytonutrient content. These losses are significant, amounting to several hundred times lower phytonutrient content. Purple phytonutrients called anthocyanins—with potent antioxidant activity—are bred out of carrots, cabbage, and potatoes. Apple varieties that are less common to North American supermarkets have a hundred times more phytonutrients than the standard fare. Older variants of organic yellow corn had 60 times more beta-carotene. It's bad enough that we don't consume nearly enough fruits and vegetables; we should really be *over*compensating!

There are also issues with what constitutes "fresh." Consider that broccoli that has been transported to a supermarket under standard conditions and sits in a tray covered with cling wrap loses up to 80 percent of its glucosinolates (the phytonutrients that have been the subject of much research in health promotion) and 60 percent of its flavonoids. Bottom line: lose the phytonutrients and lose a massive bite of what fruits and vegetables do in fighting against daily assaults on health.

Why It Matters: Getting to the Root

Much of modern medicine is directed at symptom reduction, while less attention is paid to the root causes of what might be driving the symptoms in the first place. There is also a good deal of trying to compartmentalize patients into neat checkmarked boxes to determine what they may have and not have in terms of illness. These are essential and very necessary approaches in medical practice. But they are less well suited to chronic conditions and are often ineffective when someone can check many of the boxes (e.g., fatigue, sleep problems, gut problems) but not quite enough for a specific diagnosis.

The problem is the "someone" who can check many of the boxes is actually *many* of us in modern society. The bulk of visits to family doctors are for fatigue, sleep difficulties, vague gastrointestinal complaints, frequent upper respiratory tract infections, tension, anxiety, and lowered mood states. In the doctor's office, patients are often reluctant to mention mental symptoms, preferring instead to focus on the physical.

Patients who check the boxes but don't meet the full-blown criteria for specific diseases suffer nonetheless, enduring a consistent mental and/or physical ache. They are referred to as the "walking wounded." They look good from far, but are far from good. Their basic blood tests look fairly good but deeper and more sophisticated testing would show clear signs of oxidative stress and low-grade inflammation. Like waves eroding a shoreline, these physiological changes add to a further cycle of fatigue and compromised mood.

The walking wounded often turn to quick-fix solutions that deal with the branches far away from the root causes. There are no shortage of pseudofood and psuedomedical offerings—sugar-laden energy drinks, fast food and highly-processed snacks that provide a convenient but very temporary lift, caffeine shots, antacids, and what not. They prop us up for a bit but really just contribute to the cycle, reinforcing the allure but doing nothing to address what is quite often a root cause—cracks in the foundation of nutrition.

Many population studies show that healthy dietary habits are linked with lowered risks of depression and other mental disorders. Our good friend and colleague Dr. Felice Jacka has just strengthened this evidence by completing the first randomized controlled trial to directly test whether improving diet quality can treat clinical depression. In the three-month study, adults with major depressive disorder were randomly assigned to receive either social support or support from a nutrition specialist. Specifically, the dietary group received information and assistance to improve the quality of their current diets, with a focus on increasing the consumption of vegetables, fruits, whole grains, legumes, fish, lean red meats, olive oil, and nuts, while reducing their consumption of unhealthy foods, such as sweets, refined cereals, fried food, fast food, processed meats, and sugary drinks. The 2017 study showed that participants in the dietary intervention group had a much greater reduction in their depressive symptoms over the three-month period, compared to those in the social support group. Remarkably, one third of those in the dietary support group met criteria for full remission of major depression, compared to eight percent of those in the social support group.

Fixing the Terrain: It Starts with the Gut

Taking steps to strengthen the foundations of nutrition can pay massive dividends. Properly nourishing the body is a form of love that can set up an ideal internal landscape, or what many experts call the terrain. For the most part, terrain refers to the gastrointestinal tract, the landscape which trillions of microbes call home. The foods and beverages we consume have a massive influence on the types of microbes that set up their community tents within the hills and valleys of the lining of the GI tract.

Our previous discussions have set the stage so that we can easily highlight the importance of gut microbes and their far-reaching effects on human health. We know that diversity of microbes (that is, a wide variety of community members) has been an overlooked factor in promoting human health as has, at a minimum, preventing less desirable microbes from becoming dominant players. Further discoveries should soon allow for easy analysis of stool for specific markers of what constitutes a healthy vs. unhealthy gut microbiome. Scientists are developing simple to use but sophisticated home-use microbiome test kits. These might be especially useful as a guide for users while engaging in lifestyle changes. If the available research holds true, users should expect to see increases in microbial diversity with healthy foods, especially fiber and phytonutrient-rich whole foods. Diverse microbes can enhance the value of quality foundational nutrition by increasing nutrient absorption. But you can't obtain diversity of microbes without having established foundational nutrition!

Each aspect of foundational nutrition can lead to a healthy terrain. Colorful phytonutrients, fiber, and omega-3 fatty acids can promote the growth of beneficial bacteria in the gut and, in turn, these friendly microbes transform the phytonutrients such that they are absorbed in more potent forms. Fish oil has been shown to increase the adhesion (stickiness) of beneficial bacteria to intestinal cells. But a healthy terrain is essential, too. Diversity of gut microbes may increase the absorption of essential fats and boost omega-3 levels in cells throughout the body.

Food and Therapeutics

There is a large array of natural health products that are formulated to support the phytonutrient aspects of foundational nutrition. Often these are multi-ingredient formulas with various powdered whole foods. These products are generally categorized as "Green Foods" or "Super Foods." Despite outwardly similar appearances, these products can differ significantly in their closeness to whole foods and their medicinal properties. Consumers should check labels carefully as research shows there are wide differences in phytonutrient content between commercially available products. In addition to whole food ingredients, certain products contain herbal ingredients such as licorice, ginseng, ginkgo, and other herbs with standardized levels of what scientists consider to be bioactive ingredients. Formulas inclusive of these botanicals may provide a therapeutic aspect to the whole-food-based ingredients. Health Canada's Natural Health Product Directorate allows for specific claims if a product in this category has been the subject of scientific research.

This is a fascinating aspect of nature at work. When highly bioactive forms of phytonutrients make their way through the bloodstream to their target destinations—let's say the brain or the heart—they have a beneficial effect at other terrains. From this perspective, the entire body is a single terrain or landscape that needs nourishment, not just for growth in physical stature, but growth in life itself.

The Linings Are Leaky

You'll recall that we previously placed a good deal of emphasis on one particular study in the immune chapter, the one showing if very low levels of endotoxins—the outer membrane of microbes normally restricted to the inside of the gut tube—find their way into the blood in even small amounts, this can provoke cytokine release, alter cognition, and diminish mental outlook in humans. Endotoxins can gain access to the blood through an intestinal lining that is slightly more porous than it should be, so-called leaky gut.

You can take your pick from the roulette wheel of chronic medical conditions—autism, asthma, allergies, chronic fatigue, depression, fibromyalgia, heart disease, irritable bowel, obesity, psoriasis, type 2 diabetes, schizophrenia, and many more—intestinal permeability has been reported to be in the mix. We aren't saying it's a cause, but at the very least, it is a detrimental contributor. Once endotoxins gain access to the blood beyond the most minute levels (even in healthy states there can be ultra-small levels and this is tolerable), it can lead to a cascade which increases central nervous system inflammatory chemicals and oxidative stress.

We have already mentioned several roads to a dysfunctional intestinal lining—antibiotics, stress, overwork, physical exhaustion. But the westernized diet, with its high sugar, ultra-processed fare, fast foods, and soft drinks is the express lane to permeability. Since the largest reservoir of microbial endotoxin, lipopolysaccharide endoxin (LPS) in particular, is within the intestinal tract, it is easy to see how lifestyle can put LPS into your blood. Your other brain—the immune system—responds, and mood and metabolism suffer.

When healthy adults start consuming a westernized diet high in refined fats and low in fiber, they see a 71 percent spike in blood endotoxin levels within a month. On the other hand, switching to a more traditional diet lowered blood endotoxin activity by 38 percent. The traditional diet also increases the healthy gut bacterial family

Bifidobacterium. The saturated and refined fats that find their way into ultra-processed foods can also lead to higher LPS and inflammation. What's more, when LPS gets into the blood it can compromise another barrier, one that is normally more tightly controlled than Fort Knox—the blood-brain barrier. You can imagine that our delicate neurons would normally be protected from alien substances, and the blood-brain barrier is a specialized network of blood vessels that is designed to stop a Roswell Incident before it even begins.

But it's the same old chestnut. The blood-brain barrier wasn't constructed by nearly three million years of evolutionary experience with AstroFood. Sure enough, the westernized diet increases blood-brain barrier permeability and disturbs cognitive performance. When animals consume a westernized diet and researchers inject a small amount of dye into the bloodstream—a dye normally excluded from the brain—it finds its way into an area of the brain governing memory and emotion. Remember, also, our previous discussions on the surprising lines of communication from the skin and inward to the immune system and brain. The westernized diet is no friend to the normal structures that allow the dermis to effectively filter the outside world. So, the modern diet causes leaks to be sprung all over the place. Intestinal barrier, blood-brain barrier, and the skin barrier as well.

To re-emphasize: the westernized dietary pattern isn't just a force of sugar, excess fat, and a bunch of synthetic chemicals. It's also about what it doesn't include. Inadequate omega-3, fiber, and phytochemicals are individually and collectively associated with marked shifts in the intestinal microbiota and enhanced intestinal permeability. If we look at all the animal studies showing certain culinary items are associated with antidepressant and anti-anxiety-like properties, they all have in common being antioxidant, anti-inflammatory, and able to promote the growth of beneficial bacteria. (Some human studies show this, too: for example, curcumin from turmeric has been shown to improve mood in several clinical studies.)

Examples of the dietary items shown to reduce fatigue and positively influence behavior in animals include turmeric, ginger, seaweeds, purslane, Brassica-family vegetables and sprouts, green tea, coffee, cocoa, apples, beets, grapes, plums, blueberries, and cherries. When microbes act upon these foods in the gut they transform them, changing their structures. This determines the extent of their benefits in the brain. Incredibly, it also appears that these microbially transformed phytochemicals are broken down into other structures, or what scientists call metabolites, and for reasons that remain a mystery, those newly formed metabolites can promote the growth of beneficial bacteria such as *Bifidobacterium*. One example is the phytonutrient quercetin which is found in many healthy plant foods, like kale. The fermentation of quercetin produces metabolites that act favorably upon the growth of gut microbes. Lifestyle promotes life. Good life.

Many of these same dietary items—at the whole food level and even blends of different phytonutrients—can turn down the dial on intestinal permeability and lower blood endotoxin levels. In addition, omega-3 fatty acids may play a particularly important role in preventing intestinal permeability, and these essential fats also have a favorable effect on beneficial gut microbes. In a stunning bit of research, scientists manipulated the tissue levels of omega-3 fats in animals and found that the levels of omega-3 in the cells throughout our body—distant, that is, from the gut—can control microbes in the gut. Basically, having higher omega-3 at the cellular level sets in motion a cascade of events; it signals a gut enzyme, intestinal alkaline phosphatase, to take charge. This enzyme helps coordinate the diversity of gut microbes and keeps intestinal permeability in check. So, we now have another consideration when pondering the heavy omega-6 load from all the processed vegetable oils we consume.

Dietary Acid Load and Glycation

For many years nutrition experts have suspected that there may be other, less obvious ways in which foundational nutrition might promote—and the westernized diet might diminish—the potential for optimal health. Dietary patterns are more than simply reflections of macronutrients. Since the westernized diet contains a relative deficit of fruits and vegetables, with a loss of alkaline-rich bicarbonates and minerals like potassium, scientists have started to look more closely at the consequences of dietary acid load. Fruits and vegetables represent a wonderful way to buffer the steady input of relatively acid-heavy meats, dairy, and highly processed foods.

Blood pH must be kept close to near-neutral for human survival, that is nonnegotiable. In order to maintain this steady state, we can tap into reserves for alkaline buffering. In the face of the westernized diet, the blood looks at the alkaline reserves in the bone the way a greedy oil baron views the Albertan tar sands. Except we only extract the most miniscule amounts of buffering minerals at a time. Once used in the buffering process, these minerals are subsequently lost in the urine. Obviously this could add up over time and threaten the integrity of bones as we age and some research supports this premise. However, the possible long-term threat to bone health doesn't explain why dietary acid load is now being linked to so many chronic non-communicable diseases.

Scientists are starting to figure out why dietary acid load matters so much. It's a stressor. Even though our acute survival isn't under threat, bodily systems appear to react as if they are under stress in this effort to maintain appropriate balance. At least three studies have now shown that an acid-heavy diet can elevate the stress hormone cortisol; on the other hand, neutralization of this westernized acid load with an alkaline beverage can bring the stress hormone production back to normal. It makes sense that a diet devoid of alkaline-rich plants is perceived to be a threat because from the perspective of our ancestral past, it is strange. Throughout our evolutionary experience we have always consumed plants. Meat and fish, too, of course. But we have always consumed plants.

There are enormous implications to this; continual stress hormone production can take its toll over time. As we have mentioned, cortisol is a "break glass in case of emergency" hormone that supports proper metabolic and immune reactions in times of immediate need. Keeping it elevated over time is disastrous; it then maintains a state of low-grade inflammation with higher levels of oxidative stress, disturbing immune function and increasing the likelihood of abdominal weight gain and compromised mood.

Until the last two or three years, the idea that dietary acid load (or more specifically, the absence of buffering alkaline input) is relevant to many chronic health conditions was scoffed at. Just as with leaky gut or probiotics for mental health, we have had our share of colleagues smile and roll their eyes when we broached the subject over the years. But now scientists have been forced to take a different position. High dietary acid load is now linked to type 2 diabetes, insulin resistance, high blood pressure, deterioration of bone health, and obesity. Having a higher dietary acid load can place someone on the trajectory toward kidney disease later in life.

The good news is that it is easy to remedy. The enemy isn't meat or protein (proteins are made up of amino *acids*!) or dairy, it's the absence of alkaline-, potassium-, and bicarbonate-rich fruits and vegetables. In addition to consuming more fruits and vegetables, whole-food-based supplements have also been shown to shift pH in the direction of alkalinity in just a matter of days. Since neutralization of dietary acid load can improve metabolism and keep stress hormone production in check, the alkaline buffering may help explain why diets rich in fruits and vegetables have been shown to improve energy and vitality. Alkaline-rich diets have also been associated with better exercise performance, lengthening time to exhaustion.

Vitality

Vitality is a well-known indicator of physical and psychological wellbeing. But what is vitality? Researchers define vitality as approaching life with excitement and energy; living life as an adventure; feeling alive, activated, and enthused; feeling vigor; feeling zest. Foundational nutrition extends well beyond the bones, heart, muscles, lungs, and other bodily systems—it supports the Big Picture of human life.

In a recent study Japanese scientists showed that when adults with diabetes consumed an alkaline drink they had improvements in blood sugar control. Remarkably, this seemed to be, at least in part, a product of the way that the alkaline drink changed the bacteria in the gut.

Here again, all indications seem to point back to changing the terrain. Or, if you are already a healthy person committed to keeping your wonderful health status, it means maintaining the terrain.

In addition to dietary acid load, it also turns out that the way you prepare your foods matters as well. Most highly processed foods are laden with advanced glycation end products (AGEs). The abbreviation really fits because these chemicals contribute to the aging of our body. AGEs are like the opposite of antioxidants. They are highly oxidant compounds formed when sugars react with free amino acids—and nothing provokes this reaction like turning up the heat in cooking techniques, especially those that involve no water. Like an oven. Or a frying pan. The use of water—so common to ancestral and traditional cooking techniques—keeps AGEs in check.

Steaming, boiling, poaching, and other uses of water still contribute to traditional cooking techniques in many parts of the world. According to experts, our ancestors relied upon Stone Age soups and stews dating back at least 20,000 years, but the use of steam to release necessary fat from animal bones appears to have been in practice for at least 50,000 years. Suffice it to say that water-based cooking is an ancient practice. Notable declines in water-based cooking have been occurring in Asian nations as western influences creep in.

Think of baked goods. Think of AstroFood. Sugar and a bit of protein baked in the oven. If you wanted to jack up your AGEs, that's how to do it. When you see the brown top on a loaf of bread, you are literally looking at AGEs. Same with the browning on meat, like the holiday turkey. But many times the collection of AGEs in food isn't so obvious to the eye.

Human studies show that moving the diet toward stewing, steaming, poaching, and boiling foods can lower the AGE burden by approximately 50 percent. In doing so, there are marked reductions in inflammation and oxidative stress. AGEs and high-fat foods prepared with high heat have been shown to reduce the growth of beneficial intestinal microbes in human and animal samples. Here we go again: lifestyle bearing down upon life. Higher levels of skin AGEs have recently been linked to depression and higher intakes of dietary AGEs have been associated with a decline in cognitive function.

Insurance Policy

In the context of optimal health, every effort to increase fruit and vegetable intake, fiber, and omega-3 fatty acid intake will pay off in spades. But we have discussed the realities of why that can be difficult, even for the best intentioned among us. So what to do? Can supplemental natural health products play a supportive role? The answer is, of course, yes they can.

Products designed to act as foundational nutrition insurance policies are widely available. Insurance for phytonutrient intake via natural health products typically means choosing a "green" or "superfood" formula. Scientific evidence shows that the further away from nature we move—for example, taking isolated vitamins or phytonutrients—the more we lose out on the synergy that exists in whole foods. Phytonutrients work together like an orchestra when they are consumed within whole foods. The sum of their health-promoting activities is greater than their individual parts.

Scientists who have been trying to isolate individual polyphenols as being "the one" for health promotion have not been rewarded for their efforts. The quest for the single Rosetta Stone polyphenol has proved to be largely disappointing. But when you add them together clear benefits emerge. Multi-ingredient whole-food-based supplement powders increase the odds that the consumer will ingest a wide variety of total phenolics. In many ways these are like whole foods with the water removed. The polyphenols and other phytonutrients remain.

There is certainly research on isolated nutrients and mental health. Zinc and magnesium have both been shown to be low among individuals with depressive symptoms and intervention studies using zinc and/or magnesium appear to be quite helpful. But the mechanisms are tough to tease out. The road to not having enough magnesium is paved with a westernized diet. Why? Because the primary source is green vegetables! Zinc is found in whole grains. Both of these nutrients perform countless functions that can support normal communication between nerve cells. Inadequate intake of either one of these nutrients can disturb the intestinal bacteria. Supplements can help, but getting upstream to correct the source of inadequate zinc and magnesium by fixing the diet will provide collateral benefits. Magnesium pills don't carry phytonutrients.

Fermented Foods, Natural Health Products

Thus far we have presented the nutritional problems associated with our contemporary way of living, including the common deficits. We have discussed the health implications of not giving enough nutritional attention to the terrain, and how this can influence so many aspects of metabolism, immune health, and even mood. These discussions have segued into the value of natural health products as a general form of nutritional insurance policy. We can now turn our attention to both terrain and specific foods and supplements as a way to magnify foundational nutrition.

Earlier we discussed the ever-increasing research on traditional dietary patterns. At the top level it is easy to understand why the richness of essential fatty acids, colorful phytonutrients, and fiber (in the absence of high-sugar, highly processed foods) can contribute to health. But there may be another facet to traditional diets that has been overlooked until recently: they commonly include a wide variety of fermented foods and beverages.

Fermentation is the slow conversion of carbohydrate constituents of foods into organic acids and/or alcohol. Fermented foods and beverages are an ancient tradition. According to analysis of pottery shards, fermented dairy consumption dates back close to 8,000 years and the consumption of intentionally fermented blends of honey, berries, and some grain—probably a Stone Age wine—dates back 9,000 years. Today, approximately one-third of all foods and beverages consumed within healthy, traditional dietary patterns (those that remain!) are fermented. These include wide varieties of fermented fish/seafood, dairy, meat, fruit, vegetable, and cereal dishes.

Since traditional diets have been linked with so many aspects of health promotion, it forces the question of what might be lost as artisanal foods and beverages disappear from the homogenized, ultra-processed landscape. With the encroachment of the westernized diet, some researchers are voicing concern with global declines in fermented food consumption. From our perspective, since fermented foods carry microbial diversity at far higher levels than unfermented foods—up to 10,000-fold more orally-consumed bacteria—this represents yet another potential loss of biodiversity. Canadian researchers had this to say in a recent scientific paper:

> With few exceptions, fermented foods are generally absent as a recommended category of food for daily intake in Food Guides. We believe this reflects a failure to appreciate the benefits resulting from the process of fermentation, which have been supported in numerous studies.... It would be a great detriment to human health if fermented foods were to decline.
>
> — S. N. Chilton, et al., "Inclusion of fermented foods in food guides around the world," *Nutrients* 2015

The story of food and beverage fermentation is, of course, an ancient one. There were no microscopes to guide knowledge that unseen microbes were at work, over time our predecessors crafted their techniques. They learned that fermentation could help preserve foods, make them more digestible, improve their nutritional value, and make them more palatable.

Today, the science of fermentation—using select microbes to transform foods for the better—is moving at a lightning pace. Remarkably, fermentation has been shown to support the absorption of phytonutrients and actually produce entirely new phytonutrient structures in the process of transforming healthy foods. Even more stunning is the recent demonstration that some of these newly formed antioxidant chemicals may themselves encourage the growth of beneficial bacteria. Scientists aren't sure how that happens, but it is one more indication that the terrain is set by wondrous interactions between diet and microbes. In a 2006 study published in the *Journal of Dairy Research*, researchers followed healthy adults who refrained from their usual intake of fermented foods for two weeks. The researchers found diminished immune responses indicating that the deprivation of fermented foods might reduce the capacity to respond to the latest common cold or influenza virus floating around.

In very simple terms, fermentation makes good things even better. Examples abound, including the ability of fermented citrus juice (vs. unfermented) to decrease allergy and improve quality of life in adult volunteers. When researchers examined a traditional herbal medicine with known anti-inflammatory activity, they found that it lowered blood markers of inflammation by 20 percent more when administered in the fermented vs. unfermented form. When microbial fermentation is applied to green tea, it can become more effective at helping to control blood sugar. When soy germ is fermented its antioxidant activity is over 200 times higher. When fermentation is applied to a blend of herbs with already well-known anti-inflammatory activity, these effects are magnified. Recently, it was also discovered that fermentation may amplify the active properties of echinacea in support of the immune system. There will be no putting this genie back in the bottle; the landscape of natural health products is changing, for the better.

There has also been an increased appreciation for the ability of fermentation to enhance plant-derived protein quality and ease its digestion. Legumes and other protein-rich plants contain antinutrients (e.g., phytates and lignans). You can imagine these antinutrients as a mesh that prevents proper digestion and absorption of amino acids. Some techniques, including soaking and heat application, can help. The microbes used in the fermentation process are a highly effective means to break down antinutrients that would ordinarily block absorption of essential amino acids. By doing so, they also address the bloating and discomfort commonly associated with antinutrients in protein sources.

Food scientists can easily demonstrate how fermentation can improve nutritional quality by examining the protein efficiency ratio (PER). This test is described as "real world" because it determines how a protein source can directly promote growth and development in living animals. It separates out theory from reality, meaning that a plant protein might look good on paper with bountiful amino acids when it is raw but can it lead to growth? Sesame provides a good example because it has lots of different amino acids. Fermentation of sesame increases its real-world PER by 44 percent over raw. This simple example, one among many, makes it plain to see how fermentation can amplify foundational nutrition. It also dispels quaint notions that raw is better.

It should also be noted that the large scale ultraprocessing of foods isn't simply about loss of nutrients, at least not for bread. Traditional breadmaking—the sort involving yeasts and a variety of microbes—dates back thousands of years. The careful and time-consuming techniques spread throughout the world, and bread became a truly global food. But in a relatively short period of time, production techniques have changed; not only are there added preservatives and emulsifiers, the actual preparation is far more rapid. The techniques used seem to effect the microbiome.

In a recent study published in the *Journal of Functional Foods* (2017), researchers from the Universidad de Zaragoza, Spain, fed animals one of two different breads—one manufactured by typical industrial techniques (using only a single strain of yeast) and the other traditional bread (made with seven strains of microbes and five different grains). The industrial bread was also made 12 times faster than the traditional bread. The animals fed the industrial bread showed gut microbial dysbiosis, reduced microbial diversity and elevated signs of inflammation in the blood. Separate research looking at sourdough breadmaking also found that manufacturing which used slower, more traditional techniques (with diverse microbes used for fermentation), resulted in less gas production by gut microbes and better growth of friendly *Bifidobacterium* in the gut.

Fermentation transforms foods and natural products. Benefits include:

- Increased intestinal microbial diversity
- Increased uptake of antioxidants into the bloodstream
- Increased uptake of amino acids into the bloodstream
- Production of newly-formed phytonutrients with strong biological activity
- Protection of the intestinal lining

To summarize, fermentation is both an art and a science. Cutting-edge science is validating the collective knowledge of our ancestors. Beyond fermented foods, scientists are now making innovative artisanal products a reality for consumers. These nourishing and therapeutic natural health products hold so much promise in their ability to amplify what is already known to be beneficial. Taking advantage of fermentation and the way it can change the terrain for the better will become much easier and convenient.

Get Us to the Greek

Let Food be Thy Medicine, and Medicine be Thy Food.
— Someone other than Hippocrates

Although there are many quotes attributed to the Greek physician Hippocrates, there is no evidence that he said food was medicine, or that medicine should be food. This fabricated quote started to show up in various books in the 1920s. But the more we learn about the ancient Mediterranean diet—and it's far older than Hippocrates, some 8,000 years older—the more we learn that it is very much medicinal. The early origins of the diet around the Mediterranean basin are some 10,000 years old and evolved over time with agricultural techniques and animal husbandry to include olives and their products, carrots, leeks, celery, cilantro, cabbage, peas, turnips, cucumbers, fruit (apples and plums), citrus, nuts (walnuts and chestnuts), pulses, pork, fish, chicken, eggs, dairy.

What it means to adhere closely to the Mediterranean Diet?

Cereal = no difference

Vegetables = 38% more

Fruits = 59% more

Fish = 56% more

Meat = 30% less

Fat = more monounsaturated, less saturated

Dairy = 36% less (despite perception, the Mediterranean diet is not high in dairy vs. standard diets)

At its core, the Mediterranean diet = more fruits, vegetables, fish and olive oil.

Today, the basics of the Mediterranean diet are an ideal match to the sort of diet that feeds our humicrobe entity. The human is happy, and so are the microbes living on and within. Several studies have shown that moderate to high adherence to the Mediterranean diet is associated with better quality of life and lower risks of allergic diseases, cognitive decline, depressive symptoms, and other emotional problems. Associations are one thing, but interventions with the Mediterranean diet have also shown lower inflammation. Not surprisingly, the total phytonutrient intake is a predictor of the lowered inflammation.

Foundational nutrition represents the essential aspects of food as medicine. Volumes of research have shown that the combination of phytonutrients, fiber, and omega-3 fatty acids in traditional dietary patterns are protective against many chronic medical conditions. In addition, these important constituents have been linked to a healthy mental outlook. When adults transition from highly processed, westernized dietary choices and toward traditional diets (e.g., Mediterranean), immediate improvements in mental focus and other aspects of health may be noticed.

In an era where we should be overcompensating with higher levels of fruits and vegetables, we barely scrape together a 0.5 serving of deep greens per day. As mentioned earlier, voids in fiber and omega-3 essential fatty acid intake can also be described as gross. Since these foods are such an important part of keeping the internal systems of the body working in optimal ways with the assistance of appropriate acid-alkaline balance and our friendly gut microbes, supplementation as a foundational nutrition insurance policy may be an essential consideration.

Finally, with all the attention scientists have paid to microbes and how they can work to our nutritional advantage, it has become clear that we lose our connection with traditionally fermented foods at our peril. The inclusion of fermented foods and fermented natural health products may be one of the most effective ways to not only fill in nutritional voids, but to amplify the inherent value of phytonutrients and fiber. The new science of fermentation may meet with foundational nutrition in very convenient ways.

We're Better Than This

What does it say about where we are at as a society when a clown—that's right, a clown—is responsible for cajoling our kids to eat neatly packaged fast food? You probably feel, as we do, that it's diabolical. Overpaid celebrities gulping soft drinks, taking their bucks off the backs of kids with obesity, drooling over burgers that boast over 1200 calories. Hawking pseudofoods left and right on every available medium. Executive charlatans keep on placing toys in the bag to lure the kids in. When was the last time a major animated film studio refused to pair its character toys with fast-food products? We're still waiting for the first time.

In August 2016 Australian researchers pooled all the evidence on food marketing to kids and matched it to the highest standards used within public health research to establish cause. The clown has been hiding behind the claim that there is no proof that marketing influences the dietary choices of our children. The Australian researchers concluded that the evidence was now in place. At the same time, Canadian researchers evaluated 29 different studies that provided data from more than 6,000 children. The Canadians reached a similar conclusion: children exposed to ads for low-quality foods and beverages consume more calories within the next 30-minute period compared to children who were not exposed to such ads.

It's literally a sellout to write the global production and sales of ultraprocessed foods (and their marketing tools) off as "these are part of a balanced diet if taken in moderation." This was the line of defense used by the American Dietetic Association as they informed Americans on the value of soft drinks as a form of fulfilling daily hydration needs in their National Soft Drink Association-funded "Nutrition Fact Sheet" on beverages. The document was subsequently presented to governing bodies as some sort of proof that there is no need to regulate vending machines in schools. Lifestyle impinging on life. Who is holding the marionette strings? There's the "occasional treats" line of defense. But what happens when every day has become one colossal treat?

What does it say about our culture when our external environment looks like a North American stock car race with branded logos in motion on an endless loop. The tote bag given to attendees of a recent American Dietetic Association annual meeting was emblazoned with six primary sponsors. The National Dairy Council was the standout legitimate organization offering whole foods. Two of the six—Coke and Pepsi—are the mega-purveyors of sugar-laden beverages, and a third, GlaxoSmithKline, is one of the largest drug companies. Huh? Are we missing something? Wasn't this a *dietetics* meeting? It hints at the holders of the marionette strings. Our media, politicians, and Hollywood players on the red carpet are donning designer race car fire suits. Chock-full of the logos representing the loss of diversity. Nary an inch to spare for any more logos of ultra-processing. Throw another branded toy into the bag of ultra-processed foods at the drive-thru, because it's all part of a balanced diet on special occasions.

Often in books written about nature and the natural environment there is absolutely no discussion about the marketing forces that bear down on our behavior. No mention of the juggernauts heading toward a town near the reader, which are every bit part of our relationship to the natural environment. Well, as with the proverbial butterfly flapping its wings, and Darwin's clover story, the way we see it, when the smiling clown cajoles a single child to eat fast food and collect their toy, it has impacted all of us. It has impacted our future generations.

We cannot talk of nature-deficit disorder as if it's something that is caused exclusively by not being outdoors playing. Nor, if this disorder is legitimate, can it be prevented or cured solely by going outside. Going outside and keeping eyes fixated upon a pixilated screen, with a package of Twinkies in the backpack, does little to undo the contemporary nature disconnect. How can we remain connected to nature while consuming phthalates, bisphenol, polysorbates 60/80, synthetic emulsifiers, hydrogenated fats, Red #40, Yellow #5, and undisclosed artificial flavors? Is that a Pokemon over there?

AstroFood was a disaster because of what it contained—sugar, emulsifiers, and synthetic chemicals—and because of what it didn't contain: fiber, phytochemicals, omega-3 fats, and the context of nourishment. Looking at some of our dietary realities, it would appear that AstroFood has become the way we are. An entire industry is supporting that reality. But our youth are seeing through the curtains. In our final chapter we will discuss the low tolerance among millennials and contemporary teens for the realization that they are being played. Change is in the air. Optimism. Hope.

Are Phytonutrients Essential?

Inside the walls of a dietetics classroom, phytonutrients aren't considered essential. That is, unlike macronutrients, select fatty acids, and various vitamins and minerals, it would be possible to sustain life and so-called normal physiological functions over short spans without them. But that measure of essentiality is a little silly in practical terms. Unless you are clinging to a face-painted volleyball, humans aren't looking to survive on a deserted island for just a few months. When looking at survival over the life-course—as in resistance to early mortality from chronic disease—it is possible to argue that phytochemicals are indeed essential.

First, researchers have looked at the intake of total phytonutrients, as well as groups of different ones, and found that intake is linked to lower risk of mortality. Which means lower risk of dying early. Which means survival. In simple terms, phytonutrients are emerging as essential in the maintenance of the normal physiological functions that keep us alive over time. Second, we aren't seeking to merely survive. We seek quality of life, and the ability to thrive. We seek vitality. Evidence shows that phytonutrients can influence mood and daily quality of life. Thus, we have to ask, "What is normal?" Surviving on a remote island, or thriving and reaching our potential in the urban canyons of any-metropolis, Canada, USA, or Elsewhere?

Phytonutrients are certainly *nourishing*, the dictionary definition of which means "providing sustenance, support, endurance, and strength." With each passing month, more and more research shows that phytonutrients actually help to define nourishment in this way.

Support for the essentiality of phytonutrients can be seen by examining the multitudes of different functions they perform in the body. Although best known as simply antioxidants, this is a scientifically outdated label. Yes, phytonutrients protect cells from the erosion of oxidative stress, but they do so much more. To name just a few: they turn on enzymes that influence detoxification pathways and turn off enzymes that would otherwise promote inflammation. They regulate steroid hormone metabolism and promote nerve growth factors that support cognitive and mood-regulating nerve cell communications in the brain throughout life.

They also promote immune system surveillance by supporting our front-line defenders called natural killer cells. The latter functions of phytonutrients are part of a growing body of research known as nutritional immunology. But, again, we shouldn't expect a single phytonutrient to be a game-changing nutrient.

Studies in this field, which previously linked fruit, vegetable, and phytonutrient intakes with lowered disease risk are now increasing in their sophistication. Researchers are continuing to document connections between total dietary phytonutrient intake and lowered risk of dying from various chronic diseases. These studies have been replicated many times. However, they are also finding, more specifically, that higher urinary levels of polyphenols, indicating those that have been consumed *and* absorbed, are linked to lower inflammation in the body and a decreased risk of dying early. Survival.

7

On Being a Biophilist: An Injection of Nature Relatedness

> *A simple form of biophilous behavior is that of plants leaning toward the source of sunlight.... Biophilia confirms and promotes life. The biophilous person produces rather than destroys, creates rather than hoards. He is more interested in living things, such as nature and other people, than he is "dead" things such as sports cars and spaceships.*
>
> — Michael McGrath, Scholar, 1969

The term *biophilia* is not new. Countless articles state that the term—which means attraction to life—was coined by Harvard biologist Edward O. Wilson in the 1980s, while others suggest it was coined by the sociologist Erich Fromm in the 1960s. Neither is the case. Biophilia has been in use for more than a hundred years. However, in 19th century medical textbooks the term was more narrowly defined as the drive to preserve one's *own* life. Attraction to life, yes, but the emphasis was on the self.

Animal lover Charles J. Adams had a slightly different take on biophilia, challenging its narrow use. In the late 1800s he banded with naturalists and other animal lovers and formed the Bureau of Biophilism; he defined a biophilist as someone who extended the love of life beyond self-love, and into "love of humanity, out of love of humanity into love of all sentient things. When he has so far evolved, he is a biophilist. As there may be born a poet, so there may be born a biophilist." Thus, to truly be a biophilist, a human had to love all forms of life. To move beyond self-interest and out into the sphere of all living things.

The statement by Adams clearly positions biophilia as innate. However, he went on to say that the innate love of all life had to be "drawn out" and "developed, trained" through experience. In the many years since the doors of the Bureau of Biophilism closed, the idea that humans have an innate, yet experientially driven, love of elements of the natural world has remained. As previously discussed, these concepts are part of the very fabric of life among native North Americans and Inuit, as well as Aboriginal Australians. The need for biophilism does have to be drawn out through experience; we must increase our awareness that opportunity for experience is a prerequisite for the transgenerational hand-off of biophilia from ancestor to offspring. This is top-of-mind among the Inuit and other traditional cultures.

In westernized culture we seek to understand these relationships from the reductionist perspective because we like to put everything under a single-lens microscope. However, the importance of biophilism looms larger than ever. Our nature connectedness is clinging, precariously, to the ledge of urbanization and westernization. Some are trying to throw a lifeline in the form of Pokémon hunts. But modernity isn't really reaching out a hand. We walk past oblivious to diminishing nature relatedness, an important asset, as it sways and loses its health-giving grip high above our greenless, starless, polysorbate 80-filled urban canyons.

It's difficult to know something needs saving if you can't see it. The current generation of Inuit are acutely aware of what is being lost. But what about two generations from now? Will youth recognize what is missing? Will they realize that nature connection is a form of medicine they no longer have access to? In our evidence-based world, we need proof that we are losing something of value in order to be convinced to retain it.

The greenspace images derived from NASA satellites are a wonderful way to patch together residential proximity to natural environments and link them to various aspects of health. However, they tell us little about why everyone living in the vicinity of green space *doesn't* derive the same level of benefit. Why does walking in a forest have a more profound impact on some, and not others? How does connection to nature impact upon mental wellbeing and, critically, our desire to protect disappearing biodiversity?

Yes, we need deep exploration of the tiniest ecosystems in the gut. We need far more detail on the interactions between our nerve cells and the immune system. But we desperately need further understanding of our connectivity to all life. It won't be found in the space out there somewhere beyond the biosphere. Not the stuff of Ancient Aliens or Kepler planets (some 3,700 light years away) that are purported to be Earth-like. We are talking about connectivity to life inside the delicate envelope of life. Right here. Thankfully, dedicated scientists like our good friend Elizabeth (Lisa) Nisbet are doing just that. Asking questions that help determine whether or not someone is a biophilist. And, if so, why might that be important in the 21st century dysbiosphere?

Personal Connection

Dr. Nisbet is a brilliant woman who has devoted her career to understanding the deeper connections between humans and nature, and how those relationships support health and wellbeing. She and her colleagues have developed useful scales like personality tests that would make for decent application forms to decide whether or not you would get your platinum membership card to the Biophilic Justice League. There are three scales that are often used by researchers when they try to get a handle on the degree to which an individual is connected to the natural world: the Nature Connectivity Scale, the Connectedness to Nature Scale, and Dr. Nisbet's Nature Relatedness Scale.

These paper-and-pencil tests ask respondents to rate their answers to statements such as, "I enjoy digging in the earth and getting dirt on my hands"; "my relationship to nature is an important part of who I am"; or, "I feel very connected to all living things and the earth." The nature-related statements are peppered among other questions that are oriented toward an individual's feelings regarding other aspects of the environment, like cities. They tap into personal fascination with, interest in, and desire for nature contact. But they are more than simple measurements of a "love of nature" or an appreciation for the aspects of nature that are aesthetically pleasing. The scales aim to capture your deeper awareness and understanding of the natural world. It's more than just a love of the Grand Canyon or giant pandas.

The Nature Relatedness Scale and the others like it (for convenience we will use the term nature relatedness) have proven to be very useful instruments. Convincing research now shows that high scores on nature relatedness scales is associated with high levels of overall psychological wellbeing, and lower levels of anxiety. High scores on nature relatedness are also associated with vitality and the ability to experience meaningfulness in life. These type of positive findings have been noted in various international studies, which suggests that nature relatedness has benefits that transcend cultural borders. It is also found broadly in various age ranges. One doesn't have to be a lumberjack—office workers can also score high. It's not about the constraints of the 9-to-5, per se.

We have previously mentioned the importance of vitality in daily life. That it is something you can't buy in a can of energy drink, no matter how many fluid ounces it might contain. Vitality is zest for life, doing things full-heartedly, as opposed to half-heartedly. It means feeling alive! Therefore, nature relatedness is a discussion about zest for life. Purpose and meaning in life are also worth calling out as well.

Having a sense of purpose is intertwined with meaning in life. Meaning is a psychological asset that generally refers to the extent to which an individual perceives essential aspects of their lives, particularly experiences such as time spent in nature and the feelings of love, as having significance and purpose in one's existence. That human existence is part of a larger jigsaw puzzle that transcends our fleeting moments on Earth. An individual with a keen sense of purpose is aware that these experiences and feelings are part and parcel of the pursuit and attainment of worthwhile goals, as well as a sense of fulfillment.

From the perspective of health, meaning and purpose in life are associated with better sleep, better overall mental health, lower risk of chronic disease, and increased longevity. In day-to-day living, higher meaning in life equates to quality of life. It has been shown to act as a buffer against stress and to provide a means to help cope with life stress as it happens. These associations are supported biologically, which means, as you might imagine by now, that purpose in life is linked to lower inflammatory cytokine production. Having a sense of purpose is also associated with healthier lifestyle habits such as exercise and engaging in preventive medical checkups.

Since nature relatedness is associated with greater meaning and purpose in life, we feel compelled to underscore that meaning is distinct from positive mood and happiness. Certainly, they can overlap. However, we emphasize that during times of stress, when mood takes a hit, meaning in life can compensate for lowered mood. Researchers have found that resilient individuals lean on their higher levels of meaning in life when levels of happiness and positive emotions aren't where they should be. The reverse is also true. Positive emotions bolster meaning in life when it takes a hit.

Spanning the Globe, Empathically

As mentioned, nature relatedness isn't confined to one particular group of people. Its potentiality is shared by all of us throughout the life course, regardless of occupation. However, what you actually do during the 9-to-5 may be determined by your early-life nature relatedness. For example, in a recent study at a major US university, students in the College of Agriculture and Natural Resources scored higher in nature relatedness than did students in the Colleges of Business Administration and Journalism.

It is also noteworthy that students scoring high on nature relatedness have higher levels of empathy. Recall our earlier discussions on declining empathy and the rise in narcissism in the US and elsewhere. Shortly, we will further expand on the ways in which nature relatedness can ripple outward to make for a better world. Empathy and purpose in life are a core part of the link between nature relatedness and prosocial attitudes and actions. Does empathy foster nature relatedness, or vice versa? At this point we don't know. They likely feed one another.

Extinction of Experience

Nature relatedness is really about one's frame of mind and day-to-day perspective toward the biosphere or, as Dr. Nisbet calls it, a biospheric attitude. Assuming for a moment that Charles Adams was right in that we can all be born to be biophilists, and that this needs to be drawn out through experience, it would stand to reason that nature relatedness is really an assessement of opportunity. That is, opportunity to experience the development of love for all life. Opportunity to nurture the innate capacity to love. Opportunity to cultivate the innate draw toward life within the biosphere. How can one evolve to become a biophilist when the odds of experiencing deep connections to nature are minimized by modernity?

In support of the theory that one can evolve to become a biophilist, research shows that experience in nature can lift scores on nature relatedness scales. Increased connection to nature has been observed as a consequence of relatively short bouts of walking in nature, as well as longer wilderness experiences. Forests seem to be particularly effective at provoking nature relatedness and an empathy toward nature.

Working in the opposite direction, nature relatedness itself encourages contact with nature. For example, higher scores on nature relatedness translate into more frequent visits to local green space. Moreover, nature relatedness also predicts meeting physical activity guidelines within green space. Thus, opportunity to develop nature relatedness through equitable access to natural environments seems essential. Australian research shows that among city dwellers, living in areas with greater tree canopy was associated with higher scores on nature relatedness. As the researchers from the University of Queensland concluded: "Maintaining the availability of nature close to home is a critical step to protect people's experiences of nature and their desire to seek out those experiences."

These studies have multiple ramifications; as discussed, nature relatedness is associated with psychological wellbeing, vitality, and purpose in life. These conclusions are not based on one or two studies, but on a meta-analysis of multiple published studies and several studies that have been published since the major 2014 review of papers. Thus, we should consider nature relatedness to be a strong psychological asset. Nature relatedness can be provoked by experience. Which is exciting because it means that education and experience may pay dividends. It also means that we need to ensure equity of opportunity for its development. Nature relatedness predicts other healthy lifestyle factors, such as visiting local natural environments and being more physically active when engaged with nature.

Mindfulness

The extent to which an individual absorbs the benefits of walking in a forest or within other natural environments may be determined by their awareness of the present moment—their mindfulness. Definitions of mindfulness are more plentiful than pizza shops in New York, with each one claiming to be the best. Mindfulness is often considered a soft science, and as such, it seems that academics make attempts to overcompensate for its perceived fluffiness by developing all sorts of clunky, constrained definitions. We want to avoid that trap.

In general, mindfulness is a mental state that involves purposeful awareness of unfolding experiences as they happen. This includes nonjudgmental observations of emotions, thoughts, and sensations. Meaning that the mindful person is an independent observer of the flow of consciousness. It involves suspending rumination on past events and future planning, and disengaging from strong attachments to beliefs and baggage that might otherwise dictate emotional responses.

Even these descriptions of mindfulness sound complex. In simple terms, if an individual walked into a forest alongside a creek and a waterfall or two, from our perspective mindfulness would mean entering that environment from a childlike vantage. Entering the natural environment as if it were the first time it had ever been experienced. The first time seeing the deep bark on trees and the water tumbling over a ledge into a crystal pool 40 feet below. The entire in-the-moment experience—its sights, sounds, smells—absorbed with a childlike wonder.

Despite its deep roots within complementary medicine, meditation, and traditional Asian philosophies, mindfulness—and training to foster mindfulness—are proving themselves to be as evidence-based within the westernized biomedical framework as drugs and various medical devices. Strong evidence supports mindfulness as a means to curb anxiety, depressive symptoms, and pain. Just like purpose in life, mindfulness is a psychological asset linked with biological benefits. Remember the telomeres, those shoelace cap-like ends protecting our DNA? Mindfulness is associated with lower levels of the stress hormone cortisol, inflammatory cytokines (the ones that can compromise mood), and a higher production of the very chemical that protects the telomeres against premature shortening.

Although mindfulness can be cultivated through experience and education, it shouldn't really be considered a technique per se, but rather as a way of being. A lifestyle in and of itself. In addition to psychological wellbeing, nature relatedness has also been associated with the trait of mindfulness. This might be a clue as to why being nature related is so good for us. That is, living life in a mindful way, and more specifically, spending time in nature in a mindful way, is likely to be the bridge between gaining only some health benefits afforded by nature, and receiving the true abundance of nature's bounty. Living in close proximity to biodiversity will get you only so far—it's the purposeful engagement with biodiversity that gets you to nature's health-promoting riches.

Much like nature relatedness, mindfulness can be cultivated through education and experience. The practice of mindfulness meditation—a specific focus on breathing for a period of time— may enhance both mindfulness in daily life and, if practiced while in nature, promote nature relatedness. For example, during a three-day nature excursion, half of a group of young adults were randomly assigned to a 15-minute mindfulness meditation session on two of the mornings. This was really simple stuff. Participants were asked to sit comfortably and focus their attention on their breathing so as to become aware of their experience in the present moment. They were told that if their mind wandered, they should gently bring it back to their breath. After 15 minutes of meditation, participants were instructed that they could return to mindful breathing at many points throughout the day on their own. Before and after the trip, all participants, including those who did not engage in the morning mindfulness sessions, were scored on their nature relatedness. The mindfulness meditation group reported greater increases in self-nature connection. They also had more spontaneous recollections of nature experiences after the trip concluded.

The key seems to be awareness. Both groups on the three-day nature excursion increased their scores on nature relatedness, but it was the mindfulness meditation group that witnessed the greatest change. Several other studies support the notion that awareness increases benefits. Subjects in one recent study were shown streetscapes with or without trees and had improved performance on attentional tests when trees were present, regardless of any provoked awareness of the trees. However, subjects fared even better on cognitive tests when researchers manipulated a higher awareness of plants within the streetscapes.

In another recent study, adult subjects walking in nature were asked to bring mindfulness to their stroll. The results showed that as they walked the unpaved trails along the banks of the river Rhine, the simultaneous practice of mindfulness boosted positive emotions. The emotional lift, in turn, boosted further mindfulness. The researchers refer to this as the upward spiral—the benefits of walking in nature are amplified in a synergistic way by being paired with mindfulness. Again, this wasn't a complicated task. The participants were reminded to pay attention to sights, sounds, smells, and bodily sensations as they were encountered along the walk. Meditation in motion.

It is worth noting that the practice of either mindfulness meditation or loving-kindness meditation can itself improve connectedness to other human beings and enhanced nature relatedness. Loving-kindness meditation is slightly different and involves the repetition of a series of mantras during the meditative breathing; examples include "may you be happy" or "may you be free from suffering" and the exercise may also include the visualization of light emanating from the heart. The intent is to send loving thoughts and imagery outward toward other living beings.

This matters to microbes as well. As we discussed earlier, there are plenty of studies showing that chronic stress compromises the mobile mind and your microbiome. Thus, the wealth of studies on mindfulness meditation and loving-kindness meditation as a path to stress reduction and resiliency are of relevance to this conversation. Dr. Kirsten Tillisch of the University of California, Los Angeles, caused quite a stir when she and her colleagues used fMRI scans to show that a fermented beverage with live microbes could influence the brain activity in a potentially beneficial way. Now she is studying the effects of mindfulness practice on the microbiome.

Furthermore, mindfulness and meaning in life are fuel cells for one another. Greater mindfulness equates to higher scores on meaning in life and vice versa. When researchers break it down, it appears that increased self-awareness may explain this positive relationship. Nature experience provides plenty of opportunities to foster both of these very valuable assets. Like examining the fine detail of a multicolored autumn leaf, with its textures and intricate veins. That is mindfulness, and awareness of its place in the circle of life helps to provide a connection. And meaning. Recall Thomson's forewarning: "There would be less psychopathology of everyday life if we kept up our acquaintance [with nature]." He was referring to acquaintance as practical, not textbook knowledge. He was referring to mindful contact. If we could only increase our cognizant acquaintance with nature, we would be so much better off for it.

Concern for the Environment

Actually, we can up the ante on our claim that humans would be better off by maintaining conscious awareness of nature. The entire Earth would be better off if there were more biophilists. One of the major side effects of taking biophilist pills, if there were such things, would be heightened concern for the total biosphere. We know this because higher scores on nature relatedness equates to greater concern for the environment. Even the attributes associated with being nature-related—mindfulness and meaning in life—are associated with sustainable and environmentally friendly attitudes.

Scientists generally define pro-environmental attitudes as those associated with consideration (attitudes) of lifestyle choices and behaviors as actually taking actions that have favorable impacts upon living ecosystems. If not entirely favorable, at the very least, pro-environmental choices should only minimally burden ecosystems or contribute toward climate change and biodiversity losses. Examples include reducing, reusing and recycling, decreasing use of motorized transport, cutting back meat intake, and choosing environmentally labeled products. In short, they are actions that are *more* sustainable for the biosphere.

Researchers have shown that nature videos (vs. urban built environment videos or a separate video of various geometric shapes) can influence environmental attitudes. In particular, nature scenes increase the intent to act in environmentally responsible ways. In this case the researchers set up a situation where subjects were responsible for managing a lab version of fishing where the ocean (just like in the real world) isn't a bottomless pit that can be fished to oblivion. Nature scenes increased sustainable fishing and helped determine whether or not the fish stocks collapsed.

Greater willingness to engage in environmentally sustainable behaviors was independent of elevations in mood. However, the effect was driven by nature because the group in the urban built environment and the geometric shapes group were indistinguishable from one another and did not see the increased sustainability actions of the nature group. Thus, we cannot explain away all the benefits of nature simply because it can lift our mood. There are deeper aspects to the story.

Importantly, the assets of nature relatedness, mindfulness, and meaning in life seem to help transform good environmental intentions into actual pro-environmental behaviors. The good news here is that these assets can all be cultivated. Nature relatedness appears to be strongly influenced by early life experiences. Children who frequently experience nature are more likely to develop emotional affinity with the natural world and subsequent support for protecting biodiversity. This appears to be especially so for urban children. Perhaps childhood is the best time to foster a deep nature relatedness. However, research also shows that it is never too late. Even brief experiences with nature can prod nature relatedness and a subsequent lean toward environmental concern in adults. The Earth can use all the help it can get.

Education matters as well. Later on we will discuss the oft-used acronym of STEM—science, technology, engineering, math—from a somewhat critical perspective. Spoiler: we don't like the term as used. But our message is still the need for *more* science education, not less. "Science" is a broad term. What kind of science does the "S" in STEM stand for? That will be our critical question. For now, we simply underscore that new evidence shows direct links between an individual's science knowledge (as assessed by self-perception, not objective exams) and pro-environmental behavior.

That makes perfect sense. Biology and ecology would seem particularly important subjects for bridging both nature relatedness and environmentally friendly behaviors. In fact, researchers from South Korea have shown that science education oriented toward the environment can have a significant impact on increasing nature relatedness and transforming biophobia—in this case the common fear of bees near one's home—into biophilia and an appreciation of the need for their protection. Creative arts can also increase nature relatedness in children; however, a new study in *Environmental Education Research* shows that a virtual hike is ineffective in boosting nature connectivity.

Nature, Empathy, Society

In addition to pro-environmental behavior, the attributes of being nature-related appear to extend to prosocial actions as well. There are now several studies demonstrating that experience in nature appears to foster human interconnectedness. Of course nature environments, especially parks and community gardens, can act as hubs for the development of social capital.

However, just like the pro-environmental research, there seems to be deeper ways in which nature resonates to make us more kind. More thoughtful. More altruistic. For example, viewing images of natural environments (vs. built environment) provokes intrinsic aspirations. These are drives toward personal growth, depth of meaning within relationships, and inclinations toward community value. Devaluing extrinsic aspirations was also associated with nature scenes. Extrinsic drives are those toward the accumulation of wealth, personal image, status, and fame. In a tricky portion of the study, researchers set four plants into the room where subjects answered aspiration questions. They found the same results as when subjects viewed images. Higher value was placed on intrinsic aspirations when plants were in the room vs. the same room without plants. When plants were in the room, subjects also displayed increased generosity in an economic decision-making task.

In field experiments in France, researchers pretended to lose a glove and found that the return rate was higher after (vs. before) random subjects had walked through a natural environment inside a town centre. The same research team found that when the individual dropping a lost item was holding a bunch of flowers, they were more likely to be alerted to the loss vs. holding either a T-shirt or nothing in their hands. When researchers induce awe by showing study volunteers scenes of extraordinary nature, the typical response is to feel an increased oneness to other human beings and a more pronounced willingness to do volunteer activities.

Returning to the topic of empathy and its strong link to nature relatedness, it is worth discussing the notion that empathy from the biophilist perspective is a complete merging of the self with all life. As you will remember from earlier discussions, empathy is the ability to understand or make accurate inferences based on the experiences of another; the combined cognitive and emotional aspects of empathy allow one to take the perspective of another and to experience some of their emotions in a vicarious way.

At the human level, empathic concern produces a oneness, the perspective of "we" rather than "me" and the out-group moniker "they"—a concern for others that is beyond self-interest. This perspective is in line with what Charles Adams was saying about the evolution of becoming a biophilist, except that the "we" isn't just two or more human beings. The empathic concern is a collective where more than two become one...trillions become one. We are all one.

Low empathy is associated with higher dogmatism—clinging to slim evidence, or no evidence at all, and disregarding the thoughts and opinions of others. We have previously cited examples of the cruelty of dogmatism in clinical settings. The disgusting arrogance of mid-century psychosomatics and psychoanalytics; the countless cases where those who were suffering with physical ailments being dismissed offhandedly by clinicians so strong in their convictions.

Diet and acne is a perfect example of dogma. From the 1970s to the mid-2000s, patients with acne were informed on the website of the American Academy of Dermatology that "extensive studies have shown no link between diet and acne." Which would be fine, except there were no extensive studies. In fact, there weren't any credible scientific studies at all. Just two 30-year-old, highly flawed observations. To refer to them as "studies" is a gross overstatement; they wouldn't see the light of day in peer-reviewed research today. But on and on it went. More and more strongly held the convictions, more and more textbooks exclaiming the "extensive studies." Anyone that dared mention diet and acne being related met with vitriolic opposition by those who would score high on dogma.

From the ADA website (since removed, but your authors are pretty decent keepers of historic records): "Myth #2: Acne is caused by diet. Extensive scientific studies have not found a connection between diet and acne. In other words, food does not cause acne. Not chocolate. Not French fries. Not pizza. Nonetheless, some people insist that certain foods affect their acne."

Note the blame... "some people insist." Yes, some people are just so annoying, especially as they insist upon some empathy and trust in their reality. The actual myth was the one propagated by the American Dermatological Association which suggested there *were* extensive studies. Unless "extensive" means a single chocolate bar study that simply swapped out saturated fat and replaced it with trans fat in the placebo bar. Today, the preponderance of available evidence shows that diet is very much connected with acne. We can't say that it causes acne, and undoubtedly there are strong genetic influences, but the dogma of nonexistent "extensive studies" has been crushed with a wrecking ball. Quality research has shown that diets matters.

We aren't picking on the profession of dermatology. It's a good lesson, a parable really, for all in medicine and for society at large. An empathic drought leads to doubling down on narrow-minded, Earth-is-flat perspectives. The combination of low empathy and high dogmatism is both a clinical and societal poison. It's the men in bowties telling a patient with an inflammatory condition of the bowel that it's "all in her head."

To further highlight how nature relatedness and empathy can ripple though society, we can consider some more examples of how nature can influence prosocial behavior. First, a recent study has examined the effect of nature's ability to soften the domino effect of ostracism. The feeling of belonging is a very important aspect of personal wellbeing. Being ostracized—made to feel excluded, marginalized, or ignored—can cut to the bone. Worse still, many studies show ostracism results in the victim not only experiencing depression, but also increasing aggressive behavior. Which doesn't tend to help win friends.

With this background, researchers conducted a study in which 86 adult participants were evaluated for ostracism levels and also subjected to the feeling of being ostracized. The participants subsequently viewed ten different nature scenes or ten different urban scenes. After viewing the urban scenes, ostracism was associated with a greater willingness to subject a person to painful ice-cold water, or to add more spices to the meal of someone that was known to dislike spicy food. Nature scenes diffused the aggressive tendencies.

Emotional Intelligence

Emotional intelligence is yet another important, yet often overlooked attribute in the context of nature relatedness and a better world. Emotional intelligence is a form of "social smarts" that includes the ability to discriminate between negative and positive emotions. By contemplating emotions, recognizing them in the self and others, and managing thoughts and actions, a more effective path to problem solving can be realized. It's more than just a way to win friends; the benefits of emotional intelligence can be profound.

Younger children high in emotional intelligence are more likely to be prosocial and less likely to engage in bullying behavior. It can predict academic success over time. Remarkably, higher emotional intelligence in math educators has been linked with greater achievement among their young students. Maternal emotional intelligence has been shown to be predictive of offspring emotional intelligence and adaptive behavior in children under stressful conditions. Encouragingly, emotional intelligence is malleable and programs geared toward fostering this attribute can improve mental health, wellbeing, and empathy.

In keeping with the themes of mindfulness and meaning in life, emotional intelligence has been associated with physical health and wellbeing. Positive emotions are deeply linked with the important asset of emotional intelligence; in teens, higher emotional intelligence is associated with happiness and resiliency against depression. As with all the other research on positive emotions and positive emotional attributes, there are interconnections between emotional intelligence and a healthy lifestyle. On the other hand, factors such as sleep deprivation—hallmark of an unhealthy lifestyle—can decrease emotional intelligence.

Given these connections between lifestyle and emotional intelligence, and that it can act as a stress buffer, you probably won't be surprised to learn that emotional intelligence works its way toward the cells in your immune system. The telomeres, those DNA "medical records" on your immune cells, are longer in those scoring higher in emotional intelligence. Recall, longer telomeres are associated with a longer, healthier life span. It indicates that your immune cells haven't been overworked.

Since empathy is a core feature of emotional intelligence—and empathy itself has been linked to life satisfaction, wellbeing, rich social networks, healthy relationships and workplace performance, accommodative behavior, and prosocial activity—empathic perspective-taking surely warrants a nature relatedness perspective. The developmental origins of empathy have been noted in early infancy before the emergence of verbal abilities; perspective taking expands through childhood development and adolescence. Prospective research that follows subjects over time indicates that higher levels of early life empathy predict social competencies. In youth, higher empathic concern is linked to greater connectedness to nature.

Researchers have shown that the ability to take the point of view of another person, and perceiving the natural world as important to health and wellbeing, are intertwined. There are also indications that when emotional intelligence is heightened through training and awareness, meaning in life also goes up. All this infers that connections to nature and interactions between mindfulness, empathy, meaning in life, and emotional intelligence are part of the same conversation. Each asset offers a windfall of opportunity for success in life. Yet, how often are these topics discussed in schools?

Getting a Sunny Forecast

Imagine waking up to your weather forecast and being informed it would be overcast, foggy, with multiple showers, and only a ten percent chance of sun. You are advised to bring an umbrella. But when you go outside it's a gorgeous day, sunny, light breeze with low humidity—and it stays that way all day. That would be a forecasting error. It turns out, humans make a lot of forecasting errors concerning nature, or more specifically, how good nature might make us feel.

Dr. Elizabeth Nisbet's research examines the extent to which contemporary adults value, or forecast, the psychological benefits of nearby nature. In a clever experiment, she and her colleagues recruited a sizeable portion of the student body of Carleton University in Ottawa, Canada, to engage in what they thought was a study of "personality and impressions of the campus area." An interconnecting series of tunnels connects the buildings on the Carleton campus; above ground, the campus is also home to a gorgeous green corridor.

Students took their 17-minute walk to specific locations via differing routes—through above-ground green space or by underground tunnels. Post-walk assessments queried them on their emotional state. Meanwhile, an entirely separate group of students provided a forecast—without actually walking—on how they thought they would feel after walking in the natural environment or through the tunnels. Subsequently, those students that were providing the forecasts took walks in either the natural environment or through the built structures.

The results showed that students did indeed feel more positive emotions after walking in nature vs. tunnels. Critically, the forecasters were way off. They consistently underestimated the positive benefits of walking in nature and overestimated the positive benefits of walking through the built tunnels. What's more, in perhaps one of the most striking parts of the study, the students who walked through the natural environment subsequently reported being more connected to nature. The positive emotions produced by the nature walk appeared to be the match that lit the kerosene of nature relatedness.

As temporary as this uptick in post-walk nature relatedness may be, the reinforcement value would seem obvious. In other words, nature produces positive emotions, the brain becomes more aware that nature is the provider of positive emotions, and since humans value pleasure, and since positive emotions have provided plenty of evolutionary advantages, the value of nature would be ripe for behavioral reinforcement. But lack of awareness of nature's potential benefits is the obstacle.

This was J. Arthur Thomson's big fear: that we wouldn't be aware. That we would undervalue nature's medicinal effects and overinflate the extent to which a walk in antinature—that is, a tunnel—would make up for it. He was worried that a walk would just be considered a walk. Worse still, lack of anticipation of nature's benefits fosters more time indoors, less affiliation with nature, and a diminished concern for the environment.

Interventions to remedy this and maximize the potential benefits would therefore include provoking the awareness of what's good in nature. Recently, scientists Miles Richardson and David Sheffield from the University of Derby, United Kingdom, conducted a study where they asked about 50 adults to note three factual things about the external environment for five days. The other 50 adults were asked to note three good things about nature each day for five consecutive days. The group who were asked to note just a few simple things about the goodness in nature showed significant and sustained increases in nature relatedness compared to the factual group. In turn, the increases in nature relatedness were associated with improvements in emotional wellbeing.

Nature Relatedness 2.0

Although there have been many studies on nature relatedness and mental wellbeing, as well as links between nature relatedness and empathy, mindfulness, and meaning in life, there are many unknowns concerning how nature relatedness might impact other lifestyle factors. The links to environmental concern and prosocial activity are starting to reveal themselves, which is fantastic. However, we still need to learn so much more.

At this point we can only speculate, but it would be justifiable to think that individuals scoring high on nature relatedness would be less tolerant of AstroFood. We just don't see biophilists eating up synthetic food dyes, polysorbates, and industrial food emulsifiers with glee. Moreover, we would suspect that biophilists would embrace minimally processed foods that are closer to nature. Fermented foods, full of life as they are, would likely have appeal. Animal lovers are, of course, biophilists. Among them we see an increasing concern with the disturbing, synthetic chemicals being dumped into pet food. This is a signpost for change. Sometimes intolerance is a good thing. Empathy for our pets is now bumping up against the industrial Big Food complex.

If our speculations concerning intolerance to AstroFood are true, then we would expect nature-related individuals to have a magnified diversity of gut microbes and an intestinal ecosystem slightly more reflective of our ancestors'. Furthermore, if nature relatedness places someone in greater contact with elements of the natural world, then everything discussed in our microbiome chapter should settle itself onto the skin and under the nails of those scoring higher on nature relatedness. Diversity! Thus, we would also expect a very different, and potentially healthier, skin microbiome.

Recently, a team from the School of Environment and Natural Resources at Ohio State University found that urban youth unfamiliar with wilderness often associate wild nature with the words "dirt" and "disgust." In an increasingly sanitized world where we are checking under our germicidal-appliqué bedcovers for microbes, urban youth fearing nature because it is dirty doesn't bode well. Recall our earlier discussions: skin microbes can communicate with your mobile cranium—the immune system—from the outside-in. Body-wide.

At this point researchers are now turning their attention to a deeper understanding of nature relatedness and human physiology. They are beginning to make direct connections between higher nature relatedness and reduced risk of non-communicable diseases. At present we do know that nature connection is associated with mental health and wellbeing. We also know that the assets associated with nature relatedness—mindfulness, meaning in life, and emotional intelligence—are all linked to a lower occurrence of chronic diseases and enhanced longevity. But straighter lines between nature relatedness, the immune system, biological markers of stress, and chronic disease risk still need to be drawn.

NASA's satellites can pick up colors on the Earth's surface but they can't peer into the minds of the people walking around urban areas that happen to display greater pixilated greenness. With all the research we discussed earlier relating living in close proximity to green space and healthy birth outcomes and reduced risk of asthma, we feel there is now an urgent need to thoroughly examine our hypothesis that personal connection to nature is a critical cog in the wheel that transforms merely living close to green areas to actually thriving. Not just in this generation, but in the ones that follow.

Is it far-fetched to think that prescription drug use would be lower in those scoring high on nature relatedness? Might they take less of the chemical drugs, especially antibiotics, that disturb a normal microbiome? Might they take less psychotropic medication? Although not all are as harsh as antimicrobials, there is emerging research indicating that many drug classes, including psychotropics, might disrupt the microbiome. Common over-the-counter and prescription acid-blocking medications are another example.

Being a biophilist would certainly put someone outdoors where they can make contact with nature. Being outdoors gets you vitamin D, so long as you don't overdo the sun exposure and defeat the benefits of some sunshine. Lack of vitamin D may itself be enough to disturb the gut microbiome. But might nature relatedness also be connected to less screen time and less sedentary behavior? It would make sense.

We also ponder the idea that nature relatedness might amplify the benefits of other medical interventions or therapies. It's no longer a stretch to suggest such a thing. Mindfulness enhances the benefits of many medical interventions, so we might expect nature relatedness to do the same.

In one study, researchers investigated the treatment of patients with moderate to severe depression over the course of one month. The only distinction between patients was where psychotherapy took place. Patients were assigned to once-weekly cognitive-behavioral therapy (CBT) in either a hospital setting or a forested arboretum. An additional control group were treated using standard outpatient care in the local community. The overall depressive symptoms were reduced most significantly in the forest group, and the odds of complete remission were up to 30 percent higher than what would be typically observed from medication alone. In addition, after one month the forest CBT group had normalized markers of stress physiology.

At this point there is a pressing need to know if nature relatedness can amplify the benefits of prescription medications or even nutritional agents, probiotics, and aromatic chemicals such as in aromatherapy. Research does show high levels of nature relatedness are not required in order to reap rewards from short-term nature experiences. Which is good, because if being a biophilist was completely innate, there would be no room to grow.

8

Encephalobiotics Turn the Mind

> *It is clear that many characteristics assumed to be inherent in an individual can in reality be determined by the microbial flora of the intestinal tract.*
>
> — René Dubos, 1960

In the eyes of post-mid-20th-century media, the general public, and even among fellow scientists, René Dubos was transformed from a detail-oriented microbiologist to a big picture scientific spokesperson for environmental causes. But his groundbreaking microbiological discoveries are often overlooked. Through the 1960s his work with rodents, including pregnant mice and their offspring, demonstrated the far-reaching effects of intestinal microbes, including on the mind. For Dubos, microbes, especially early exposures to them, were at the heart of what he called "Biological Freudianism." But this descriptive term wasn't in support of Oedipus complexes or the completely unacceptable blaming of "schizophrenogenic" mothers for serious biological mental illness in their sons or daughters. Nor did it refer to the homeopathy of mental health treatment, psychoanalysis. As he described his animal studies to journalist Anne Chisholm in her 1972 book *Philosophers of the Earth*:

> All the mice possess the same genetic constitution, but we are showing that depending on the conditions of their earliest development, you end up with giant mice, or tiny little mice,

> skinny animals or fat animals, with all sorts of different physical characteristics, and even different brain development. That is crucial. It is a demonstration of what I call "Biological Freudianism," to convey my view that all the characteristics of animals and human beings are, if not determined, at least conditioned during the very first period of life.

Dubos demonstrated the links between early life conditioning and later life health outcomes. This included not only disease risk, but also the observable characteristics that represent an interaction between genes and the total environment. Put simply, Dubos was demonstrating that contact with microbes and the manipulation of microbes by environmental experiences and "lifestyle" (insofar as mice can have a lifestyle) could have dramatic effects in offspring. He connected microbes to the degree of sanitization within the environment, nutrition, and the parental grooming of pups. All of these variables were intertwined; they could even influence the characteristics of subsequent generations of pups.

Dubos was demonstrating that the holistic environment matters. While interactions with caregivers—grooming and nurturing behaviors—are important, they are part of a complex environment that gets absorbed by offspring. Sole blame for mental and physical disease cannot be pinned on a mother or father because so many aspects of these interactions are a product of societal pressures and priorities.

Since then, remarkable advances in nutritional psychiatry and radical new ways in which we view microbes have illuminated the half-century-old findings of Dubos. With each passing month, volumes of new research demonstrate that healthy dietary patterns and select microbes can support optimal mental health. And so, the emerging research provides plenty of room for optimism, plenty of room for hope that we can fix the diseases of modernity at points closer to their origins.

In this chapter, we more closely examine specific microbial and nutritional interventions, and the prospect that we might regain health by correcting dysbiosis and manipulating microbes. The microbiome revolution has changed everything. The dawning of truly holistic medicine is upon us because microbial ecology in the intestinal tract, on the skin, and within other anatomical locations is a central hub of unified discussion. The microbiome is influenced by all aspects of lifestyle and, in turn, it might influence all aspects of living. Thus, virtually all aspects of science and medicine are being seen anew. Mental health is being viewed from an entirely different angle. Biological Freudianism lives.

The New Psychotropic

The term psychotropic is derived from its Latin and Greek roots which mean "mind" (*psyche*: soul) and "turning" (*tropikos*: pertaining to a change, or turn). Thus, it means turning the mind. Since the term tends to be reserved for mood-lifting, stabilizing, and sedating drugs, where expectations and the placebo effect run high, the turn in the mind is anticipated to be a good one. Given that westernized nations are literally megadosing on psychotropics—residents of the United Kingdom consume over four times more antidepressants than they did just two decades ago—the collective mind of nations *should* be better off.

But it isn't. At the national level, overprescribing of antidepressants is the equivalent of trying to heal a lesion of the brainstem with a couple of drops of homeopathic Arnica. Expectations are high, but a placebo will only get you so far when the root cause is being masked. To be clear, we are not against antidepressants. They can be extremely helpful, even life-saving, for those in whom they work. But we need alternatives, desperately. Adverse effects are common and treatment outcomes far from good.

Convincing research by Drs. Irving Kirsch and Joanna Moncrieff shows that 89 percent of depressed patients are not receiving a clinically significant benefit from antidepressants per se. The placebo matches the antidepressant in its efficacy to an extremely high—virtually indistinguishable—percentage in most cases. As Dr. Moncrieff points out in a recent article in the journal *World Psychiatry*, the very name "antidepressant" is a misrepresentation of what these drugs might actually do. It's not that the pills are mere sugar, like homeopathy. They can induce physiological changes, but their effects (they might even be called "side effects") are not specific to mood. Sedation, lowered libido, and gastrointestinal side effects are considered signs that the drug is "working." Participants in antidepressant drug trials can easily guess beyond chance whether they were in the "real" drug group or the placebo. Moreover, serious side effects, such as increased aggression, may have been underreported by drug company scientists. Gastrointestinal side effects, including altering the rate at which food passes through certain sections of the intestinal tract, may impact the microbiome. Indeed, new research in *Gut Microbes* suggests that antidepressant use is associated with changes in the gut microbes, including the Bacteroidaceae family of bacteria.

Although the significance of possible antidepressant-induced changes to Bacteroidaceae are unknown, it is worth pointing out that stress and synthetic chemicals like polyethylene glycol have also been shown to alter this family. In a recent study involving university students on the lead-up to major exams, consuming a fermented beverage with live *Lactobacillus casei* strain Shirota maintained gut microbial diversity, but the percentage of Bacteroidaceae was far lower than it was for students consuming the placebo beverage. Moreover, the students drinking the fermented beverage with bacteria also felt less stressed, experienced fewer gastrointestinal complaints, and maintained healthier immune functioning through the exams and in the three weeks following.

Returning to medications, even *if* antidepressants are mildly to moderately effective in mild-to-moderate depression—simply by creating an altered state, the way that alcohol might alter the mental state to quell social anxiety—the root cause often remains. Relapse is high. Put simply, many things are broken in the biosphere that create modern mental health and its disorders, many excesses and deficits in the contemporary ecosystem, but one thing is for sure—humans are not suffering from a physiological lack of Prozac or any other selective serotonin and/or dopamine reuptake inhibitor!

The fresh light of mental health and disease viewed in a clinical ecological context is bursting through the clouds of a 50-year focus on serotonin drugs. We are living at the dawn of a new psychotropic era. Progressive psychotropic interventions need not involve synthetic chemicals. As "new" psychotropics, phytonutrients and probiotic microbes have the potential to manifest their benefits over a course of a lifetime. In particular, the early origins of adult health can be traced right back to how the psyche was—or was not—fueled by nature contact, nutrition, and microbes early in life. So, what can we do to keep the psyche turning toward fulfillment of potential, rotating elegantly on its ideal axis, supported by its natural evolutionary requirements?

Probiotics, Prebiotics: The Back Story

Amidst the modern microbiome revolution, the commercial category of probiotics has sprawled from the back corner of the local health food store, to the shelves of Walmart. This change reflects the commercial barometer for advances in science surrounding the application of microbes for the promotion of health. And there has been a good deal of growing scientific interest. To illustrate, a 2016 study published in the *World Allergy Organization Journal* combed through journals looking for any and all published papers examining probiotics specific to pediatrics. They went back two decades to 1994, finding a 90-fold increase in research publications since then.

More than twice as many North Americans consume yogurt on any given day than they did in 2000, and probiotic supplement consumption has increased by 156 percent in the last decade. However, an inflated probiotic category doesn't necessarily represent science—in many cases the cartful of product claims are put well before the scientifically proven horse. Many products are mere pixie dust, and it is still buyer beware.

Before delving into the commercialized category of probiotics, we would like to hold our big picture vantage for a while longer. It is important to take stock of the very word probiotic and reclaim its original usage. "Probiotic" (pro-bios: life-promotion) has been in English parlance for well over 100 years. Interestingly, the word probiotic was actually used to describe antibiotics in their early days because their use was considered to be life-saving, hence probiotic. This is captured in the 1953 Proceedings of the International Veterinary Congress; the animal growth-promoting effects of antibiotics were described here as potentially ushering in a new "Probiotic Epoch." In hindsight, the overuse of antibiotics has indeed ushered in the "Need for Probiotics Epoch," but more on that shortly.

Contrary to what is often stated in various dogmatic textbooks and pop nutrition books, the term probiotic was not coined by any known scientist or physician. However, in 1943, the Harvard-trained electrical engineer turned philosopher Percy A. Campbell helped to refine its use with more precision. He used the term probiotic to describe the conditions favorable to life over the millennia and the functioning that represents bliss. Antibiotic conditions, stated Campbell, are a threat to survival: "Sooner or later even the very fittest of living things upon the Earth will no longer succeed in surviving. A new order of conditions will have arrived which will be definitely antibiotic, as the conditions of the past billion years have been definitely probiotic. [Antibiotics] condition the opposite of probiotic functioning, [probiotic is] the functioning which eases life along... [the] full-bodied tone of contentment which the organism automatically wears when it comes to the end of a perfect day."

Campbell's elegant words—"full-bodied tone of contentment"—provide an apt description for optimal health. Recall Sir Arthur Thomson's conviction, now shared by the World Health Organization, that health is not simply the absence of disease. Thus, probiotic is a term that equates not only to living. It is beyond mere survival. Probiotic existence is a complete state of vitality. Living, yes, but living with zest.

In 1954, naturopathic medicine expert Ferdinand Vergin took to the pages of the German holistic medical journal *Hippokrates* to provide a fresh take on probiotics (in German, Probiotika). He moved the term more directly in line with beneficial microbes, pushing it closer to the way in which it is used now. Dr. Vergin suggested that antibiotics were taking their toll on the beneficial microbes in the gut, and that culling these important microbes with antibiotics was essentially turning off the flow of their life-giving nectar; hence, commensal microbes in the intestinal tract and the products produced by them should be considered *pro*biotics.

Vergin was not alone. The massive increases in antibiotic use, including in animal husbandry, necessitated a counter to prodeath chemical applications that were causing lots of collateral damage. Moreover, when agriculture industry scientists figured out that they could rapidly fatten animals with low-level antibiotic use in the 1950s, it started raining antibiotics from above the fruited plains—where animals used to roam—and into their confined feedlots. In the 1950s and 60s, antibiotics were literally called rations in the feed of animals, and textbooks were written on the *Nutritional Effects of Antibiotics*.

Few seemed to wonder whether or not the tons of antibiotics being poured into feed and upon the land could be fattening more than the animals used for human food. But the United States Food and Drug Administration (FDA) at least started to consider their effects on antibiotic resistance. Long-time chemical industry scientist Thomas Hughes Jukes tried his best to quell any fear of antibiotic resistance. An anti-holistic medicine crusader, he told us all in the pages of *Advances in Microbiology* (1973) that there was nothing to see. Keep moving along folks. Antibiotics applied to animals pose no risk, said Jukes, only benefit for consumers.

Perhaps feeling bruised by the FDA's unwillingness to take him seriously concerning his claims on the safety of antibiotics, Jukes conjured up nine reasons why they may have been pressured into scrutinizing antibiotics and legislators caving in on the use of possible carcinogens. In his 1973 position paper on antibiotics and his proclamation that they are safe, referring to potential dangers as "alleged," he blamed the lack of confidence on :

- The rejection of the establishment by the counterculture.
- The success of Rachel Carson's *Silent Spring.*
- The success of books similar to *Silent Spring.*
- The fact that scientific advances allow for the identification of trace residues on foods—so for a public that doesn't know better, just knowing that small amounts of chemicals are on foods sounds scary In other words the public can't handle the truth.
- Legislative bans on known carcinogens; to the public this sounds scary, even though from the perspective of chemical scientists, they might be used by industry at levels below which they might be capable of causing cancer.
- Alarmist media.

- High priests of quackery.
- The ignorance of urbanized Americans who don't understand the importance of chemicals used in agriculture.
- "Agri-business" exemplifies capitalism and thus profits are perceived to be placed before consumers.

Immediately following these propositions, in the very same 1973 document used to support his claim of the harmless benefits of antibiotics used in enormous quantities, Jukes summarized and pinned the potential problems with antibiotics on medical clinicians: "I conclude that the responsibility for the spread of [antibiotic] resistance in human clinical medicine rests primarily with those who are using antibiotics in clinical practice, not with the feed industry.... The continued efficacy of antibiotics in animal feed shows clearly that there is no significant resistance problem at the level of the farm."

Jukes then offered up the words of famed 19th century scientist Claude Bernard. He poked at the FDA scientists who disagreed with him, perhaps not realizing the irony of using Bernard's words: "If an idea presents itself to us, we must not reject it simply because it does not agree with the logical deductions of a reigning theory." Yes, indeed. Like the ideas that presented themselves in *Silent Spring*. Like the ideas that antibiotics, pesticides, herbicides, combined with synthetic non-nutritive chemicals added to food might end up wreaking havoc on microbial ecosystems and health. Those ideas were a challenge to the logical deductions of reigning theory. They challenged the reigning theory of Jukes.

At that time, naturopathic doctors, or high priests of quackery, as Jukes called them, were talking about how important yogurt and fermented foods are in the new antibiotic epoch. They were talking about probiotics. But while Jukes was extolling the benefits of antibiotics, a few of his colleagues in the animal feed industry were also talking about what they termed, in uppercase, The Probiotic Concept. Richard Parker, chairman of the Department of Microbiology at Oregon University and vice president for research at NuLabs had been studying an alternate way to develop what they called the ideal intestinal flora in animals. Put simply, they were seeking alternatives to antibiotics.

Dr. Parker and his team performed many studies showing the benefits of applying non-harmful bacteria, including various strains of Lactobacilli, the sort that would be killed off by antibiotics, to animal feed. With consistency, the results showed that adding friendly bacteria improved the growth and health of animals. Parker called his alternative approach the Probiotic Concept and a commercial product, Probios, was born. In 1974 he formally defined probiotics in a way that works well to this day: "organisms and substances which contribute to intestinal microbial balance."

However, Parker's Probiotic Concept didn't begin and end with animals and intestinal tracts. It was really an ecosystems concept, one that could be applied to any situation in which microbial balance, and hence the promotion of life, was in jeopardy. For example, he considered coating seeds with beneficial bacteria to make them more resilient to harsh conditions. This is, in many ways, what beneficial microbes might do for us humans, as we try to increase our own resilience in the hostile physical and mental environment of the dysbiosphere.

Notwithstanding the focus on the intestinal tract—as important as that might be—Parker's definition of probiotics in the microbial context is ideal because it does not confine probiotics to live microorganisms, and does not exclude other substances and agents that might contribute to microbial balance. Many have attempted to modify and massage the definition of probiotics, one might even say co-opt the term and pigeonhole it into specific, commercially profitable microbes. When it comes to probiotic definitions, it's always good to ask who is doing the defining. Are these players holding patents? Are they working for an industry that might favor the word "live"? Are there vested interests?

Prebiotics also suffer from convoluted definitions. From our perspective they are uncomplicated; we view them as any dietary constituents that selectively promote the growth of friendly bacteria. That is, the microbes that confer benefit to you. Which is pretty much most bacteria. Since there are countless dietary constituents that promote a healthy microbiome, from apple fiber to omega-3 fatty acids, countless are the food components that can be considered prebiotic agents. The entire Mediterranean diet is a prebiotic special agent! Hiding in plain sight.

Commercial Probiotics Remain Undefined

In 2002, a small group of five scientists and two representatives of the World Health Organization settled on their own definition of probiotics as "live microorganisms administered in adequate amounts that confer a beneficial health effect on the host." This definition has stood the test of time, and although we agree with it, many others don't subscribe to this interpretation. Although this definition is often presented as an official and ironclad World Health Organization affirmation, it is not. It was merely the opinion of the group of seven. The document states quite clearly: "The opinions expressed in this report are those of the participants of the Working Group and do not imply any opinion on the part of FAO and WHO."

The problem with this oft-repeated opinion/definition is that beneficial bacteria in the promotion of life (ie. pro-bios), commercially available or otherwise, need not be living. That much we know. Heat-inactivated (nonliving, nonviable) strains of beneficial bacteria have been shown to provide a number of health benefits in humans, including the promotion of healthy immune status and even mental health.

For example, studies of heat-killed (HK) *Lactobacillus brevis* strain SBC8803 have shown it to protect liver cells, influence the release of the neurotransmitter serotonin, decrease oxidative stress, and improve sleep. Many studies on HK *Lactobacillus plantarum* strains have also been published, including one recent animal study showing that a strain isolated from a coastal tree can inhibit *Salmonella* colonization in the intestinal tract. This study highlights that bacteria for the promotion of life are all around us, and their benefits are part of a collective interaction with nature. Making contact with the coastal tree—being near it, touching it—means making contact with a microbe that might protect you.

This is not to say that live and heat-inactivated microbes equate to one another. Undoubtedly live microbes are much more likely to impact health. However, the need for viability in order to confer *any* benefits is grossly overstated.

Other critics of the narrow definition of probiotics say it presents other problems; it is virtually impossible to determine what the threshold level of "adequate amounts" of bacteria might be—dead or alive—to confer those beneficial health effects. By this definition, we would need to do clinical studies on ever-increasing colony-forming units (cfu = the number of viable microbial cells), one by one, and then determine what level of cfu was the elusive adequate amount. That is, the threshold amount that provides benefit, let's say in acne or anxiety, for *most* people. Simply showing that x billion cfu helps with acne in one study does not rule out the possibility that a slightly smaller level of cfu may be helpful. It may be even more helpful!

The other problem with "adequate amounts" is at the opposite of the minimal-dose end. What about the upper limits of adequate amounts? This is left out of the equation, a void to be filled by a zealous industry. Products with higher and higher cfu are pushed on the public. More must surely be better! If two billion cfu works, why not 200 billion? So suggest the marketers.

But at what point do ever-increasing cfu levels compromise the best interest of gut ecology? Especially if restoring balance is a central goal. If we spent a year consuming 400 billion cfu of some strains of bacteria, even if they look good when served up at 2 billion cfu, could there be untold consequences of the very high cfu? This won't be addressed by the almost exclusively short-term studies on probiotics. Nature provides plenty of examples where some is good, but more is far from good. So-called "antioxidant" vitamins are a prime example.

Drug companies are often required to show dose response studies in order to be added to the FDA drug formulary. These studies examine the safe and effective ranges between the smallest doses with a useful effect to the maximum dose beyond which no further beneficial effect is seen. The so-called "adequate amount" in the probiotic industry definition is well intended, but there is virtually no evidence to support what this might mean from the dose-response formulary perspective. For example, a study might show that 50 billion cfu provided various health benefits, but it could be that 45, or 35, or 15 billion cfu might perform as well under various conditions. Could 100 billion cfu be detrimental?

In sum, many studies would be necessary just to confirm a single health benefit for a particular amount of a single strain of bacteria—the Goldilocks level of enough, but not too little and not too much.

Then, just like proving the essentiality of phytochemicals in human health, we have the issue of time frames. It is very likely that relatively small amounts of cfu consumed via fermented foods and even heat-inactivated probiotics would provide benefit over long periods of time.

Like Dr. Richard Parker, we consider beneficial microbes to mean organisms and agents which contribute to balanced microbial ecosystems and the conditions favorable for optimal health. In this way, the healthy microbes we make contact with, the ones we touch, breathe, consume with fermented foods, whether they be living or not, all collectively make a contribution to health throughout life. These are the probiotics that make up a critical part of *vis medicatrix naturae.*

Probiotics: The Marketplace

As much as we may be concerned with "wild-west" claims associated with probiotics, we also acknowledge the tremendous benefits, some already realized, some only hypothesized, of many commercially available microbial formulations. That is, the specifically selected probiotics sold in bottles, cups, capsules, and so on. Yes, we are both concerned that contracted narrow definitions of beneficial microbes will obscure the fact that they are all around us. However, we are also picky consumers of commercial probiotics ourselves, and we are convinced that many times, especially when travelling, they have conferred health benefits upon us!

There is little doubt that, overall, commercially available probiotics play a role in the promotion of animal, human, and planetary health. If the only benefit were their ability to diminish the use of antibiotics and antibiotic-induced adverse effects, that would be a good enough selling point. Indeed, a Canadian study published in late 2016 showed just how effective they can be in that regard. Based on published meta-analyses (review of all existing, quality studies) showing that probiotics reduce respiratory tract infections, the researchers estimated that probiotics are responsible for the avoidance of 52,000 antibiotic courses. Plenty of animal studies also show that probiotics can reduce reliance upon antibiotics in animal rearing.

But their promise is even more than that. Probiotic products can help to defend against the dysbiotic forces we must confront in the modern environment. They can play a role in helping to recalibrate microbial ecosystems in the intestinal tract, on the skin, and perhaps even in the air we breathe and the soils from which we harvest much of our food.

In fairness to narrowly defined specifications of what constitutes a probiotic in commercial settings, these are well intended efforts to at least *try* and help regulatory bodies and consumers cut through the nonsense in a very crowded marketplace. Label claims are outrageous, with some products presenting themselves as a direct route to eternal youth and beauty, the map that Ponce de Leon was missing. Others provide research citations right on the box, inferring monumental scientific discoveries. But then you look closer and it's a single test tube study. Confusion abounds. If nothing else, it is definitely safe to say that all commercial probiotics are not alike.

Discussions surrounding commercial probiotics involve some relevant nomenclature: genus, species, strain. These words are important because they often help to identify specific, research-based expectations related to products that sit on the shelf. Take, for example, the internationally famous fermented dairy drink Yakult. It contains *Lactobacillus* (genus) *casei* (species) Shirota (strain). Align, a well-known probiotic from Proctor and Gamble, contains *Bifidobacterium* (genus) *infantis* (species) 35624 (strain). Sometimes the strain is named in honor of a scientist, but mostly it's a number.

It is impossible to evaluate the quality of a probiotic if the strains are not identified. Any decent probiotic should have its strains listed on the bottle and identifiable within an Internationally Recognized Culture Collection. This way, consumers know exactly what they are consuming. Identification also helps, at least to some degree, with the ability to bypass the wild west label claims and research what the actual expectations might be. The preponderance of scientific evidence suggests that different strains do different jobs and may play very specific roles in health. Some strains may be helpful for urinary tract infections and others for irritable bowel syndrome.

If you are picking up a Greek yogurt in the dairy case strain identification might be less important, but for pills, chewables, potions, gels, etc. it is essential. Still, some fermented dairy products in the

Probiotic Genus and Species with Potential Health Benefits

- Lactobacillus: acidophilus, rhamnosus, plantarum, casei, bulgaricus, brevis, johnsonii, fermentum, reuteri, gasseri, helveticus
- Bifidobacterium: infantis, lactis, bifidum, longum, breve
- Lactoccocus: lactis
- Enterococcus: durans, faecium
- Saccharomyces: boulardi, cerevisiae
- Streptococcus: thermophilus
- Pediococcus: acidilactici, pentosaceus
- Leuconostoc: mesenteroides
- Bacillus: coagulans, subtilis, cereus

These are the names you are likely to see on a probiotic product. For more information on claims for specific species, see Table 1 in the elegant review by Dr. Sabina Fijan. "Microorganisms with claimed probiotic properties: an overview of recent literature," *International Journal of Environmental Research and Public Health* 2014;11:4745–67. The chart developed by Dr. Fijan is freely available to view and print.

refrigerated display case present themselves as extraordinary probiotic formulations, specially packaged to sit apart from run-of-the-mill yogurts. Here, it helps to know if indeed these refrigerated foods contain well researched microbial strains. An example of this would be *Bifidobacterium animalis* DN 173010 found in Activia (Danone) or *Lactobacillus plantarum* 299v in Goodbelly (NextFoods).

It is helpful to consider strains as the details in the microbiome, and here's why they matter. At a top-level glance, looking only at the phyla/genus and species level, two people might appear to have very similar gut microbiome profiles. But upon closer inspection, they can have broad differences in their strains. And those differences might translate into good health or ill health. Strains do a lot of very personalized heavy lifting in the gut. They vary in their genetic make up, which means they vary in their capabilities, the bioactive peptides and metabolites they make, the ways in which they metabolize phytonutrients, and the interactions they have with microbial neighbors and your immune system. Put simply, strains have consequences.

Post-mid-20th century, probiotic supplements have tended to focus mostly on *Lactobacillus acidophilus*, and to a lesser degree, *Bifidobacterium bifidum*. The products on shelves and those used in research were often single species formulations. Over the last two decades there has been a rapid expansion of species and strains used in probiotics. Recent trends have been to include a wide variety of different strains in formulations as a way to hedge bets toward beneficial outcomes. After all, the gut ecosystem isn't controlled by *Lactobacillus acidophilus* and *Bifidobacterium* alone. There are several hundred species and untold numbers of strains residing in the intestinal version of Gotham. So we would expect multispecies, multistrain probiotics to perform well. Indeed, while there isn't a ton of research where single strain probiotics have been pitted against multistrain, in most cases where they went mano-a-mano—about 75 percent of the time—the multistrain emerges as the champ.

Many commercial probiotics, both single strain and multistrain, are showing themselves to be scientifically and clinically remarkable. For others, the golden dust of pixies has a greater chance of fulfilling the expectations of consumers based on label claims and marketing literature. The current situation must be a source of extreme frustration for industry scientists and ethical companies that spend years researching specific microbial strains. Tremendous amounts of energy are expended attempting to prove or disprove the potential value of various probiotic strains. Too often, well-researched probiotic formulas end up on the shelf right next to a bottle of encapsulated white powder with plenty of claims and very little, if any, research.

Despite the lack of regulatory oversight, things are improving as companies tighten up quality control in response to consumer expectations and education. Clinicians in family medicine, pediatrics, gastroenterology, and even psychiatry are now taking a close look at quality and research in species and strains. Smart clinicians need to know when they are being asked to be marionettes on strings, dancing to the tune of marketing and tissue-paper-thin evidence. They are passing this knowledge on to their patients.

Beyond picking the right species and strains, new encapsulation techniques are allowing for survival of probiotic bacteria for up to 24 months at room temperature or below. In addition, capsules using plant-derived materials are now being used to bypass the harsh conditions of the stomach, protecting the microbial contents for delivery to the intestines. Processing of probiotics now includes the removal of potential allergens and materials on which the beneficial microbes were actually grown. You can think back to high school biology and the Petri dish, and imagine large-scale versions of that growth medium. In addition, blister packs are available, providing an extra perimeter wall around the precious microbial cargo. Moisture is the enemy for most strains of probiotic bacteria in capsules; no matter how science based a particular strain might be, humidity will quickly put a dampener on viability throughout the production process and while on shelf.

Probiotics: Friends *for* Life

The range of medical conditions for which probiotics have been studied runs the gamut from acne and Alzheimer's to schizophrenia and urinary tract infections. While there are plenty of studies that demonstrate little to no benefit of probiotics, the vast majority show at least some health benefits. The best evidence, however, is probably in respiratory tract infections, allergic skin conditions, the prevention of antibiotic-associated diarrhea, and intestinal conditions in general.

Experimental studies in animals have shown them to do everything from helping gastrointestinal detoxification of heavy metals and environmental plastics to improving the behavior of unruly animals. In 2014, researchers from the University of Toronto had a look through the published evidence and concluded, from a general perspective, that probiotics can be recommended for prevention of diseases that are associated with altered intestinal ecology. From our perspective this includes just about every chronic, low-grade-inflammation medical condition we have described in previous chapters! Importantly, they provided reassuring words regarding the broad applications of probiotics: "Because probiotics are generally recognized as safe and can be removed with antimicrobial agents, their use should be considered in patients of all ages."

There are several important considerations regarding the potential health benefits of commercial probiotics. The microbes have the potential to be your best friends for the promotion of life, but for the most part, if you are an adult, that friendship will only last so long as you take the product...for life. Yes, probiotics can act as facilitators, acting as the cables to help jump start a cold battery. In general, however, commercial probiotics mostly confer health benefits in association with consumption. Once you stop being a consumer, especially in the context of ultra-processed food consumption, stress, and antibiotic use, the bling is gone.

The big exception, however, might be early life. There is already evidence in place showing that probiotic administration in infancy can have long-lasting effects on reducing allergic conditions and lowering the risk of behavioral problems much later in childhood. These studies should be followed up with vigor. One of the greatest possibilities for commercial probiotics to be truly life-transformative is their potential to help establish a healthy immune system—thus, mobile intelligence and balanced emotion—in early life. All experiences with microbes that aren't out to harm us can help form a relationship with nature, an outside-to-inside connection that can pay dividends over a lifetime.

Whatever the returns on consumer investment might be post-consumption, they are not likely to be predicated upon the probiotic microbes establishing their own little community within the gastrointestinal tract. Researchers can't find much in the way of remarkable changes in the main phyla (major family groups) after probiotic consumption. Many, if not most, probiotic microbes are likely to be transitory and provide their benefits in ways that don't shift the major ruling classes within the gut.

But even if they are only transitory, it seems very likely that they shift the dinner tables of the smaller groups of resident species and strains, and, as described earlier, the strains residing in the gut can make or break our health. Since the idealized, health-equating gut microbiome profile is as elusive as Sasquatch, and scientists may never truly find out what it once was, or is now, you can only determine the grades of probiotics based on whether or not they work—for you. Commercial probiotics won't transport urbanites back to the Paleolithic era in gut microbial ecology; but if they can help offset modern dysbiotic pressures, and improve quality of life along the way, that is a pretty good investment.

There is so much to celebrate about probiotics in the promotion of human health. We've outlined that wonderful research with the words of expert scientists. Next, we will examine the use of commercial probiotics in mental health. Here, we get a bit more critical.

Unbridled enthusiasm of products by their purveyors and patent holders can be dicey, most especially when we are discussing mental health. It's one thing to expect a probiotic to help with traveler's tummy or combat some of the side effects while on antibiotics, but it's another thing entirely to walk up to a register with a bottle of probiotics and think that they represent a solution to the 21st century mental health crisis. Cajoled by the content of some websites and soothsaying blogs, consumers—especially if they are patients suffering from emotional difficulties—might be excused for having lofty expectations concerning the use of probiotics in psychiatry.

Encephalobiotics Defined

In the autumn of 2016, Japanese researchers published three remarkable studies that made us sit up straight and double down on our appreciation for the confusion surrounding the WHO definition of probiotic, not to mention that of its ill-fitting cousin, the psychobiotic. More on the latter term shortly, but first the new studies: the first involved 118 healthy participants. Over the course of three weeks they consumed a pasteurized milk beverage with or without 10 billion cfu of *Lactobacillus gasseri* CP2305. Post-pasteurization the microbes were obviously nonviable. The results showed normalization of bowel movements, reduced production of potentially harmful stool metabolites, increased numbers of *Bifidobacterium* species in the gut, and increased parasympathetic nervous system activity (the branch associated with relaxation) in the group that consumed the milk plus nonliving *L. gasseri*.

The second study examined the consumption of a citrus juice containing heat-killed *Lactobacillus plantarum* LP0132. The researchers showed that it has beneficial effects on symptoms of perennial allergic rhinitis, having previously shown it to be helpful in atopic dermatitis. The third study showed that heat-killed *Lactobacillus brevis* SBC8803 could improve sleep in healthy adults; this after an animal study had shown the same heat-killed strain added to the diet can help regulate circadian rhythms and promote voluntary wheel running. That basically means the rodents slept well, got off the couch—without cajoling from the proverbial "You ate too much during the Holidays, now set a New Year's Resolution!" marketing campaign—and went to their local gym.

In sum, the first study dismantles the notion that beneficial bacteria need to be living and clearly shows they can even shift the gut ecosystem toward the family group most often linked to a healthy state—that is, *Bifidobacterium*. The demonstration that HK microbes can shift species in the gut is entirely remarkable. Collectively these studies add to others in showing that nonliving bacteria (still very much in the business of pro-bios) can influence the rhythms of health. Although these studies did not involve massive numbers of subjects, and all require confirmatory follow-up, they add to an existing body of research. The finding that HK probiotics can exert an effect on the nervous system was ever so close to our interests.

As we described earlier, among the very first conversations that we had with one another almost 15 years ago in Japan was around the realistic possibility that probiotics could influence mental health. The fact that Japanese researchers went on to show that nonliving Lactobacilli strains could turn up the dial on the parasympathetic nervous system and help sleep made us smile. It validated exactly what we had been thinking all those years ago. It would have been exciting enough if it were a viable, living probiotic, but the fact that these observations were made with a nonviable microbe made it a pure *nature* story. Not necessarily a *product* story. The key message here is that the microbes around us, even if they are nonviable, can influence our own microbial ecosystems, our immune systems, and in the process, our mental wellbeing.

The sheer numbers of studies on heat-killed probiotic microorganisms forces researchers to start splitting the hairs of definitions. The whole thing starts to look silly. "Probiotics can only be living!" exclaims one researcher. Then what of that very active *Lactobacillus plantarum* in my food, beverage, or supplement? Well, it's a *para*biotic, of course! No wait, it's an *immuno*biotic! Not at all, it's a *post*biotic argues another. Throw some prebiotics in and it's a "synbiotic"?? Or wait, is that a "symbiotic"? Cue the George Gershwin music and "Let's Call the Whole Thing Off."

But it gets worse. Probiotics have recently had the prefix "psycho" added. It is seemingly not enough to just let probiotics be probiotics when used for potential benefits in emotional support. So-called psychobiotics were needlessly defined in 2013 as adequate amounts of living microorganisms to be used for the benefit of *patients* with psychiatric *disorders*. The emphasis on patients and disorders is ours. Medically focused, the definition did not use the more generic terms of *individuals* and *mental health*. This is not semantics. Readers should understand that for many, if not most developed countries, any orally consumed products—chemical, natural, or otherwise—that specifically target patients with mental diseases and disorders are considered drugs. Period. Full stop.

For example, in Canada, claims related to depression, panic and acute anxiety, psychosis, as well as symptoms *associated* with these conditions are included in Schedule A of the Food and Drugs Act. They are strictly prohibited for natural products. Put simply, by definition, psychobiotics are not to be found in natural health products, dietary supplements, yogurt, or any over-the-counter remedy. They might, however, be found in prescription drugs one day. Be that as it may, the length of time it takes to do the rigorous clinical studies required to enter the prescription drug formulary, especially to treat classified mental disorders, suggests that psychobiotics as currently defined would be years in the making. What's more, writing in the *Annals of General Psychiatry* (February, 2017), a group of researchers from Queens University, Ontario, Canada, suggested that the term psychobiotics will only add to the stigma of depression. Individuals with depressive symptoms might be more inclined to broach the subject of depression with healthcare providers if offered an opportunity to do so through the door of nutrition or probiotics.

Amid all the confusion and ever-expanding laundry lists of definitions, we decided to coin our own term and define the whole shebang. We didn't convene under the auspices of the World Organization of a Sciencey-Sounding name or anything; we simply decided by our Committee of Two, a sample size similar to the group of seven that is said to have defined probiotics for the WHO. We wanted to formulate a simple term that everyone will surely remember. We came up with *encephalobiotics*. Biotics for your brain—the encephalon! We defined this term recently in the *International Journal of Environmental Research and Public Health* as follows: probiotics, prebiotics, postbiotics, microbes, microbial parts, and/or agents that influence the microbiome for improved cognition, overall mental wellbeing and brain health.

Of course we are jesting about it being a simple term, and it is more likely that it will be easily forgotten. Even we can hardly recall it. Encephalon sounds more like a well-known brand of pots and pans available in Target. But our point is more serious. We really want to provide an alternative for the view concerning *living* microbes as probiotic *drugs* for mental *disorders*. We are emphasizing potential cognitive benefits, stress resilience, generalized emotional health, vitality, and a sunnier mental outlook that might be mediated by a manipulation of the microbiome. Thus, we defined natural products that might operate toward that aim.

We are concerned with ludicrous headlines that suggest patients with depression "forget Prozac" and turn to psychobiotics. When we first posited that lactic acid bacteria (that is, microbes found throughout nature) and probiotics could help with fatigue and depression, writing in the pages of *Medical Hypotheses* over a decade ago, the key word was "help." At best, we considered them to play a role as an adjuvant to established care. As much as we may question the degree to which prescription antidepressants are overprescribed, the magnitude of their benefits in comparison to placebo, and the extent to which their side effects have been minimized, there is a great danger in suggesting they should be forgotten in favor of bacteria. The same is true for prescription antibiotics; there are legitimate concerns in the ways they insinuate themselves through the crevices of our existence, but they can save your life.

Neither one of us hold patents, nor do we have a company that sells natural products, probiotics or otherwise. So we really don't care too much about directing the conversation toward narrow definitions. Our view is that all of nature is connected. It could be dietary fiber, living probiotics, heat-killed probiotics, prebiotics, a yogurt, fermented foods, fish oil, or a blend of phytonutrients—they are all encephalobiotics. If one of these or any other agent impacts the microbiome in a way that works toward a better brain for you, if it turns the mind via the microbial ecosystem, then we have the only definition that matters.

Silent Lucidity: Microbes and Mental Wellbeing

It's taken almost 20 years for us to get from the rekindling of old ideas under new light, and forward to the place where beneficial microbes are once again being taken seriously for mental health. The origins of these ideas were sparked in the early 1900s. But what do we really know about this topic from an intervention perspective? Are commercial probiotic microbes really beneficial for mental health? Are they helping us "see it through," are they protecting us in the night? What is the state of the science, the human evidence, supporting the contention that microbes are silently watching over us? Can all those encouraging rodent studies really translate to humans?

> **I will be watching over you**
> **I am gonna help you see it through**
> **I will protect you in the night**
> **I am smiling next to you, in Silent Lucidity**
> **— Queensryche, "Silent Lucidity," 1990**

As we write, there have been three meta-analyses (examination of the collective pool of controlled studies) directed at the current human studies involving the effect of probiotics on psychological health. Although each of these 2016 reviews critically outlined the shortcomings of the existing research, not the least of which is the generally small size of studies, the overall conclusions of the three separate research groups are very favorable. Dr. Natalie Colson and colleagues from Australia's Griffith University concluded in their meta-analysis that "probiotic consumption may have a positive effect on psychological symptoms of depression, anxiety, and perceived stress in healthy human volunteers." The salient word here is "healthy"; this highlights that the studies so far have generally avoided patients with diagnosed mental disorders.

It's also important not to get too carried away by the favorable conclusions of these primary reviews. It's not like these meta-analyses were taking on the job of evaluating dozens and dozens of clinical studies. Just to keep things in perspective, the reviews only pooled together a half-dozen or so controlled studies examining probiotics and psychological wellbeing in human adults. This means that if a couple of negative studies were to be published in the next little while, the pendulum could quickly swing back the other way. Then the media headlines would be "Forget psychobiotics, get back on Prozac". Put simply, we need more research.

With the appropriate cautionary language in place, we point to some of the studies behind our optimistic view that probiotics can play a role in general wellbeing. The studies that interest us most involve early life. In 2015, researchers from the Department of Pediatrics of the University of Turku, Finland, reported on a study with significant implications to childhood behavior. It was a study they initiated almost two decades earlier. Supported by the emerging research on beneficial microbes and lowered allergy risk in the 1990s—the hygiene hypothesis—they set out to investigate probiotic use during pregnancy and for the first six months of life outside the womb, with the primary aim of determining if it could lower the subsequent allergy risk as the infants grew into childhood and beyond.

It was a randomized double-blind, placebo-controlled study where moms received ten billion cfu of *Lactobacillus rhamnosus* GG or placebo for a month before birth. The probiotic or placebo administration continued for six months after birth—mothers consumed the capsules if they were breast-feeding, and if formula-feeding, capsule contents were given directly to the children. The probiotic did substantially lower the risk of allergic skin disease in children when evaluated at various points in their development. But as the research on other possible benefits of probiotics started to mature—including the once implausible idea of behavioral health—the researchers seized on the opportunity to see how the kids who had been supplemented in early life actually turned out.

It turned out there were some differences in these burgeoning teenagers; those who entered the world having made contact with *Lactobacillus* GG had lower risks of attention deficit hyperactivity disorder and Asperger's syndrome. This was a relatively small study, and the findings need to be substantiated, but it provides suggestive evidence to support one of our important themes in *Secret Life*: there may be critical windows of opportunity in early life, periods of time where training with nonharmful microbes may pay off in spades later. Like the piano lessons at age two that get the kid to Carnegie Hall later.

We are also looking with keen interest at a well designed, placebo-controlled study that is just wrapping up in New Zealand. In a study involving almost 400 women, scientists from the University of Auckland have been examining the effects of *L. rhamnosus* HN001 when consumed at about 15 weeks gestation and onward to about 6 months postpartum. They are determining if the probiotic can influence postnatal mood, including symptoms of depression and anxiety. Given the seriousness of depression during pregnancy and postpartum, the implications of this study for maternal mental health and offspring health could be magnificent. It is also the largest probiotic-mood study yet performed.

As far as studies in adults, the various controlled studies demonstrating psychological benefits from probiotic supplementation have included strains within the following species: *Lactobacillus casei*, *Lactobacillus acidophilus*, *Lactobacillus brevis*, *Lactobacillus gasseri*, *Lactobacillus helviticus*, *Lactobacillus salivarius*, *Bifidobacterium lactis*, *Bifidobacterium bifidum*, *Bifidobacterium longum*, and *Lactococcus lactis*. These are smaller studies, about 50 subjects or less in each trial. Taken as a whole, the results suggest the benefits can manifest as less tension, anxiety and aggression, and a generally happier outlook. Remarkably, a University of Toronto study published in *Nutrients* (2017) showed that the benefits of probiotics extend to reduced food cravings and depressive symptoms in women with obesity. Some, including a recent Irish study evaluating the benefits of *Bifidobacterium longum* 1714 and a Japanese study looking at *Lactobacillus gasseri* CP2305 have reconciled the psychological benefits with improvements in human stress physiology. The latter study, involving medical students in the midst of a stressful period of course work, also showed improvements in sleep quality compared to placebo.

In addition, the administration of 100 milligrams of heat-killed *Lactobacillus gasseri* OLL2809 was shown to improve tension and anxiety in university students compared to placebo. The one-month study also showed that the microbe—yes, it's still a microbe from our perspective—prevents the usual reduction in immune functioning that otherwise occurs after a significant bout of strenuous exercise. Intriguing research published in 2016 also shows that 500 milligrams of dry *Saccharomyces cerevisiae* tablets—in this case a strain of yeast used to make Japanese sake—improves sleep quality vs. placebo when consumed for four days. Previous animal research had shown that Japanese sake yeast improved sleep in animals.

Further support for the effect of probiotics on daily emotional health and wellbeing can be found in the studies that assessed various aspects of general wellbeing and quality of life. While not directly studying mood stress and anxiety per se, quite a few of these studies reported improved quality of life and overall wellbeing in association with the benefits to the primary goals of addressing medical conditions. For example, when probiotics are found to improve skin conditions or various gastrointestinal diseases, it shouldn't be surprising that quality of life improves. Although there hasn't been as much attention to prebiotic fiber, a 2015 study by researchers from the University of Oxford, England, showed that three weeks of consuming a galactooligosaccharide prebiotic lowered cortisol and decreased attentional bias toward negative information. An attentional vigilance toward negative information—threatening or aversive stimuli—compared to positive information characterizes anxiety and depression.

Finally, given the links between mental disorders, obesity, and metabolic syndrome (the cluster of elevated blood sugar, increased blood pressure, abdominal fat accumulation, abnormal cholesterol, and/or triglyceride levels), we have been keeping a close eye on the emerging clinical studies using probiotics to help with weight management. Probiotics have been shown to lower the stress hormone cortisol, body-wide levels of inflammatory cytokines, and oxidative stress, as well as helping with blood sugar regulation. Since these are the metabolic dysfunctions that encourage abdominal fat deposition, it would make sense that probiotics might be helpful. Although it's too early to draw firm conclusions, several new controlled studies leave room for optimism for the use of probiotics and prebiotics in weight management. Reductions in energy intake and waist circumference in these small clinical studies have indeed been linked to lowered intestinal permeability, inflammation, and oxidative stress.

Candida: Old Wine in New Bottles

> **Orthodoxy means not thinking—not needing to think. Orthodoxy is unconsciousness.**
>
> **— George Orwell, 1984**

Not to be overlooked is a pair of studies from Johns Hopkins University, Maryland, that have revived discussions on possible connections between fungal overgrowth and mental health. Specifically, they reported that men with schizophrenia or bipolar depression had higher levels of antibodies to *Candida albicans*, indicating exposure to the fungus. Compared to individuals without such history, women with schizophrenia or bipolar disorder and a history of Candida infection had lower scores on aspects of cognition, especially memory. In addition, the researchers have been publishing their preliminary studies using *Lactobacillus rhamnosus* GG and *Bifidobacterium lactis* BB12; so far the results suggest that this combination might improve leaky gut and some of the symptoms of schizophrenia.

Often gut microbial dysbiosis refers to bacteria, but it can involve fungi as well. The connection between fungal-type dysbiosis and mental disease is not new. This is an old concept that was first described in the 1930s. However by the mid-20th century, just as surely as the topic of beneficial microbes *for* mental health became taboo, even uttering the word Candida in conjunction with any condition other than easily detectable thrush was enough to raise questions of competency to practice medicine. It was like the Bond movies where the eyeball in the framed oil painting was moving, looking around for anyone who dare speak of *Candida albicans* at a gathering of physicians and scientists.

Naturopathic doctors, however, had been long testing for stool *Candida albicans* overgrowth and connecting it to general fatigue, emotional difficulties, and poor cognitive focus. Just prior to our meeting in Japan, a group of general practitioners from Norway did dare to speak in the mainstream journal *Family Practice*. Before describing their study, they reminded readers that about half of all visits to family doctors include a shopping cart full of unexplainable symptoms that don't fit into cookbook diagnostic criteria. Patients get a pat on the head and told it's all in their head. Psychosomatic.

Their study involved 116 adults with vague, chronic symptoms. Fatigue, low-grade anxiety and depression, muscle aches and pains over the body, constipation, abdominal pains, and insomnia. Ah yes, ripe, low-hanging fruit are the sufferers of these symptoms—just waiting to be plucked by bow-tie-wearing experts. Rigid in their thoughts, some clinicians see the world with designer glasses *by Dogma!*; this functionally-fashionable eyewear has been around since the time of Freud. One lens bends the light into sweeping diagnoses of psychosomatic disorders for individuals, and the other lens provides a steady stream of conversion disorder at the population level. Except in this study the safe and well-tolerated anti-fungal nystatin produced marked improvements in these symptoms compared to placebo. What's more, a group of patients that incorporated a low-sugar diet had even more pronounced benefits over drug alone and placebo.

The Norwegian research wasn't intended to suggest that nystatin should be added to drinking water or that fungal-type dysbiosis is at the root of 50 percent of the business of family practice. It was a call to clinicians and scientists to look more deeply. Look to lifestyle and microbiota. Let's explore this, there might be something here. But the response was silence. Advocating for a diet low in sugar to see if it might help unexplained medical symptoms was, after all, such heresy.

It would stay that way until the microbiome revolution turned everything on its head, including the oil painting with the eyehole.

Those behind the once vigilant hunts for the insolent anti-Candida diet proponents now have less to say. Petrified—in both outdated thought and fear of watching their dogma dissipate—they still find channels on sectarian science blogs. But they are on the wrong side of history now. It's getting harder to belittle the minority perspective and dismiss the minority patient. The future is bright, and the much-maligned individuals with food sensitivities and medically unexplained symptoms are marching closer to real solutions. Science is making it so.

Phytonutrients, the critical component of healthy low-sugar plant-based diets, have been shown to have anti-Candida properties. In addition to sugar, the excess soybean oil characteristic of westernized diets has been linked to Candida proliferation in animal research. If Dr. Heiko Santelmann and his colleagues in the Department of Family Medicine in Oslo, Norway, published their nystatin study today, it would make international headlines. He would be invited to do a TED-something talk. How about a TEDx talk on why alternative views shouldn't be silenced?

In sum, there is now plenty of animal research demonstrating clear mechanisms whereby probiotics could potentially provide benefit to mental health. There is also just enough human research to justify a momentous movement toward the *study* of probiotics (and by that we mean encephalobiotics) for mental health. Not the *selling* of probiotics for mental health, but their study. There are plenty of other legitimate reasons to promote the sales of probiotics, but for the moment, the treatment of mental *disorders* should not be oversold. Nay, it should not be sold at all.

As much as we would like to say that commercial or prescription probiotics can enter the psychiatric armamentarium—not only because we hypothesized they could play such a role a decade and a half ago, and it feels good to be right when virtually everyone else was saying we were wrong—we can only say that there is hope for prevention and intervention. With large studies demonstrating the safety of probiotics for emotional health during and after pregnancy, and in early life for children, the probiotic as defined by Campbell in 1943—that is, functioning which eases life along—seems possible. In other words, probiotics in early life might actually lead to a functioning which eases life along, at least to some degree.

The Mobile Brain Fans the Flames

We must underscore that the secret life of your microbiome is also being controlled by the immune system—and its privileged position as holder of the nuclear codes of inflammation. Sometimes this gets overlooked in the biotech positioning of future products and storytelling blogs that translate the exciting microbiome research. We don't want to unwittingly fall into the same trap of inference by omission. Yes, microbial dysbiosis can contribute to the burden of low-grade inflammation and oxidative stress directly. But low-grade inflammation, especially in the gut, can itself lead to dysbiosis. Microbes that are normally kept in check in states of health can start flexing their muscles when the fires of inflammation start burning. They see opportunity and seize on it.

You can think of it as being like an algae bloom in a body of fresh water. In that case, chemical phosphates and nitrogen from agricultural or household products enter the water and lead to massive growth in algae and a resultant suffocation of other plants. The balance is tipped and the entire system goes toxic. Except in the human gastrointestinal tract, the phosphates are inflammatory chemicals.

In an interesting parallel, in chemically induced blooms, the algae often produce high levels of neurotoxins with detrimental effects on wildlife. This may be happening in humans as well, where inflammation-induced dysbiosis ramps up the production of potentially neurotoxic chemicals produced by microbes as they interact with food. For example, so-called "putrefactive" chemicals like cresols that are formed by microbial action on proteins (and linked to autism, anxiety, and multiple sclerosis) may be ramped up.

The cresols, the best known being p-cresol, can get a free pass right from the blood to the brain. Although miniscule levels are normal, the brain runs a tight ship and even small increases can increase oxidative stress, the promotion of inflammation, and, hence, the suspension of normal physiological operations. Writing in the *Journal of the American Medical Association* in 1909, Philadelphia physician Judson

Daland suggested that cresol, although not overtly toxic, could be linked to irritability, fatigue, and depression. Recent studies suggest he might have been correct in his thinking.

The good news is that this might be modifiable. In a Japanese study that compared standard yogurt to one with added *Bifidobacterium longum* BB536, the version with added probiotic decreased p-cresol among its adult consumers. Similar results were found with the fermented dairy beverage Yakult. Foods that are known to act favorably upon the gut microbiome, like cocoa, have also been shown to normalize p-cresol in association with reduced cortisol.

The take-home message, however, is that keeping the immune system operating in optimal ways is essential to preventing dysbiosis in the first place. When the mobile mind is thrown for a loop by chronic stress, when it doesn't like what it is touching and feeling—think environmental toxins, synthetic chemicals, AstroFood—it fires up the inflammatory pathways and stress hormones. Thus, lifestyle habits that support normal immune functioning will inevitably be in favor of a healthy microbiome.

Until now, the term *psychotropic* has been reserved for mood-regulating and sedating drugs, but in reality there are many ways in which microbes and nutrition can intersect to turn the mind for the better. Progressive psychotropic interventions need not involve synthetic chemicals and they may manifest their benefits over a course of a lifetime. In particular, the early origins of adult health can be traced right back to how the psyche was—or was not—tropic, or turning, beautifully in early life.

There are untold benefits to the consumption of probiotics, prebiotics, and encephalobiotics, as long as we invest in the good stuff, and as long as we maintain our adherence. But it is important to underscore that these commercial products are supplemental to dietary choices, physical activity, proper sleep, and other lifestyle choices that help maintain a healthy microbiome. To what end will probiotics help when our 3-squares-a-day-plus snacks are all in the form of ultra-processed food with synthetic, non-nutritive chemicals?

From the thriving ecosystem comes a state of upmost bliss—beatitude.

Beneficial Microbes: Wanted Dead or Alive

Despite perceptions that friendly microbes are only good for us while living, the science shows otherwise. There are volumes of experimental and human research showing that heat-killed or otherwise inactivated strains of lactobacilli and other friendly microbes can have numerous beneficial effects. Human research has shown that these inactive microbes are particularly beneficial to supporting immune function and research has even shown that they may reduce stress and tension. Among the microbes with lots of research in the inactivated form is *Lactobacillus plantarum* (the name means "from plants") a microbe found on edible plant foods and in wild nature. This microbe is commonly used in fermented foods and natural health products. Even though probiotic microbes may not be living when consumed in fermented foods or supplements, they can still shift the living microbes in the gastrointestinal tract in healthy ways.

9

Biophilic Science: A Brighter Future

Science for life implies research motivated by a profound vision of harmony, or ecology in its broadest meaning, of rational inquiry into the interrelation of things.... The difference between science for life and other science is not one of method, but of openness to a larger universe of variables. To the "biophilic" scientist, a problem does not end with the suppression of a symptom or the discovery of a relationship; the central question is how the cure or discovery affects a wide balance of forces.... The "biophilic" scientist is not so intent on conquering nature as he is on discovery or on learning how to employ the forces of nature to create a greater harmony for man within his environment.

— Michael Maccoby, PhD

In 1964, the social scientist Michael Maccoby described the need for "biophilic science," a science directed at life, rather than one heavily weighted toward destruction and the lifeless. Half a century on, his words are even more profound. So much of applied medical science today is about suppressing symptoms and skewed towards the "anti-bios."

Science is applied as a technology and blended into marketing because, as marketer extraordinaire Steve Jobs famously stated in *Businessweek*, "A lot of times, people don't know what they want until you show it to them." But people *do* know what they want. They want happiness and quality of life. However, marketers have long understood that the mirage of happiness could be presented in aesthetically pleasing brand names; *marketed* gadgetry technology sends out a shiny spinner lure into the stream of humanity.

If packaged properly, our genus will bite at the decoy prey of vitality in the form of convenience, speed, and of course, entertainment. The result is a 24-hour-daylight Las Vegas-like world of continuous screen-based divertissement. It looks like the Eiffel Tower and a Venetian Gondola, but it's really some girders of steel and a chlorinated canal on a desert strip. People enjoy the illusion but still leave knowing what they really want.

The acclaimed microbiologist and environmentalist René Dubos stood up to the nonsense of marketed technology. In his Pulitzer-Prize winning book *So Human an Animal* (1968), he called it straight on artificial wants: "All too often, science is now being used for technological applications that have nothing to do with human needs and aim only at creating new artificial wants. Even the most enthusiastic technocrat will acknowledge that many of the new wants artificially created are inimical to health and distort the aspirations of mankind. There is evidence furthermore that whole areas of technology are beginning to escape from human control."

Dubos also expressed his frustration with a modern body of "science" that was losing its big picture context. Scientists make scientific discoveries, or work to strengthen, verify, or nullify existing data, but the discoveries worth chasing, or considered scientifically virtuous, are at the mercy of human culture. Said Dubos in 1965:

> Despite our pathetic attempt at objectivity, we as scientists are in fact highly subjective in the selection of our activities, and we have goals in mind when we plan our work. We make a priori decisions concerning the kind of facts worth looking for; we arrange these facts according to certain patterns of thought that we find congenial.... A more disturbing aspect of modern science is that the specialist himself commonly loses contact with the aspect of reality which was his primary concern, whether it was matter, life or man.

Dubos was foretelling a time when generations of specialist scientists would have incredible working knowledge of the structures and functions of deep cellular apparatus, and yet have little appreciation for what it means in socioeconomic and ecological context. That day has arrived. Even though all the king's horses and men came up short on rebuilding Mr. Dumpty, the microbiome revolution provides a guidebook for scientists of all stripes to put the cellular pieces back together again. More than ever, it brings us back to the greater need for a holistic vision for a better world.

What do these cellular biology discoveries mean to someone walking down the street being lured in by the latest in marketing science to consume fast food at the sight of French-Fry-Yellow arches? Who profits from the pursuit of science? What is the endgame? Another symptom reducer, one more marketing app that acts as a layer of concealment over an inflamed lifestyle? It gets even more knotty when we consider what to teach youth regarding science. Put simply, science is a very broad term, and although it is certainly beautiful to pursue anything for the sake of discoveries alone, it gets tricky when humans are funding the choices of what is actually worthy of pursuit.

Maccoby wondered, as we do now, why so much publicly funded science is directed at destruction (the war machine) and *leaving* the living Earth behind (space exploration)? Like us, he acknowledged that military and space programs have done much for the advancement of science—and, indirectly through trickle-down effects—human comfort and convenience. Hard to imagine we ever lived without mobile phones. However, when you are doing one thing—in this case, the nonbiophilic science of weapons and rockets—you aren't doing another. What if the same colossal resources were directed at mental disease, inequalities, suffering, and ecology?

It would seem we are long past due a Manhattan Project for the 21st century, one that is determined to bring greater understanding of mental health and ecological justice. The disturbingly skewed direction of scientific interest, where funding dictates interest, is exemplified by the staggering amounts of money spent on the study of weaponry and the vast tools of military might. The lack of attention paid to our wounded veterans represents a system of scientific study that is devoid of equity, if not morality. Crumbs are devoted to the veteran whose brain, heart, immune system, and mind have been shattered by war. The six trillion dollar and counting war effort spanning the last 15 years seems little interested in throwing more than a few coins toward fixing the broken humans left in its wake. Nonbiophilic science is discarding veterans like a Sherman tank in a Texas salvage yard. Sorry veteran, there is a new rocket to build.

Maccoby finished his 1964 piece with a haunting two-sentence paragraph: "The scientific community has the knowledge, the ability, and the prestige to reorient the nation toward 'biophilic' goals. Unless science wakes up to the dangers of acquiescence and objects to priorities set by fear and political expediency, there is little hope of reversing a process which at best leads to waste and stagnation and at worst to moral disintegration and suicide."

Fifty years after Maccoby described the urgent need for biophilic science, the acquiescence he spoke of abounds. Scientific exploration of mental health and trauma is minimal. Rockets and the bronze busts of the great scientific men that built them remain as towering monuments to a science that is led by politics and fear. Meanwhile, more than two dozen US, Canadian, and Australian veterans commit suicide—each day. Biophilic goals remain low on the list of priorities, while the families of more than two dozen veterans receive news that is too much to bear—each day.

The multimillion-dollar rocket to Pluto flies on, and on...and with each passing day mental health is swept under the synthetic, pixilated rug. Preventive medicine is ignored, AstroFood is produced, and we become more nature disconnected. Each day. The system is broken.

Mars Can Wait

To be clear, we are advocating a pro-science message. Scientific exploration is vital, but science for whom? No disrespect to celebrity astrophysicists, but on the earthly plane where mental health and non-communicable disease epidemic needs are at crisis levels, we line up behind the "Mars Can Wait" sentiment expressed by a minority of vocal scientists, including Geoffrey Goodman and colleagues. The universe within the human neurons, immune system, and microbes is in much more urgent need of exploration. We concur with microbiologist René Dubos, who stated in 1970: "The man of flesh and bone is not likely to remain long impressed by the fact that a few of his contemporaries can explore the moon if the planet Earth becomes unfit for his everyday life."

Dubos underscored that the greatest aspect of manned space exploration was enjoyed by astronauts who turned back from lifeless, bleak, grey-colored rock to see the colorful Earth rise before their eyes. On Wednesday July 16th, 1969, he addressed the World Health Assembly in Boston, Massachusetts, stating: "What surprised [the astronauts] and charmed them is that the Earth, our Earth, is like an oasis, a paradise, in the midst of the emptiness of space, and in comparison with the drabness of the moon".

In a world where empathy is on the decline, and narcissism on the rise, the idea that 20-billion-dollar checks should be written to explore and establish a colony on Mars, simply because it inspires youth, just doesn't hold weight. We will touch on this a bit more in the final chapter. In addition to his concerns about an earthly environment losing its ability to sustain life, Dubos recognized that while a few humans would enjoy the privilege of seeing the Earth from space, with polluted skies the other 99.99 percent of humanity will not even be able to see the stars. Pictures delivered from space are no substitute. We've seen Earthrise. That is the photograph we need to focus on. He worried whether even basic needs would be fulfilled, writing in 1974: "But everywhere and always human beings have longed for a perception of space which would incorporate the sun, the moon, the planets, the stars, a sweep of landscape or seascape beyond the horizon. The visual apprehension of the cosmos seems to be an essential psychological need. It nourishes the imagination and gives larger significance to life."

Amazing, isn't it? We long to see stars, but as they disappear from the night sky, we are somehow palliated by more pixilated images of Pluto. Politicians basking in the strobe-light glow at podiums celebrating rocket trips to Pluto, when citizenry can barely see any stars due to light pollution. The much-ballyhooed Pluto rocket passed through its close-to-Pluto encounter for six days. Never mind better images of

Pluto—and it will always be a planet in our hearts—those who cannot realize the fullest potential in life due to mental ill-health represent the neuron-immune-mind universe we should be exploring with our publicly funded micro-rocket ships.

Nonbiophilic (one might say, necrophilic) science is the autocratic status quo. We are living in a dysbiosphere and yet we invest heavily in knowledge of lifeless technology and speculations of a fantasy of distant life, as on Kepler-1649b and the like. Why? We set aside the fact that chronic pain affects 100 million Americans and costs society 600 billion dollars. We set aside the fact that mental health disorders, when assessed even at a superficial level, cost well over a trillion dollars per year in Australia, Europe, Canada, and the United States alone.

Even that trillion dollars doesn't capture the true cost of mental illness. Its vampire squid-like arms suck the quality-of-life blood out of individuals and society. They wrap themselves around substance abuse, and a bloated criminal "justice" system that seeks to treat by incarceration. Veterans are the poster children of this system. Lack of care, higher rates of incarceration. Defense spending in United States exceeds $824 billion but the costs to veterans and their families are truly untold.

But wait, there's more. We haven't even added in the ways in which mental health disorders and subthreshold conditions contribute toward other burdensome non-communicable diseases. Recall, low-grade depression and anxiety increase the risk of cardiovascular disease and a basket of other conditions. And that's not all. The emerging research on transgenerational outcomes of psychological distress shows us that our children are indeed a product of biological, psychological, and social ecosystems of parents and even grandparents.

So, we will say it one more time. Sorry, celebrity astrophysicists and pretend-scientist spokespersons for planetary exploration, we remain unconvinced of your priorities. Twenty billion dollars for Mars? Biophilic scientists get crumbs to study the human mind and immune system from the perspective of purpose, mindfulness, empathy, nature, and emotionality. It's actually amazing that the study of mindfulness has paid such dividends, despite limited government support. Peanuts have been spent on the search for emotionally intelligent life here on Earth. Findings that might actually save us from ourselves.

Commandeering STEM

The spokespeople of science take on a stereotypical ethos. Commonly it is dressed up as an older Caucasian man in a bowtie, perhaps overcompensating for the lack of an advanced degree in the sciences. He selectively drones on about "science." Whose science? What kind of science? The science of mindfulness, empathy, cultural understanding, ecological injustices, inequity and/or manipulation of public perception? The science of marketing deception that allows children to consume more ultraprocessed foods? The science of contentment that allows our children to settle into the glow of a pixilated screen with images of the once-visible Milky Way, while remaining detached from nature? The science of inequalities? The science of "safe" chemicals that turn out to cause intestinal dysbiosis and adverse epigenetic changes? The science of poverty?

We could drone on as well. But we will simply return to the Latin root of the word science, that is, *scientia*. Acquaintance with *knowledge*. Not opinion. There are two parts of this word root of science—the fact itself, which is the subject of the knowledge, and then actually knowing it. Acquaintanceship. Somehow this relatively simple, yet elegant and broadly applicable word, *scientia*, has been co-opted by the stereotypical: characters with test tubes in a lab-coat pocket, holding small rockets, and plastic models of atoms. Anything phenotypically distinct from this branding isn't really considered science.

The greatest example of this can be seen in the mad rush toward the trendy and very well intended fields of science, technology, engineering, math (STEM). As most parents know, even in kindergarten an educational focus is placed on the so-called STEM subjects. Especially in dominant westernized nations where global ranking means everything. Never mind celebrating that many transitioning nations are lifting up their STEM educational rankings, it's assumed to be a ghastly thing when G8 nations aren't ranked neatly one through eight on the medal podium of STEM rankings. Fear abounds. Quick-fix solutions appear out of the digital-o-sphere.

There are, of course, multiple economic reasons to support STEM in the general sense. Skills under the STEM umbrella are an asset when seeking a well-compensated job, and that isn't likely to change. The problem, however, is that when STEM is privileged within academia, and when purveyors of screen media education and their proverbial app-hawking uncles claim we can digitalize kids to academic success, two things happen by default.

First, we relegate so-called soft subjects, humanities and the arts, to near if not complete irrelevancy. Second, the entire educational environment becomes a means to palliate what is viewed as Brainier Baby DVD Deficit Disorder. Which means more screens in more places. Even though, as German neuroscientist Dr. Manfred Spitzer pointed out in *New Scientist* (October 17, 2015), the evidence supporting the effectiveness of screens in the classroom is thinner than bargain-brand tissue. But on it goes. Companies name their must-see early-life DVDs after Nobel Prize-winning scientists and market them as pure genius. Clearly, there must be some very lofty expectations on the part of parents.

The problem sits with the belief that STEM stands for success. Some find the acronym offensive simply because it translates as "Not the Humanities." But it's more than that. The S in STEM is symbolized by Bunsen burners, beakers, and the prospect of a job at Boeing. It's the dissection of the biosphere and its geological theatre into bits. Which is good except that those bits aren't put back together again. One can become a university student, then a medical doctor and/or a PhD scientist, all without having ever studied the unified science of ecology or ecosystems. As currently used, the S and the entire acronym do not symbolize the interconnectivity of life.

It is an S that breaks apart the unity of life, of living organisms. Self-appointed science guys in bowties talk to our children O'Plenty about scary germs. They do one hour shows on them, say "eewwhh" when touching surfaces, tell the kids we are covered in them. You can check out these videos on YouTube. Nary a mention of beneficial microbes and their positive connections to all life. They talk about wonderful plastics in the space shuttle, but not about the ones found in fast food, denaturing our food supply. The S doesn't represent the science of empathy. It doesn't represent the science of positive emotional assets that will serve us—individuals, society, and the planet—over the course of a lifetime.

T for technology? That technology even makes the list is absurd. It can't be defined, but this much we know: there is no shortage of dreamers in Silicon Valley looking to make it big with the next app that serves up fast food even faster, or turns kids into geniuses. Cash out before 30 and get the latest Tesla.

The actual *Tech Times* headline screams: "The Latest and Greatest Fast Food Apps for 2016." The article goes on: "In a world where we're all trying to get everything done at a lightning-fast pace, it only makes sense that we need to eat faster, too. Luckily, there are several fast food chains to satisfy this need, but that creates a problem in itself—which one should you buy from? Furthermore, how do you narrow down all of the delectable options? Here are all of the fast food apps you need to quell your hunger, whether you want a burger or a giant soda."

Who is asking what *kind* of technology? What does it even mean? The inappropriate placement of the word *technology* in STEM is reflected in the inability of economic scientists to actually quantify whether or not we actually have a surplus of STEM jobs. Many experts legitimately ask if the so-called STEM crisis is a myth.

René Dubos routinely pointed to the attitudes conveyed at the 1933 World's Fair as a way to elucidate the runaway enthusiasm toward the T of technology. He would pull out quotes from the pages of the World's Fair Guidebook to highlight a system that was only getting worse. Technology was shaping humans without pausing to consider the unintended consequences. Here are some passages, often cited by Dubos, from that famous Guidebook.

> Science discovers, genius invents, industry applies, and man adapts himself to, or is molded by, new things...industry accepts its findings, then fashions and weaves, and fabricates and manipulates them to the uses of man. Man uses and it affects his environment, changes his whole habit of thought and of living. Individuals, groups, entire races of men fall into step with the slow or swift movement of the march of science and industry.

The Guidebook sets the stage for visitors in describing how nasty North American life was in 1833, one hundred years earlier:

> You live roughly, in your own tiny, lonely world, bound in by forest or houseless prairies or towering mountains. No means of quick communication have been contrived to overcome natural barriers or to break, for months at a time, the solitude. You wear crude dress, ill-fashioned.... You eat foods that must be indigenous to the territory in which you live.... Electricity has not come to work its wonders, even the kerosene lamp is in the future...
>
> Come back to 1933. You hurtle through the air over mountains and plains on motorized wings, or speed along the ground in luxurious trains, or over smooth highways in motor-powered cars. You live in a home, made of materials created by the genius of man anticipating the vanishing of forests. Electricity is your servant to give you your light and do your work. You whisper and your words wing their way across the seas to be heard by listening ears.

The euphoria associated with technological advancements was spilling over its golden chalice:

> Consider the architecture of the buildings. Wonder, perhaps, that in most of them there are no windows. Note curiously that these structures are for the most part unbroken planes and surfaces of asbestos.... Windowless, these buildings assure, by virtue of the advancement in the science of interior lighting... they exemplify too, the advancement which has been in healthful, controlled, filtered ventilation.
>
> And science has achieved a brilliance and skill of electric lighting which, as exemplified in the buildings of the Fair, render windows and skylights no longer a necessity in buildings.

And in one of the exhibits, in order to demonstrate how disease and illness are becoming a thing of the past: "You can watch the antics of an Indian medicine man, practicing his primitive medicine, in the exhibits of the Milwaukee Public Museum."

Let's reflect on this: bound in by a tiny, lonely world? Really? Celebrating the genius of man who can make plastics because the forests have vanished? The "antics" of traditional healers? Windowless buildings? Surfaces of asbestos? Technology that changes the habits of thought and entire races falling lockstep into what industry decides? The arrogance of it.

Yes, STEM is where the jobs *may* be. Yes, science *can* be cool. But trying to make it the new cool, as in the new high school jock or another variant of designer-label clothing, isn't going to lead to a better world. On the contrary, exchanging one in–group for another isn't cool at all. Science with *heart* is what is needed. The science of empathy shows us that success in the career of living will be magnified by an empathic perspective.

A STEM education might lead to opportunities for a higher paying job, but research from Belgium shows that fostering *emotional* intelligence increases employability. Moreover, empathic education will allow that career to be *savored*. It will allow the fruits of discovery and application to be enjoyed by all. STEM needs to break out of its little cubby hole of simply being the route to highly compensated employment.

Here's why we are commandeering the acronym and staking the claim that empathy is the new E in STEM. Empathy is linked to life satisfaction, personal wellbeing, rich social networks, healthy relationships, accommodative and nonaggressive behavior, workplace performance, and prosocial activity. Individuals who are empathic do not bully. Empathy is all wrapped up in symbiotic tentacles with an understanding of nature. Research shows that being able to take the point of view of another person, *and* at the same time perceiving the natural world as important to health and wellbeing, are intertwined. This, from our perspective, is the coolness of science. The awesomeness of *scientia*—knowledge.

Jonas Salk: Nature and Coalescence

Jonas Salk, the scientist who cured polio by preventing its occurrence through vaccination, is a premier example of an empathic scientist. He grew up with two brothers in a four-room apartment in New York, the son of garment workers. While many know of his monumental achievement in polio eradication, he left a legacy of how biophilic science can work in the era of the microbiome revolution.

In April 1955, when the success of his polio vaccine was announced with great fanfare, he did what few on the modern TED Talk circuit do. He downplayed a colossal personal triumph by reminding viewers in a national broadcast that no person can really lay claim to sole ownership of scientific ideas and advances. Almost all are built upon the work of others. He also turned down massive financial gain. When asked on US national television who owned the patent to the life and quality-of-life-preserving vaccine that bore his name, Salk responded: "Well, the people, I would say. There is no patent. Could you patent the sun?"

During his nationally broadcast interview with Edward R. Murrow, when city bells were still ringing in celebration of the end of polio, Salk was asked, "What is the next big thing in preventive medicine? In what direction should he and fellow polio-conquering scientists turn their attention?" His answer was simple. The turn should be toward mental health. In the midst of the Mad Men era of pseudoscientific psychosomatics and psychoanalysis, he made his case for the biological understanding of mental suffering: "The area of mental disease is one of largest by far, and particularly desperately in need of quantitative measures that tell us precisely what we are dealing with."

In the 1960s, Salk set up his own research institute in California to study chronic diseases. He believed that the pathway to solving the mysteries of chronic disease, including mental disease, would roll through the immune system. However, he viewed science in a holistic way. He knew that the immune system was an extension of the cultural policies and practices, educational systems, and way of life in which it is surrounded. Chronic diseases were distinct from neatly defined polio, and therefore scientific solutions required an expanded outlook. Solutions, according to Salk, required the recognition and appreciation of biology and metabiology (the latter meaning biology in humanistic and cultural context).

In an interview with Bill Moyers, Salk described his vision in attempting to make for a healthier existence: "We have to learn to think like nature...to allow the fullness of the potential of the individual to be expressed, to flower...we must acknowledge ourselves as part of nature, not apart from nature." Generating the right kinds of scientific questions, according to Salk, is a prerequisite to a healthier, peaceful, and more equitable world. To get to this place, to truly evolve in a healthy direction, mere cooperation and coexistence wouldn't be enough. Biology and metabiology required, in his words, balanced coalescence. Scientifically, this means combining droplets into one. From a human perspective, it means understanding that we are part of the single droplet known as the biosphere—the big beautiful droplet seen by the astronauts on Christmas Eve, 1968.

The bench scientists at Salk's Institute worked with the perspective of a team that included mathematicians and humanists. Molecular and cellular biologists at his Institute could uncover life's processes, but their relevance to complex, chronic disease in modernity could only be illuminated by asking the right biophilosophical questions. This was how the prestigious Royal Society once worked. Scientists and philosophers shared membership and ideas. Salk's larger aim was to understand humans in their physical, mental, and spiritual complexity so that each individual would be free to cultivate their full potential.

Like René Dubos, who discovered the first clinically studied antibiotic, and then later devoted his career to the unification of silo-sequestered bits of science, Salk pulled the lens back from his reductionist microscope. Writing in his book *Anatomy of Reality* (1983), he stated: "I think we ought to stand back and say 'What is it that the world needs? And what is it that people need to experience in order to fulfill themselves and their potential?' We need to address ourselves to questions such as meaning and fulfillment, satisfaction. And I'm not talking about satisfaction of myself as an individual without respect to all others. I'm thinking of the notion of mutualism... recognizing that my fulfillment helps in your fulfillment, your fulfillment comes in my fulfillment and we are all part of humankind together."

He advocated for a collective effort in understanding human thoughts and behaviors as they pertain to the future health of the planet, and humankind within the biosphere. Said Salk in the Australian publication *The Age* in 1979: "When we value a planet better balanced in economic differences, when we value healthy, and civilized individuals more than their opposites, and civilized behavior more than pathological and violent behavior, when health becomes at least as worthy of attention as disease, and when we value peace more than war, we will be on our way into the new Epoch."

Oneness. That is what Salk envisioned. He made brilliant discoveries through detailed scientific work, and was well aware of the larger ecosystem of life and lifestyle in which his reductionist toil had been taking place. The test tubes were always connected, in his mind, to a larger purpose. He also wrote on the transformation of education. In particular, the famed immunologist felt that traditional education needed to immunize children in ways beyond the needle: "We talk of immunizing children...but what about immunizing, say, against the kind of behavior that leads to drug abuse? Or immunizing that leads to [youth being] more responsible for themselves and responsible to society?"

Salk declared that an exclusive focus on analytics would not lead to a better future or accomplish this mental immunity. The educational environment must support creative, out-of-the-box thinking in order to provide a context to science and technology. He was part of a coalition known as the Universal Movement for Scientific Responsibility (UMSR). Note the word *Movement* vs. *Institute* or *Organization*. As he said, "This is not an elite. It is essential that scientists, humanists and managers of society work in concert." The keyword *Responsibility* meant understanding not just causes and mechanisms, but also what the effects of scientific findings might be when applied to the commercial T of technology.

The UMSR was the antidote to the World's Fair Guidebook. How are societies and humankind influenced by applied science? Was the march of science and technology leading us in lockstep to a cliff, or was it going to highlight the interdependence and interconnectedness of life? Salk envisioned the movement toward scientific responsibility to be one guided by "large measures of wisdom, altruism and generosity." In 1992 he implored fellow scientists to see the big picture: "We must see ourselves as part of the ecosystem. Where we once were a product of evolution, we are now a part of the process."

Today, many scientific role models for our youth are Teddy-Talkers who have done little in the way of original research. Many ask for exorbitant speaking fees. They symbolize the chum for the shark-feeding frenzy that occurs in the cultural seas of the Kardashianism of science and technology. When packaged speakers pretend that theories and hypotheses don't already exist and claim prior art and ideas as their own, science starts to become a mirror image of what is happening in society at large with all its rising narcissism and dipping empathy. Audiences golf clap to a superficial satiety provided by the bait.

Salk was concerned about the narcissism in science, so much so that he turned a lecture, "Narcissism and Responsibility in Science and in Man," into his book *Survival of the Wisest*. To what degree do money, patents, and fame become the driving force of "science"? When scientific career success is quantified by your number of Twitter followers and YouTube downloads—the Kardashian Science Index as genetic scientist Neil Hall calls it—to what end is S making for a better world, or simply providing a surrogate marker for what ails it? In what ways are we teaching children, teens, and young adults that empathic science matters? And that the science of empathy matters?

Mathematics is an obviously important skill, the basics of which are vital to everyday personal economic decisions. But again, the positioning of M as simply being a road to a higher paying job privileges mathematics as a subject out of its context. Knowledge of M is more than managing personal finances and preventing runaway credit card bills. In the transition to adulthood, a grasp on the M in STEM, if it is delivered in the context of social sciences, can allow individuals to see how a living wage can lead to a better world.

Even biological sciences can lend critical, real world color to economic decisions and personally vital mathematics. Recall the mobile mind. Your immune system. In a recent Japanese study published in *Neuroendocrinology Letters*, subjects in whom low-grade inflammation appears to be persistent—that is, higher blood levels of the inflammatory cytokine IL-6—made responses to economic questions in ways that indicate they are less averse to inequalities. In other words, low-grade inflammation in the dysbiotic world may be driving us to care just a little less about the plight of others. If this research is replicated, it does not bode well for ongoing global westernization that seems to be setting the stage for higher inflammatory cytokines by default. The dysbiotic lifestyle brought to us by concentrated wealth could actually be making us less caring of inequity!

While this might be far-fetched, the topic of greater scientific understanding of social and ecological injustices is critically important. Education on a dominant, yet narrowly-defined STEM platform, continues the disconnect from the oneness of nature and interferes with the ability to truly fathom what might be missing in the contemporary dysbiosphere. When this type of education occurs coincident with the relegation of topics such as emotional intelligence, humanism, and biophilosophy to the irrelevant, it increases the likelihood of division and inequalities and diminishes adaptive methods of coping with social challenges and interpersonal conflicts.

This is what Salk, the scientist who transformed the lives of millions with a microbial intervention, was ultimately concerned about. He believed that regardless of the degree to which genetics might play a role, all children have the capacity for great empathy, kindness, and recognition of the interdependence of life. However, society has a responsibility. Speaking to Arianna Stassinopoulos (better known as Arianna Huffington of the Huffington Post) in 1984 on goodness and nobility as a formula for the future he said: "They [children] need to be taught...by example and experience. They can't be taught passively. They must be taught actively—in the same way we speak of active immunization as distinct from passive immunization. Passive immunization transfers antibodies from one person to another, but it doesn't last. In active immunization, you form your own antibodies; you evoke your own potential for nobility."

Vaccines like those developed by Salk are a prime example of active immunization. We develop lifetime immunity against the likes of polio. Experience is also a vaccine. In other words, nobility doesn't belong to an upper-class elite, it is a product of the classroom. Every single child has the potential to feel that they belong to an exalted, equitable, and loving class. They see the goodness in each student reflected in every other classmate because they have experience and examples. Even the notion that Jonas Salk and the celebrated creators of fast-food apps could be considered together under the umbrella of this superficial educational acronym is beyond disturbing. STEM, indeed.

Superficiality vs. Experience

> **Marshall McLuhan writes about the global village in which all of us are interconnected through the marvels of electronics. This kind of relationship, however, affects our lives only in a superficial way... in fact, the global village concept is of little importance for daily life because it exists only in the form of cerebral abstractions.**
>
> **— René Dubos, *Beast or Angel?*, 1974**

Dubos worried that although access to screen-based information might add to awareness and general knowledge, it also has the potential to become a purely voyeuristic, non-organic means to accumulate facts dissociated with emotion. He was right. Screens are everywhere. Even at the gas pumps screens are being installed, the latest cable news fuel injected to our retinas, lest we suffer a few minutes of entertainment deficit disorder. Horrific scenes play out live on the seat-back screens of aircraft travelling at 30,000 feet; a passenger watches real life victims of a tsunami while ordering up another diet cola. Cerebral abstractions and distractions.

As Dubos said: "Learning about the world through news reports, talk shows, television broadcasts gives the artificial thrill that comes from the illusion of proximity to events without the necessity of being involved in them; it does not elicit an organic interaction and therefore gives at best a trivial quality to the experience of the global village.... Telephones, radios, television sets, and other gadgets enslave us to a nonhuman and often antihuman environment."

For Dubos one of the primary ways in which to recapture the fundamental satisfactions of our distant past was through mindfulness of the "richness of sensory experiences that can be derived from direct contact with nature—human values that are being eliminated by technological civilization." Dubos further argued that economists and futurologists consistently ignored these values derived from nature contact, even though one of the limits to growth (no matter how one may choose to define the term) may turn out to be lack of opportunity to experience such sensory pleasures: "In order to retain their sanity human beings must try to recapture these satisfactions because the need for them is an unchangeable part of human nature.... Certain situations are naturally more conducive than others to the *joie de vivre*. From a purely biological point of view, the best environment for the human species may have been the pre-industrial, pre-agricultural world based on the hunting-gathering ways of life. This may be the biological environment to which we are still best adapted, the biological paradise we have lost."

Holistic Medicine in the Microbiome Era

At what point did *holistic* become a dirty word in the context of healing? As the microbiome era unites, some non-scientists and scientists who strive for a better world by chemistry become the mouthpieces for division. They crave attention, hiding behind the dressed-up ethos of scientific rhetoric—bowties and Clark Kent glasses—waving their swords of "ethics" and "science." Yet ignorance and pharmaceutical-selective referencing undercuts their messages. Partial messages are not scientific. Nor are they ethical or just. They pedal what we call Sectarian Science. Only a bit more filler for the culture wars, is all.

Sectarian Science: Biased, partial science dressed up as authoritative, legitimate, and canonical. Often applied to medicine, it is weaved into the fabric of financial and institutional power, prestige, and privilege. It shouts from these pulpits, selectively using terms such as "evidence-based" to defend dogma, maintaining that all the answers are in. Thus, by definition it is pseudoscientific. It is a sect of science. The new McCarthyism, its purveyors shout out "anti-science!" in all directions except inward. In medicine, sectarian science is little interested in evidence-based optimism, cultural competency, and ecological justice. Based on the biases of its brokers, it is less concerned with outcomes and selectively focused on mechanisms. Rachel Carson, the author of *Silent Spring*, was one of the first notable victims of the sham of Sectarian Science. Carson lived by the scientific truth *Nullius in verba*: "Be not obliged to swear allegiance to a puppet master, wherever the storm may drag you." Her sectarian-scientist detractors did not. Nay, the authoritarian puppet masters smeared Carson while pretending environmental degradation and species loss was a fantastical notion.

For science, no matter its cultural, commercial and, ultimately, anthropocentric translation, demands a holistic context. We have seen time and time again in these pages that the microbiome cannot be pulled apart from its lifestyle context. In Chapter One we described how Darwin's clover really isn't just clover; it's clover, bees, mice, cats and feline-feeders, cattle, and an entire biosphere. As we will discuss in the final chapter, many trees communicate with the assistance of nearby fungi. The fungi have "telephone wires" called mycelia and the trees uses these lines for signalling. But the fungi are also assisted in this by the microbes that massage their roots and work *with* the fungi.

So too with medicine. It cannot be pulled out from its lifestyle context. Conventional medicine, alternative medicine, complementary medicine, allopathic medicine, functional medicine, traditional medicine, integrative medicine, orthodox medicine—all are applied as team sport labels pitted against one another. Bring your own helmets, mouth guards, and shoulder pads as required for the offensive and defensive gamesmanship that ensues. Diminished, if not lost, in the fierce rivalries is the commonality of the desire to *heal*.

There can only be one medicine—holistic medicine. Holistic medicine is ecological medicine and it is not in the domain of one particular type of doctor. It is the promotion of health and the healing of the *entire person* in their ecological habitat. There is no way around this. In fact, it is not just the individual person, it is the promotion of health and healing at the community level; public health is holistic medicine. If there were any doubting this before, the microbiome revolution proves that all aspects of life, including medicine, must be viewed in its own biosphere. Once again, we go back to the Latin and Greek word roots for perspective.

Medicine is derived in part from the Latin *medeor* which means to correct, restore, and to heal. But it is also derived from *medicina* as in the *art* of healing. Somehow the art of healing, *medicina*, has been diluted, and even corrupted from its original purpose. The body has been dissected into separate pieces, more often considered separately than as part of a whole and commercialized beyond belief. Catchy jingles in direct-to-consumer drug ads are splashed across primetime television. A carefully staged actor posing as if with a complex, chronic, non-communicable disease, does an Irish reel dance while a trusted anonymous voice urges us to "ask your doctor about Curitall-Rx."

In the real world things aren't so simple. Real suffering individuals must be viewed in the complex holistic context of their environment. Acute situations are one thing, but for most chronic diseases, there is little to dance about with pharma treatments alone. The statistics on stress, mental health, and chronic disease certainly suggest otherwise. Biopsychosocial medicine was initiated in the 1950s by social workers pleading for a more holistic outlook. The term was subsequently popularized by some psychiatrists and formulated into a model of care. It was an approach that tried to emphasize simultaneously attending to the biological, psychological, and social dimensions of patient care.

By and large, however, the biopsychosocial movement has never looked at holistic ecosystems; in fact, it has never lived up to its well-intended expectations. For someone with depressive symptoms, biopsychosocial medicine often translates as "Here's your drug (bio), you can see the side effects listed in our ad in the back of *Golf Gazette* magazine; here's the name of a therapist doing cognitive behavioral therapy (psycho), the latest of many large studies by the University of Amsterdam's Pim Cuijpers shows that such therapy is really good, but really good means it's really only marginally better than placebo; hopefully you will be able to get out of that unhealthy work environment soon (social)."

Medical Holism in the Microbiome Era

This is of no fault of clinicians. The system is broken. It is dominated by biomedical influences and neoliberalism. The former posits that pharmaceutical drugs and technology can fix most problems and the latter works from the framework that an individual can assume responsibility for personal health problems by simply adopting what biomedicine has to offer. Neoliberalism doesn't consider that a broken socio-ecological system might be the driving force for the need of biomedicine in the first place. The social in biopsychosocial doesn't translate as advocacy; it fails to live up to the dictum of Rudolph Virchow, the brilliant public health physician of the 19th century:

> Medicine is a social science, and politics is nothing else but medicine on a large scale. Medicine, as a social science, as the science of human beings, has the obligation to point out problems and to attempt their theoretical solution: the politician, the practical anthropologist, must find the means for their actual solution.... Science for its own sake usually means nothing more than science for the sake of the people who happen to be pursuing it. Knowledge which is unable to support action is not genuine—and how unsure is activity without understanding.... If medicine is to fulfill her great task, then she must enter the political and social life.... The physicians are the natural attorneys of the poor, and the social problems should largely be solved by them.

Salk and Virchow had one thing in common: they both emerged from working class environments. It is interesting to note how the mayor of New York, Robert F. Wagner, acknowledged the ecosystem of equality and underscored the system that supported the polio cure when he bestowed an award on Jonas Salk on February 19, 1958:

> We in New York City are proud of Jonas Salk because we think of him as a product of our schools, our educational system... in a way Jonas Salk is also a product of the labor movement. His father, Daniel Salk, was a member of Local 142 of the International Ladies Garment Workers' Union; in fact, in the early days of [the] National Recovery Administration, he used to organize this neckwear local of ILGWU. I wonder if Jonas Salk would have been able to go through high school, on to college, and through medical school to become the great doctor and

> scientist he is, if there had not been a garment workers union to raise the wages, to win security, to raise the living standards of the family of Daniel Salk.... Had Daniel Salk, without the protection of a fine, clean union like the Ladies Garment been forced to work under sweat shop conditions, for sweat shop wages, would his son Jonas Salk have been able to get a medical education? Perhaps—but it would certainly have been far more difficult.

Salk's upbringing no doubt had an influence on his vision of optimal health. His San Diego Bio-Philosophy Institute was concerned with the minute workings of the cell—molecular biology. But as reported by Howard Taubman in the *Pittsburgh Post-Gazette* (Nov 24, 1966), Salk had his microbiologists sit down with both union and industry executives in order to nurture the intellectual Petri dish of what scientific findings might mean to humans in their workplaces and social contexts. Salk wanted his scientists to hear and understand what union leader and Presidential Medal of Freedom recipient Walter Reuther had to say to them: Reuther told the Institute scientists that labor and industry should join with scientists and humanists to think about what we wish to make of man and our society.

Rather, the current "social" in clinical biopsychosocial medicine is largely viewed as "pull your socks up" and get out to the local sports club, or keep hunting for that job that pays 50 cents an hour more than minimum wage. Personal responsibility. Of course, there is no doubt that personal choices are critical to health; but think of children and pregnant mums. We must acknowledge that there is a societal womb surrounding, and ideally supporting, the mother carrying a child to term. Social medicine stands as an advocate when the societal womb is unhealthy and unjust—when it reinforces a dysbiotic lifestyle, or deprives a pregnant woman (and her partner) of access to resources, education, natural environments, or to any other environmentally driven factor of fairness. We can talk of the microbiome, but in order to remediate distressed terrain we must also consider the inequities in dysbiosis—that is, the marketing and socioeconomic forces that promote dysbiosis in its general and more microbially defined aspects.

Mayor Wagner had a point about the decent wages and working conditions enjoyed by Salk's dad. From our perspective, these apply to dysbiosis. Several studies have shown that relatively small increases in wages can improve depressive symptoms in socioeconomically disadvantaged workers. Given the well-known connections between parental mental health and offspring health, the implications are enormous. Two separate 2016 studies published in the *American Journal of Public Health* show that lifting the minimum wage in the United States would have a dramatic effect on reducing infant mortality. That's *minimum* wage! But humans need a *living* wage, anything less is anti-bios.

Living wages are probios, their effects ripple through society; we can talk of wonderful ways to target the microbiome, but for whom? Anything less than a living wage increases the magnetic lure of the dysbiotic lifestyle (for example, unhealthy dietary choices) and diminishes opportunity to engage with others and spend time in natural environments due to the time constraints of working additional hours or multiple jobs. That's before we even start discussing the multiple studies showing that natural environments and access to them are far from equitably distributed.

Society must realize its own collective role in the connections between health, the microbiome, biodiversity, and nature connectivity. Research demonstrating that a living wage increases mental outlook is no less important than research showing that prebiotic fiber can lower cortisol. Both sets of research will influence the microbiome. Except the former will provide not only the means to seek out healthy, prebiotic-containing foods, but also the time required to seek out nature.

Victory Lane

Strangely, we'd like to offer special thanks to direct-to-consumer pharma advertising and ultra-processed food-driven dysbiotic drift for working in tandem to promote knowledge. The former industry drives incredible awareness: a public now so familiar with type 2 diabetes drugs and the A1c test, to the point where we can sing along to their Madmen-formulated tunes. Pharma ad spots that are often followed by the jingle of the ultra-processed food purveyor. The latter drives the need for the former. Why are we not asking what is wrong with our environment, and why is it that traditional populations living in nature are not only unaware of A1c, they don't need to be?!

In the Divided States of Western Nations, individuals take to social media and the comment sections under op-eds to make their case about the opinions of others. The whole world starts to look like an op-ed piece. Groups set up websites with sciencey-sounding names that are really only fronts for pharma or a façade for what is merely a small sliver of science with an agenda. They dress them up with legitimate names like "Center of Excellence" and "Science-Based," "Evidence-Based," and so forth. Sometimes they have even have university affiliations.

But mostly, if the resumes of these folks behind the websites were adorned like NASCAR firesuits, it would be a parody of *Talledega Nights'* Ricky Bobby, with the logo of every pharma/chemical/synthetic company peppered over the driver of "science." Except the big difference is that NASCAR hides nothing. There is something elegantly transparent in the Anthropocene about cars that get five miles per gallon making continuous left turns with logos for ultra-processed foods, Big Pharma and Big Agriculture emblazoned from front to back. Antacid logos chasing those for fast food, diabetes drugs chasing soft drinks and allergy meds on the hood while the oval is full of airborne particulate matter...and AstroFood will take the checkered flag in the Dysbiotic 500! Winning drivers ease out of their cars through the window and take a big slug of their sponsored soft drink before uttering a word in the post-race interview. But it is all so refreshingly honest.

Maybe science should be that way, too. Perhaps those standing at the podium of "science" should wear the corporate-emblazoned firesuit. Why not display their backers' names, the patents they hold, the financial investments in their own work? Then we would know exactly the brand of science and technology on the track. Is it science with an aim toward the betterment of the total environment? Sectarian science is often wielded for vested convenience. Objectors to this are painted, nay, tarred and feathered, as part of an "anti-science movement." The solutions to complex problems are being presented in the teflon names of science and technology, while the upstream things we can do right here, right now, are being ignored. That is, taking steps to curb the need for the downstream palliative "advances" in the first place.

René Dubos had a name for this: intellectual dishonesty and escapism. Frustrated by seeing the same old schemata at medical and scientific conferences, he ripped up his planned speech at the International Design Conference in Aspen, Colorado, in June of 1969 and let rip:

> Technology cannot cure the intellectual dishonesty that burdens its channels today.
>
> Even more dangerous and more universal is intellectual escapism, the attempt to escape the real problems around us. Typical is the endless discussion among humanists as well as scientists about the prospect of manipulating our genetic endowment. This is wonderfully entertaining, titillating kind of science fiction. We organize meetings about it in all sorts of pleasant places to talk about this, and that saves us the responsibility of walking across the street, where 100,000 children are being poisoned every day by lead in paint.... Something can be done immediately about this problem, but it is not being done because it is not of sufficient interest or as exciting intellectually as talking about changing the genetic nature of man. I have rapidly improvised my speech this morning in the faith that despite power structure, soulless technologies, and the mechanization of life, we will prevail because we can remain the architects of our future.

The relationships between defenders of a certain brand of science and their products are often the equivalent of subprime mortgages packed up into complex derivatives, very difficult to see. We are particularly focused on this because when the yarn gets untangled, the groups donning said firesuits appear to be losing sight of how healing actually takes place. It takes place in the context of the ecosystem. This holistic variant of science isn't about knowing whether the world is round or flat, or appreciating incredible patent-free advances such as the sharing of refractory lenses that allow people to see. Salk never wore the firesuit. He said you couldn't patent the sun.

Naturopathic Origins of Biophilic Medicine

Holistic medicine in North America took root in the early 20th century as a German import known as naturopathic medicine. Its uncomplicated principles were virtually identical to those now described in Harvard Medical School's 2017 continuing medical education course entitled Lifestyle Medicine. During this blossoming time, naturopathic medicine focused its attention on the lifestyle of patients—en-

Does Naturopathic Medicine "Work"?

The websites of so-called excellence centers lambast the principles and practice of naturopathic medicine while making no reference to the studies that show it works, and works well. Sidestepping inconvenient science that doesn't fit with the pre-formed agenda is an act of pseudoscience. What an inconvenience McMaster University's study in the *Canadian Medical Association Journal* (Gordon Guyatt, Dugald Seely, et al., 2013) must be, showing as it does that holistic medicine offered by licensed naturopathic doctors reduces cardiovascular disease risk and the frequency of metabolic syndrome. So, too, the RAND Corporation research in the *Journal of Occupational and Environmental Medicine* showing that the major disease risk reductions via holistic medicine provided by licensed naturopathic doctors in a workplace resulted in an average savings of $1,138 in societal costs and $1,187 in employer costs per employee per year. These authoritarian websites don't seem to understand that there is a cultural context to healing, and research published in the *Canadian Journal of Public Health* (2010) shows that licensed naturopathic doctors are particularly effective at delivering culturally sensitive care to the Canadian Aboriginal community.

couraging a healthy diet, time spent outdoors in nature, exercise, sleep hygiene, stress reduction through mental relaxation techniques, and hydro- (water) therapies.

In its early coalescence in North America, the naturopathic profession was guided by its primary Hippocratic principle: *vis medicatrix naturae*. It was a biophilic perspective wherein the patient was encouraged to connect with nature to find healing through its many outlets. Obviously this wasn't something that was going to help with more serious forms of acute illness but it found followers, especially in urbanized environments increasingly disconnected from nature.

The origins of the term *physician* are a Greek-Latin mix which actually translates as *naturalist*. René Dubos underscores this when he

The solution to these inconveniences is to not mention them. Given that pseudoscience is defined by the systematic presentation of selective information as science, the warriors of ethics and sectarian science-based medicine who claim that naturopathic medicine doesn't work, and worse, that it's dangerous, become the pseudoscientists. Thus, they are actually of the same fabric as those at the other end who claim that vaccines should be avoided because they cause autism and/or that homeopathy can cure cancer. Pseudoscience is like coincidence; there aren't big coincidences or small coincidences, there are only coincidences. Sectarian science is pseudoscience.

Ronald Reagan was famously quoted as saying the nine most frightening words in the English language are "I'm from the government and I'm here to help." We disagree. It's up to governments, insulated against lobbyists, to help curb the wild west that is fueling the dysbiosphere. Time and time again industry has shown it just can't stop itself from pushing dysbiotic choices on our children. Far more frightening is when someone shows up to say: *"I'm a scientist, well at least I play one in this video, and I'm here to help you see why integrative medicine doesn't work."* If the person is wearing a bowtie, your flight instincts are correct.

quotes Hippocrates: "Every practitioner of medicine was to be skilled in Nature and must strive to know what man is in relation to food, drink, occupation and which effect each of these has upon the other." Moreover, physicians were required to understand that a disease identified in a single organ was a disturbance of the whole person: "To heal even an eye, one must heal the head and even the whole body." Thus, naturopathic doctors were essentially fulfilling the definition of the physician in this way. At least that was the noble intent.

However, its mostly laudable roots were co-opted. Homeopathy, which was not a part of naturopathy's origins, found its way onto its rolls. It is the notion that diseases can be cured by substances which would otherwise produce similar symptoms in healthy people, if only the "medication" can be delivered in infinitesimally low doses (diluted to the point where no molecules of the original substance remain). Thus, homeopathy isn't aligned with the fundamental concepts of physics and chemistry. It is, however, an elegant example of the science of the placebo.

Eventually homeopathy was joined by bogus food allergy tests, iridology, machines that can diagnose anything by your pulse, applied kinesiology (not even worth explaining), psychic healing, the vagaries of energy healing, and, most alarming, antivaccine zealots. Like a giant magnet picking up paperclips of puerile thoughts and writings, everything started sticking to the term naturopathy. In their polarized battle against allopathy, adherents lost sight of the marvels of surgery, technological diagnostics, drugs (yes, drugs, too!) and other undeniable benefits in curbing the sufferings in society.

But in the midst of all the nonsense, many once scoffed at naturopathic notions, especially those centered around the microbiome and the immune system, have actually been shown to contain more than shreds of legitimacy. Leaky gut is a prime example. Probiotics, another. Food sensitivities, diet and acne, nutrition and mental health, essential fatty acids, fermented foods, and many other topics that were once laughed off were mainstays of naturopathic practice. For all their baggage, naturopathic doctors held steadfastly to the idea that total health is only as good as gut health. *Vis medicatrix naturae* ran through the intestinal tract and the immune system. If there is any profession that has been targeting the microbiome by advocating for healthy dietary practices and using fermented foods and probiotics, it's been naturopaths.

And that's the rub. People are suffering in the here and now. They don't need to wait 20 more years until a collection of studies says yes, it's okay now, you can go ahead and give probiotics and a diet low in fermentable oligo-, di-, and monosaccharides and polyols (FODMAPs) a try in patients with a microbiome-mediated condition called irritable bowel syndrome; you know, those patients we used to label with the spastic colon that we thought was all in their heads!

We have seen it time and again: what is considered today to be based on robust evidence can become tomorrow's folklore. On the other hand, what is not taken seriously today, may be in full evidence tomorrow. The point is that we need to be open to questioning dogma, especially our own. We all benefit from expanding our horizons, and building on each other's ideas—symbiotically.

The fantastical fake news inferences of the authoritarian, culturally insensitive brokers of fear suggest that North American emergency rooms are filled with patients harmed by practitioners of integrative medicine. Not a shred of evidence supports this alarmist nonsense. Like any form of medicine, there are undoubtedly cases of practitioners using high-risk procedures outside normal training, gross negligence, and incompetence. Those are the exception, not the rule. Naturopathic doctors and their medical colleagues within integrative medicine have been successfully helping people with chronic gut problems for decades. As long as a diet isn't overly restrictive, as long as it avoids megadosing a macronutrient and/or missing out on essential nutrients, experimentation with diet through elimination and re-introduction is not unsafe. Years ago many naysayers had a good chuckle when naturopathic doctors experimented with gluten-free diets in patients with irritable bowel syndrome, depressive symptoms, and unexplained symptoms. Research published over the last several years has supported that premise.

Although it might appear that the entire state of California is on a gluten-free diet, the research shows that wheat sensitivity is legitimate, more common than originally thought, and probably plays a decently sized role in irritable bowel syndrome. Naturopathic haters emerge from their various crevices when they see these studies validating wheat and gluten sensitivities, running to "science" blogs to try and defend their old dogma. Cornered, they say we need more evidence. Sure. We always need more evidence. But it's getting harder and harder to say that food sensitivities don't exist. Rather than defending dogma, its time to start asking what on Earth has changed in our environment to be spiking autoimmune diseases and allergies, however we define the latter term. The microbiome revolution is turning the very definition of allergy on its ear.

Beyond the studies showing that trials of gluten and wheat-free diets can be helpful in some people, separate research is validating patients by other techniques. Dr. Antonio Di Sabatino of the University of Pavia, Italy, showed in a placebo-controlled study that very small amounts of gluten could provoke a host of abdominal symptoms and depression in those without celiac disease but with suspected gluten sensitivity. Researchers from Columbia University Medical Center in New York reported that individuals with wheat sensitivity have significantly higher blood levels of inflammation and immune hyperreactivity to bacterial endotoxins. This means that gut microbes are likely making their way through a leaky gut. The entire cascade may be mediated by dysbiosis of gut bacteria.

New research published in *Science Reports* (2017) by our colleague Susanne Brix shows that relatively small amounts of gliadin—small protein units found within gluten—added to a high-fat diet (vs. high-fat diet alone), can cause marked changes to the gut microbiota and promote intestinal permeability in animals. This research indicates that dietary context matters. Perhaps our ancestors tolerated gluten well because they weren't consuming it in the framework of a westernized diet. Results showed that the gliadin-induced microbial dysbiosis and leaky gut led to disturbances in the immune system, blood-sugar-insulin control, and lipid metabolism. Meanwhile, the amount of gliadin used in the study matches the percentage typically found within mass-market wheat flour.

There are few patient groups that have been more maligned by doctors than those with food sensitivities. Another group of the walking wounded in society. The so-called hypochondriacs with their wheat and dairy issues, dismissed as nonallergenic and thus with a neurosis of some sort. These are people that end up in the offices of naturopathic doctors, as they have done for years, because no one else was listening to them. Certainly not the folks running "science" blogs.

As naturopathic medicine enjoys its 21st century renaissance in North America—with increased recognition and regulation of its practitioners—we can only hope that the resurgence is one in which the magnetized paperclips get removed. In fairness to the profession's educational standards, many of the pseudoscientific practices are not part of naturopathic training at accredited universities and colleges, homeopathy being the colossal exception. Medical doctors are moving toward clinical holism to embrace lifestyle medicine (otherwise Harvard wouldn't be teaching it) and it would seem that naturopathic doctors are poised to get back to their ecological origins. It's never too soon at this point for the profession and its governing bodies to cast off the homeopathic paper clip and to take an official position on the importance of vaccines.

In this era of epidemics of chronic, complex, non-communicable diseases, we need to work together toward the common goal of healing dysbiosis. The healing professions are united by the desire to heal, yet so many factions point at each other's cures as placebos. Antidepressants really aren't much better than placebos in most cases of depression, and neither is cognitive behavioral therapy. Pim Cuijpers' research shows that scientists pushing antidepressants haven't disclosed their conflicts of interest, and that might be a factor in why they suggest medication is better than talk therapy.

On the other hand, he has also shown that cognitive behavioral therapy (CBT) is massively overblown as an evidence-based treatment, especially when so many of the supportive studies use what are called wait-list controls. The wait-list refers to folks that didn't get any therapy. Think about that for a moment. It isn't really a control at all. It's not like the wait-list group was assigned to active treatment such as gardening therapy or a nature intervention, had experience with personalized dietary interventions, or engaged with an exercise physiologist. Interestingly, sectarian science remains quiet on these discussions of placebos involving therapies that are broadly accepted as science-based. Cuijper and colleagues could only conclude that CBT is "probably effective" in major depression from an evidence-based perspective.

But that doesn't mean it doesn't work. CBT most surely does work. Patients with depression are clearly better off from their experiences with a qualified professional. Should we cease and desist all CBT for depression because it only "probably" works? Of course not. We just have to face up to the fact that much of medicine is governed by past experiences, cultural context, expectations, and belief systems. That is, the placebo. If anything, the placebo effect shows us just how powerful our mind and our beliefs are in influencing our health. It is a very real effect. Placebo groups are only included in clinical trials to subtract this often significant effect to determine the "real" effect of another treatment. We would do well to understand, ethically utilize, and even develop the power of this beneficial effect, especially as most placebos don't have side effects. Ignoring or dismissing this because we don't understand it, or because it has less commercial potential is short sighted. We strongly suspect that history will make this point for us one day.

So, although neither one of your authors consider homeopathy to have any inherent biological activity, we must also check ourselves against judgment of its cultural use; we cannot call for holistic unity here and at the same time appear judgmental against anyone who might lean upon homeopathy for "dis-ease" in daily life. Dr. Ben Gray is a brilliant physician who teaches Cultural Competency to medical students at Otago University in New Zealand. He does not use homeopathy but has recently given us, the medical profession, and society at large some food for thought:

> Culturally competent care is about having insight into our own values and beliefs, and an ability to work with people who have values and beliefs different from our own. We need to understand that the concept of culture applies widely, not just to ethnicity but to sexual orientation, education level, lifestyle, age

and perceived economic worth. We need to respect the views of others, which does not mean agreeing with them. Rather than asserting what we think is right we need to embrace diversity and learn how to live with difference.

Dr. Gray, a 25-year veteran family doctor, reminds us that if serious illness is ruled out, and no exaggerated claim of efficacy/cure is presented, a patient's use of a remedy like homeopathy that is unlikely to cause harm should not be judged. We need to understand, and respect, that cultural belief systems can have a powerful effect, *because* the patient believes in them. This is the foundation of health in traditional societies. We cry out that practices different from our own are not ethical, yet it is not ethical to deny or refuse people their culture.

Yes, issues arise when such remedies are used in lieu of available medications in cases of serious illness, or when proper medical evaluations have been bypassed in favor of homeopathy, or other practices. Yet, in daily practice physicians often prescribe placebos in the form of antibiotics, antidepressants, paracetamol/acetaminophen—because they feel pressured to give something. And often these drugs have no real benefit in the situation. But all have side effects. They can all cause dysbiosis. And each one is also linked to health problems down the line. Many considerations have to be made. More than ever, this underscores the need for health care professions to work together, to learn from each other, and all take a more holistic view that includes one another.

Disrupting the Dysbiotic Status Quo

Rather than lambasting Hollywood A-listers for writing books about the "detoxifying powers" of beet-and-kale smoothies and kombucha tea, maybe we should focus our attention on their celebrity colleagues who hawk soft drinks, fast food, polysorbate 80, and sugar-laden cereals to you and our kids. Especially our disadvantaged communities.

We need a united voice in advocating to correct the forces of dysbiosis. Yellow-gloved hands emerging from the creepy red-and-white ringed sleeves of a clown are directing our kids toward their dietary choices, targeting their gut microbes. Meanwhile at medical conferences we've attended, the participants laugh it up as some nonscientist provides pseudo-entertainment about the silliness of veggie-detox drinks and gluten-free diets for those without celiac disease. Spinning a good yarn speaking *against* the "misinformed" advocates of holistic medicine, while remaining hushed about cultural competency, the microbiome, socioeconomic inequalities, and their connection to the institutionalized systems they are speaking *for*. The same systems that are driving dysbiosis. The same systems that paid the entertainer's travel expenses and fine meals. René Dubos, the microbiologist who discovered antibiotics, saw right through this. He had the guts to chastise his fellow scientists. He called it intellectual escapism. We agree.

Holistic as a prefix to medicine is not a dirty word. It is Hippocratic medicine. Avoiding food with polysorbate 80 and synthetic food dye Red #40 accords with the beauty of its simple origins. Suggesting a homeopathic sugar pill in lieu of a life- and limb-saving vaccine is not. Routine childhood vaccines are essential to protecting the herd—that is, looking out for one another. They are in line with *vis medicatrix naturae*. Encouraging mindful contact with nature, yes. Maintaining a whole foods approach to nutrition, check. Select botanical and nutrient supplementation, sure. Overselling isolated nutrients as cures and promoting fad diets is at odds with holistic medicine's foundations. The roots of holistic medicine, as the only ecological medicine, are uncomplicated and in alignment with Hippocratic medicine; individuals who are either well or ill are viewed as a whole person in their total environment. Viewed as if the larger aspects of nature matter.

Dubos continued: "Human crowds per se are not responsible for the nervousness of life in modern agglomerations. We suffer less from contact with people than from exposure to the unnatural stimuli generated by the machines that accompany them everywhere in the industrial world. Motor cars, motorcycles, telephones, radios, television sets, and other gadgets enslave us to a nonhuman and often antihuman environment."

In the future Dubos suggested that humans might finally dismantle the ironclad notion that progress and quality of life equate to technologies of greater complexity. That through rediscovery of "the satisfactions that come from direct contact with reality through the senses and from sharing emotional experiences with a few people" there would be hope. "These satisfactions have long been and still remain the surest approach to wellbeing and happiness, but they are the ones most threatened by the ill-conceived utilization of science and technology in the Western world since the Industrial Revolution."

What nourishes our interconnectivity from one generation to the next? How does the malnourishment of "B vitamins"—that is, "B" as in Biodiversity deficiencies—on this planet compromise the fulfillment of human potentialities? How does human arrival on a red planet covered in iron dust help us to coalesce on this planet? It should be enough to recognize that a trip to the moon or Mars is a trip to a planet incompatible with human life without first donning billions of dollars of the most finely cut designer space suits.

The two largest purveyors of soft drinks on *this* planet are deeply embedded in sponsoring leading health and medical organizations. Yet at almost every turn, they block legislation that would be in the interest of public health. It's time to turn the light toward the pressure placed upon our microbes by the reckless actions of various corporate conglomerates. The University of Melbourne's Professor Rob Moodie and colleagues examined these international entities and published their findings in the prestigious journal *Lancet* (2013). They didn't mince words: "Despite the common reliance on industry self-regulation and public-private partnerships, there is no evidence of their effectiveness or safety. Public regulation and market intervention are the only evidence-based mechanisms to prevent harm caused by the unhealthy commodity industries."

This takes us back to antibiotic defender and antiholistic medicine crusader Thomas Hughes Jukes. The microbiome revolution and advances in the study of toxicology have placed Jukes on the wrong side of history. His dogmatic positions on antibiotics and chemical residues in the food supply and in our soils are now as untenable as claiming that major depression can be cured by sorting out the superego. The chemicals that Jukes defended are showing their dangers one-by-one, most especially dichlorodiphenyltrichloroethane (DDT) which is linked to a host of problems in fetal development, including lower IQ in girls. Jukes once claimed that any attempts to ban DDT were akin to genocide. Almost a half-century later it is banned in most countries, the science against it so clear, but we will never know the cost of dogma. Sectarian Science, indeed.

10

The Concert of Life: Holistic Harmony

Do plants feel? One cannot give a direct answer to this inquiry without being a plant. A man who loves his dog believes that it has feelings and emotions which are almost human; but when he goes hunting, he refuses to believe that the hunted creature has any feeling when it is being done to death. Some would extend consciousness and feeling to the quadruped, but not to the fish. At what point does consciousness enter the domain of life?

— J.C. Bose, *The Century*, 1929

Now, in our final chapter, the twists and turns of our story unite. Here, we travel full circle back to where we started, to *vis medicatrix naturae,* with new eyes. Biodiversity, as we will see, *is* consciousness. It is life. We will also see that the story of interconnection, of Darwin's clover—cats meant less mice, more bees, and thus more clover—is everywhere. Until recently, humans have only seen the tiniest portions of these ecological interdependencies. But now we know otherwise. Massive communities are in action everywhere, every day. Unseen microbes and tiny insects make their own versions of the great wildebeest migrations in Africa. Even the plants around them, around us, are alive and connected in ways we had not realized. They, too, ebb and flow with microbes.

On closer inspection of the previously unseen activities of microbes and plants, we can see the world anew. Casting light on this plant-microbe-animal-human interconnectivity will allow each of us to enter a supermarket and pick up an apple as never before; we can hold it in our hands, look at its deep colors, feel its textures, inhale its aroma, and feel the interconnection of the lives that provided that life in our hands. It represents millennia of evolution, a storied past and a more recent history of the microbial actions in soil and on leaves, colluding with the plant to shape those colors and aroma. It is the present. The apple is very much alive, teeming life-giving microbes. And so are we. More aware. More connected. More *alive* in our own awareness. The apple also represents the future. It will nourish us, providing the fibre and phytonutrients that feed our microbiome. Collectively, we bite the knowledge in that apple. Ancient wisdom and molecular biology merge into their holistic context.

We can determine more than ever before if we want to resuscitate the biodiversity that sustains life.

Consciousness and a Kind of Life

We have posited that biophilic science is that which seeks to understand the biosphere, and to do so by applying that knowledge toward the promotion of life. Promotion of the life that humans want can ultimately only be one that ensures survival of the life that sustains us. Thus, biophilic science is also concerned with how we *think* about life. How we *describe* life. Quality of life. Harmony of life. Happiness of life. Commensal sustainability of life. Respect of life. And, despite the inevitable competition of evolutionary forces, Symbiosis of life. *Joie de vivre* for all. Holosphere.

Fundamentally, we, your authors have been so bold, so audacious, as to answer the question posed by renowned physicist and Nobel laureate Erwin Schrödinger: What is Life? The answer is consciousness; being fully aware of, and responsive to, one's surroundings. More specifically, awareness of the environment, in any present moment, and the capacity to respond to it. It has nothing to do with being the possessor of a human brain. Consciousness permeates the biosphere; from aloe plants to zebras, from acidophilus to zeaxanthinibacterium, it is an attribute that unites us all.

A better question is not, what is life, but rather what *kind* of life do we want? The noble crusade of biophilic science can help us answer that question. Here on Earth, we humans have found ourselves at a place of great responsibility. As Percy A. Campbell, the Harvard-trained electrical engineer turned philosopher who truly defined probiotics, stated in 1943: "Man lives largely at the expense of other animals, yet he shows little inclination to apply to them the golden rule.... To the naturalist, who envisions the world of life as a unitary evolutional process of nature, this high disregard of [other life, is] often indefensible self-profiteering.... Homo sapiens occupy the evolutional place of honor with respect to the whole terrestrial drama of a billion years. Noblesse oblige."

With the culmination of biophilic science also comes the obligation to take responsibility for a world in which there will soon be nine billion humans exerting intense pressure on ecosystems large and small. This brings a strong call to promote scientific perspectives that unite us in our humanness and in kinship with all other forms of life. For example, it allows us to see that antibiotics are not simply chemicals with biological impact. They spill into political, social, economic, environmental, and ecological injustices. The delicious apple we might use as our surrogate for taking hold of nature must be fresh, widely available, affordable, and free from the antibios pesticides that tear up the microbiome of life.

Biophilic science allows us to understand how and why the human microbiome can be influenced by the lack of a living wage. Key word: *living*. We might call it a probiotic wage. It allows us to understand how a manipulated environment might be pressing down on the microbiome of health; whatever the utopian microbiome profile might be, it's becoming fairly clear that the one carried by most of us today, the one essential to the soils that grow our food supply and dominating the indoor air we breathe, is probably not the ideal.

As we work toward these goals, it would help if we shed our anthropocentric pelt. Humans have always pondered the existence of some easily drawn line defining sentient forms of life residing within the biosphere. We have a tendency to apply our anthropocentric view of consciousness to the macro forms of life, the beings that move like us. Especially things with eyes. But not houseflies, they have too many eyes. Plus, they carry *Campylobacter*, that nasty microbe that can cause food poisoning, and as you may recall, the one that can also fire up the emotional centers of the brain, producing anxious behavior even when administered in miniscule amounts to rodents.

Think of the popularity of Animal Planet's series entitled *Whale Wars*, in which brave crews are filmed chasing down the hunters of whales. Despite massive species losses, series entitled Butterfly or Amphibian Wars probably wouldn't have done as well during sweeps week. The means of human assault on all these animals may differ, but in the era of the sixth mass extinction, the destructive ends are the same. Harpoons on the one hand, neonicotinoid pesticides on the other. Labels such as "lower" forms of life were acceptable in the time of Darwin, but today they ignore the complex mutual interdependencies within ecological systems.

Formal Sentire

From its Latin root *sentire*, which means "to feel, or to perceive," we can easily argue that all life is sentient. Even microbes, the smallest among us, are certainly capable of sensing and perceiving the conditions of their environment. Seriously capable! For example, microbes can detect a 0.1 percent change in the concentration of sugar in an environment, a change equivalent to a single drop of sugar in a pool of 1,000 drops. They make biochemical and biophysical responses in kind. These decisions about how to live—where to set up camp, what to be attracted to and what to defend against—are based on such ridiculously small environmental changes.

It would be really difficult for microbes to maintain their range of existence from the depths of the ocean to the upper ranges of the atmosphere if they were incapable of swift responses to their local environments. The microbiome revolution has uncovered vast networks of intracellular machinery, signaling capabilities, and all sorts of regulatory chains within microbes, all of which allow them to survive and appropriate their needs from so-called higher life forms like us. They work together socially, and even, in a sense, altruistically, to achieve those ends. Are microbes actually super-intelligent? The affirmative answer is not only turning healthcare into clinical ecology, it is transforming the way we view the living world around us.

Our anthropocentric view of *sentire* is being undone by remarkable biologists such as Marie-Claire Cammaerts of the University of Brussels. For many years it has been known that various creatures, including our own dogs and cats, can react to the presence of food some time before its effective delivery. That is, they are able to acquire spatiotemporal learning over a relatively short period so they can anticipate the time and the place at which an event—feeding—regularly occurs. Astoundingly, Dr. Cammaerts has shown that ants can do this as well. In other words, if she were to create a new ant-themed park in Orlando, after a few days she could have the ants show up center stage knowing, from experience, that she would throw them a fish-shaped candy at a certain time.

If Dr. Cammaerts shifted the feeding time by just 20 minutes, the ants caught on fast. What's more, her work shows that if the candy fish (she used sugar, just not shaped like candy fish) were only available for 15 minutes and then removed, ants quickly learn that they need to chow down faster and become more aggressive in the face of impending food removal; after only two rounds of experience they knew the time frame in which the food would be removed. This research reveals that we are only scratching the surface of the ways in which tiny creatures develop expectant, intelligent behavior. Scientists are also just tapping into the remarkable ways in which insects develop symbiotic relationships with microbes, and how fungi may be the life blood of many insects. For example, certain insects break down plant material into food for the growth of fungi; insects at the larval stage are protected by fungi's ability to inhibit the growth of harmful bacteria! The more we look, the more it seems the default position of nature is cooperation for mutual benefit—far more complex than the simplistic assumptions made in ignorance. Seeing nature (and even human nature) only in terms of "survival of the fittest" and "kill or be killed" is to underestimate the power and the potential of nature. Including our own.

That brings us to the other end of the size spectrum; from tiny ants we can visualize the largest living creature (as measured by volume), the giant sequoia. Tourists may look up in awe at their vastness, and perhaps because of their mammoth size and ancient existence, just like the giant whales, they pull on the heartstrings more than a pine tree in suburban Pittsburgh, or a white gum tree in Perth. But whales are mammals with eyes, so it's still difficult to imagine massive evergreens—let alone trees that aren't 760 meters tall—as being sentient, social, and loaded with intelligence.

Trees, of course, have branches. But because we often can't see them, we forget that they also have hands with fingers that touch and feel—roots. In the case of the giant sequoia those hands and fingerlike projections can be a hundred meters long. Trees do more than sway in the wind; their once secret underground movements in association with microbes are a secret story no more.

The fixed nature of plants belies their unseen operations below the surface, elegant performances that include rich dialogue, nurturing, sharing, and even what looks to be altruism and self-sacrifice. Dr. Suzanne Simard of the University of British Columbia has published groundbreaking studies on the symbiotic association between trees and below-ground fungi. The latter can't photosynthesize but they can get the sugar made by photosynthesis through the generosity of trees. In return, the trees take up the extra nutrients—water, phosphorus, and nitrogen—provided by the fungi. The fungi are brilliant hunters, sending out a web of long thin strands of mycelia through the soil to collect these nutrients.

The fungal mycelia play the part of an underground network of nourishing tubes connected to tree roots, connecting all the trees to each other. Literally, a "wood-wide web!" Remarkably, trees actually cooperate with each other by sending much-needed nutrients to one another through this tubular network. Even more impressively, Dr. Simard's work shows that the trading goes on even with trees of different species, including when there are seasonal differences forcing higher needs by particular tree species. What's more, trees linked together by these networks behave altruistically to support the whole ecosystem. In the face of mortality, the elder trees, which have the largest links throughout the community, dump their nutrients—carbon—into the fungal pipelines so it can be taken up by younger trees for their survival.

But the fungal network also acts as a conduit for warning signals or defensive communications which help other plants, even those of completely different species. For example, when Douglas fir trees are injured, they can produce distress signals that alert neighboring Ponderosa pines to arm themselves in preparation for the threat. This is seen in an uptick in the activity of the Ponderosa antioxidant defense systems. The key is that the Ponderosa pines are connected to the Douglas fir trees via the fungal pipeline. If the mycelia are cut off, the warning signals can't make it through.

Microbial Shane *and* a Pot of Chili

Diving into the fascinating story of the ability of trees to communicate below ground, it is easy to lose sight of the role of microbes. It turns out that the sorts of microbes that gather around the mycelia can dictate the health and growth of trees. Gram-negative bacteria, often living up to their negative label, can stymie growth. Mycelia are clearly the underground telephone lines for many trees, but life in the plant kingdom also uses wireless methods to communicate with one another topside. Plus, as abundant as mycelia might be on planet Earth, they aren't everywhere and many plant communities thrive without them. Bacteria, however, are everywhere.

Above ground, plants do their talking via interactions with insects and, yes, you guessed it, microbes. As discussed earlier, plants release volatile chemicals into the air. These are the phytochemicals that we inhale, which in turn can provoke beneficial effects on our immune system, and, by extension, our mood. Emerging experimental evidence also suggests that these minute concentrations of airborne phytochemicals might have a favorable influence on our gastrointestinal microbiome. But we anthropocentrically digress. The chemicals secreted by plants also serve to let other plants know that they might need to defend themselves against pests or other environmental stressors.

Microbes are facilitators of communication. The ones hanging on to the phyllosphere—the aboveground portion of the plant, its stems, leaves, flowers, bark, and branches—can help determine the volatile chemicals manufactured by the plant. In addition, friendly microbes respond to the call of the plant when a predator is on the scene.

Here's how it works. We can use the example of the phloem-sucking whitefly as it shows up to poke holes in a pristine leaf of a pepper plant. In response to the little plant vampire, the plant sends out chemical signals down into the soil. These chemicals are the equivalent of organizing a posse. New bacteria show up around the roots, like ranch hands responding to a cattle-driving intruder on the homesteads of Shane's Wyoming.

The recruits include the class of Gammaproteobacteria. Recall, it was the family linked to allergy-protective effects within the green space surrounding European homes. This enlistment shores up the defenses around the roots as well. It is as if the pepper plant knows that while it is contending with an assault above the ground, at its leaf, it is ill-prepared for an assault from the soil below. In addition, since many leaf pests begin their life cycle in the soil, the belowground defense is a nice example of concern for future consequences.

The pepper plant provides an excellent example of the interconnectivity of life because a recent human study showed that chili pepper consumers had a 13 percent reduced risk of all-cause mortality. In order to take advantage of the edible peppers, to allow them to benefit our ecosystem, we need the entire plant to maintain its healthy growth in its own environment. Here we can see so many parallels with Darwin's clover story of interconnectivity. The recruitment of helpful bacteria to the root also enhances the plant's aboveground defensive work, helping tackle the whitefly on multiple fronts, to reduce the threat to the chili. Your chili. The very fruit destined for your delicious meal. This fruit has been shown to have beneficial effects on the gut microbiome and the immune system. Not to mention, it's the same culinary additive shown to extend lifespan!

Obviously pepper is one of countless edible plants that work to promote health. Even in using it as an exemplar of the clover-style interconnectedness of life, we are reluctant to elevate one particular food to superfood status. This happens all the time in the media and it ends up diminishing the synergistic way in which dietary phytochemicals work. But the pepper does provides a superb example of how personalized precision medicine may be on the horizon vis-à-vis the microbiome. In a study published in the *Journal of Clinical Endo-*

crinology and Metabolism (2016), researchers showed that the extent to which an individual might benefit metabolically from peppers—lowering obesity, type 2 diabetes, and cardiovascular disease risk—is dictated by their microbiome enterotype. These so-called enterotypes are clusters of microbial families in the gut.

The implications are many, including tailored diets and interventions that might target the microbiome in a way unique to an individual. Even though you may be healthy today, the dominance of a particular enterotype may determine what you get out of chili, or other foods, over the long haul. So really, the studies showing that chili is associated with reduced mortality are very complicated and deeply embedded in the microbiome. Indeed, you may need to send out a call to recruit some friendly bacteria from your doctor's office to help your own root microbiome fight the whitefly of modernity. Hopefully, that day of what is now increasingly known as "personalized medicine" will be here sooner than later.

Climate Change; Probiotics for Plants

In general, the bacteria in the soil surrounding plants differ from those in the bulk soil that might exist nearby. Below the surface of towering giant sequoia or tiny clover, this diverse set of microbes hanging out in the rhizosphere, the area around the roots, get what they need. And they give back in kind. In addition to all their defensive work, they also provide plants with a useable form of vital nitrogen. Just as our commensal bacteria get what they need from us, and we them.

When the ecosystem is healthy, things work like a charm. But as is the case with humans, stress can tip the apple cart, and that is when it becomes clear that certain microbes are no friends of plants. In fact, there are microbes that can actually *increase* the susceptibility of your tomatoes, cabbage, not to mention urban-foraged foodie plants, to attack from rootworm, whitefly, and other threats.

Scientists aren't sure exactly what to expect concerning the effects of climate change on environmental microbiota, but new research from Dr. Vanessa Bailey and her colleagues doesn't bode well. Recently, she evaluated soil samples that had been transplanted 17 years earlier from cooler moist sites to warm dry locations. There were little signs of adaptation to the new, warmer digs, even after nearly two decades. Soil microbes are also an essential consideration in carbon sequestration. Microbes are the most adaptable organisms on earth, but if bacteria can't readily adapt to new climate change conditions within areas where most humans live and grow food, that is cause for concern for the wider ecosystem. The time for adaptation is passing us by. NASA reports 2016 was the hottest year on record, and the area of the world's oceans covered by floating sea ice is the smallest recorded since satellite monitoring began.

In agriculture, continuous monoculture is to a healthy soil microbiome what eating a steady diet of fast food is to the human microbiome. Microbial dysbiosis sets in, followed by disease in the crops. In addition, the research of microbial ecologist Eva Figuerola demonstrates that lack of good agricultural practices is putting the diversity of soil microbes at risk, perhaps in the same way that we are doing by continuing to reduce our own fiber intake. Several studies have shown marked reductions in soil microbial diversity based on farming techniques; differences between organic and so-called conventional farming in their effects on soil microbiota have been noted. "Conventional" in this case means using herbicides, fungicides, and pesticides. The results are similar to monoculture—loss of microbial diversity!

Between climate change, monoculture, and the absurd amount of chemicals we throw on top of the Earth, there is legitimate concern that essential microbes—bacteria that are instrumental in promoting the health of edible plants and, by incidental contact, humans—may be going the way of the woolly mammoth. For all the talk of how genetically-modified crops are going to feed the world and save us from ourselves, a detailed *New York Times* report ("Doubts About the Promised Bounty of Genetically Modified Crops," October 29th, 2016) shows that they have not lived up to their vast promissory notes. Massive amounts of pesticides and herbicides are being used despite colossal expansion of such crops in North America. The fears aren't of the genetic modification itself, it's the chemicals that could work their way through the ecosystem. *Silent Spring* lives on.

A 2003 study in the journal *Ecosystems* indicates that once unique soils in certain parts of the United States are gone forever, with many others close to the edge of extinction. We should be very concerned about this possibility because the soil is a treasure of biodiversity, one that is being encroached upon. Nearly one-third of terrestrial lands have agricultural crops or planted pastures as the main type of land use, and in a few short decades there will be nine billion people to feed. The treatment of soils in that mission should be done with microbes in mind.

But there is cause for hope. We can find optimism in research showing that we can inoculate the soil and suppress disease by the application of beneficial microbes. Since select bacteria and fungi appear to work for the promotion of plant life, why not use them in probiotic mixtures as a way to improve crop health, whether in your backyard hothouse or on the large scale? This was Parker's vision when he coined his version of the term probiotic in the 1970s. Modern research is demonstrating the tremendous value of such applications, and with emerging technical knowledge of the microbiome, hopefully we are entering a new era of sustainable agriculture. The application may also lead to atmospheric greenhouse gas reductions.

Emerging studies show that, much like with human probiotics, combination formulas of friendly bacteria and fungi added to soil provide synergistic value above and beyond a single-microbe formula. For example, when two root-associated microbes were applied together as soil probiotics, they reduced arsenic uptake into rice, and strengthened resilience against a harmful microbe that otherwise causes global losses of enough rice to feed 60 million people a year. Topside, on the leaves, stems, and branches the application of probiotics are also showing value. Microbial diversity is known to enhance community resistance to pathogen invasions, so why wouldn't it work for crops? Jie Hu and colleagues from Utrecht University's Institute for Environmental Biology used a combination of experiments to show how the diversity of multiple strains of applied probiotic bacteria can protect tomato plants against disease.

Plant microbiome experts from the University of Warwick recently weighed in on the synergy of probiotic soil microorganisms, writing in the journal *Plant Molecular Biology* (2016): "This synergism serves as an indication that experiments focusing on the effects of single microbial species may overlook important multipartite interactions of more naturalistic microbial communities. The study of microbe–microbe synergism might provide valuable models to decipher underlying communication and validate these findings in more complex microbial communities."

Bose and Plant Music

The point of our discussion of how plant communication is facilitated by fungi and bacteria has been to underscore, once again, that so much of life and its glorious music is transmitted in ways that are not audible to the human ear. Microbes are at the very heart of these communications, and are thus representative of the interconnectivity of life. When we use language such as "trees talking" and "altruism" it is obviously for convenience, and the descriptive term isn't literal in the human sense of the words.

In 1880, Darwin provided some fuel to the notion that plants have their version of a brain in those "hands" we referred to—the root tips. In his words: "It is hardly an exaggeration to say that the tip of the radicle thus endowed, and having the power of directing the movements of the adjoining parts, acts like the brain." Following this, several publicized articles by distinguished authors, including psychiatrist R. F. Hutchinson, positioned plants as sentient beings. Mostly, however, these were speculative. But Sir J. C. Bose was an exception.

Our own discovery of J. C. Bose was by pure chance. Early in our research for *Secret Life*, we scoured many older magazines foraged from eBay, primarily because they contained the writings of J. Arthur Thomson; while thumbing through copies of *The Century* to find the Thomson article we needed, a particular title jumped out from the Table of Contents of the February 1929 issue: "Are Plants Sentient Beings? Recent Discoveries Have Revealed the Unity of All Life, By Sir J. C. Bose." Seeing it for the first time together, we looked at one another, speechless. Herein, Bose described elegant research in which he showed that plants are electrical beings. They perceive and move based on electrical impulses.

Up to that point, plant communication was thought to be exclusively hydromechanical, like water pressure moving in pipes. He proved otherwise by showing that the movement of impulses in plants could occur without altering the flow within its vascular system. Importantly, he stated that "the pipe will not lose consciousness and stop the flow of water if it is chloroformed." He was emphasizing that upon chloroform anesthesia, plants can lose consciousness—their normal movements and responsiveness suspended—just like humans. But just like us, if the level of anesthesia is carefully monitored, our blood continues to flow. When chloroform is removed, normal movement and responsiveness return. He concluded that plants have their own unique nervous systems, albeit different from the neurons that comprise the human nervous system.

He also argued that consciousness is not a feature exclusive to the human nervous system. We agree. Our more specific thesis is that the immune system is a mobile mind; hence it is easy for us to suggest that the immune system is a form of consciousness.

Sir J. C. Bose knew a thing or two about energy and electrical communication. He was a biophilic scientist cut from the same cloth as Salk. Pure genius, he was an acclaimed biophysicist who shared his tremendously novel ideas with the public. Freely. The very epitome of what STEM should mean.

While best known for inventions associated with wireless communications—by some accounts, well before Marconi—Sir Bose was famous for his distaste of patents. He described why he was interested only in biophilic science in *The Century*: "There has been a feverish rush, even in the realm of science, for exploiting applications of knowledge, not so often for saving as for destruction.... He forgot that far more potent than competition, is mutual help and cooperation in the scheme of life." Like Salk, Bose also opened a research institute "where scholars devote their whole life to pursuit of knowledge for the common benefit of humanity." Contemporary portraits of Bose make little mention of his groundbreaking work on the secret, sentient life of trees, but there is even less public appreciation for how this incredible scientist viewed nature as the facilitator of scientific creativity. Some clues emerged in *The Century* piece as he describes the working environment at the Bose Institute:

> The Garden [is] the true laboratory for the study of life and its marvelous manifestations. There the creepers, the plants and the trees are played upon by their natural environment—sunlight and wind and the chill at midnight, under the vault of starry space...sheltered by trees, the student will watch the panorama of life. Isolated from all distractions, he will learn to attune himself to nature; the observing veil will be lifted, and he will gradually come to see how, throughout the great oceans of life, community outweighs apparent dissimilarity. Out of discord, he will realize the great harmony.

There have also been many pseudoscientific articles and books describing plant behavior published over the years. They capture the attention of the public, but when the truly amazing life of plants is embellished into articles suggesting that they can read minds, it becomes a spoon-bending fiasco. The end result is damaging. It frightens serious researchers off the scent of a legitimate scientific path. This is what happened with the pioneering science of the gut microbe-to-mental-health research in the first third of the 20th century.

It's occurring now in the realm of the massive introduction of terrestrial, human-generated electromagnetic (EMF) radiation. Far-fetched stories on EMF dangers make it too easy to laugh this topic off as tin foil hat stuff. Researchers doing legitimate work dare not speak out of fear of ridicule. But some are speaking. Research published in the reputable journal *Science of the Total Environment* (2016) shows the technology we have come to depend on may be less than healthy to trees. Experimental studies show that EMF in the wireless communication range can influence the growth of microbes. There are countless reports of nonthermal effects of EMF on biological tissue. Nonthermal means not related to heat and thus nothing to do with microwave ovens or bogus videos showing that you can pop popcorn with two mobile phones!

From a biological perspective it would actually be shocking if EMF had *no* effects on the smallest, most sensitive creatures among us. Among many studies, evidence published by Alfonso Balmori in *Science of the Total Environment* and Marie-Claire Cammaerts in *Elec-*

tromagnetic Biology and Medicine shows that seedlings, shrubs, trees, birds, and insects may be far more vulnerable than we are to the effects of EMF. Those effects include diminished reproductive capacity. Ironically, the bucket-loads of pseudoscience concerning the effects of EMF on humans has served to diminish this fact. Dr. Amparo Lazaro's 2016 research in the *Journal of Insect Conservation* shows that EMF effects on flying insects appears to be species dependent. Some are more vulnerable than others. We suspect that J. C. Bose, the early expert on electromagnetic influences in biology, would agree.

Comfortably Numb

The culture of science isn't very kind to those who may question its widely accepted certitudes. But some scientists forge on, and when armed with solid, reproducible data, they climb into the cab of the Bantam Dragline crane and swing that wrecking ball so hard against the masonry of falsehoods. Many times the true heroes of science are never acknowledged, having suffered just like the patients with "spastic colon" and other so-called psychosomatic illnesses who were patronizingly told it's all in their heads.

In recent decades Dr. Stefano Mancuso has picked up where Bose left off, completely transforming how we view the plants we often walk past as if they were bricks and steel. Dr. Mancuso established the field of plant neurobiology. Yes, plant and neuro! His recent book *Brilliant Green* with Alessandra Viola is truly exceptional. Never mind five human senses, Dr. Mancuso and colleagues in this area of research show that plants have over a dozen additional senses. These senses are used to juggle behavioral responses to what they feel in the total environment, including but not limited to humidity, gravity, electromagnetic radiation, chemical gradients, and gases.

Since the release of *Brilliant Green*, Dr. Mancuso has published several more papers, including one on the coordinated movements of corn plants. There are many examples of this in the animal world, including flocks of birds, animal migrations, and especially those schools of fish that move in unison just before a Great White shows up. The roots of corn seedlings growing in a group appear to coordinate a plant's motion in association with its neighbors. Mancuso's team showed that young corn plants display coordinated motions that are unrelated to the light source as in the classic motion of sunflowers following the sun. As corn plants were separated from each other, their coordination declined. The researchers theorize that plants may interact through touch in the form of electric field detection, or by the release of chemical substances.

Two other papers caught our eye. The first concerned plant ocelli, which is not a form of gluten-free pasta. It is a plant's visual apparatus. Its eyes. *Cyanobacteria*, or blue-green algae, have been shown to use a system of micro-optics set up like a retina and lens to sense light direction. But several new studies show that plants gather and use information from light in their environment to recognize the shapes and forms of plants growing nearby, including their own kin. This information is processed which in turn leads to movement of leaves, shoots, and roots. The ocellus may indeed be the eye that sees you.

Finally, we make note of another interesting research paper by Mancuso and colleagues on anesthesia. The precise mechanisms by which anesthesia induces loss of consciousness are still scientifically foggy. Claude Bernard, the renowned scientist who basically said to never trust dogma, was known to have performed volumes of studies on animals, so when he concluded that "all life is defined by the susceptibility to anesthesia," people listened. This is turning out to be correct, as even microbes are susceptible to anesthesia!

Microbes, algae, and plants also make massive amounts of chemical anesthetics, especially when they are in stressful situations. Plants also produce pain-relieving chemicals. Think back to our friend the pepper plant. One of the chemicals it manufactures when it is being ravaged by whitefly and others is salicylic acid. Aspirin! So, too, the anesthetics are made in plant fruits; it is possible that these chemicals make them comfortably numb as they are literally being eaten alive by various creatures. This act allows for the reproduction and spread of seeds of flowering plants, but the chemicals might act as an epidural, or have some effect on their predators we have yet to understand.

Credit goes to progressive scientist Dr. Frantisek Baluska of the University of Bonn, Germany, for bringing these ideas on anesthesia, microbes, plants, and consciousness to life. Literally, to life. Breaking dogma, the key point in these discussions is that anesthesia kicks in with extreme speed without any direct involvement of genes. Restoration of consciousness is typically rapid as well, which is truly amazing. Thus, our susceptibility to anesthesia unites all forms of life. Consciousness, which is what anesthesia takes away, allows us to recognize danger and makes responses that safeguard our survival.

Germ Theory for the 21st Century

The quote to the right from Dubos displays tremendous foresight and truth; he had a knack for going places where others wouldn't dare tread. He was making a stand that microbes would be proven to be connected to vague symptoms, tension, fatigue, stress, and the emotional states that detract from quality of life. He had the foresight to link microbes to chronic disease *before* the epidemic of non-communicable diseases had even started, by making the case that the ecology of the person, their "climate" or terrain, would dictate how the road to non-communicable disease is travelled.

> **Ubiquitous microbes rarely cause death, but they are certainly responsible for many of the ill-defined ailments—minor or severe—which constitute a large part of the miseries and "dis-ease" of everyday life. They establish a bridge between communicable and non-communicable disease—a zone where the presence of the microbe is the prerequisite but not the determinant of disease, a situation in which the fact of infection is less decisive in shaping the course of events than the physiological climate of the invaded body.**
>
> — René J. Dubos, *Scientific American*, 1955

The title of this piece in *Scientific American*, "Second Thoughts on the Germ Theory," was particularly gutsy in its time. In mid-century modern America, the germ theory of disease wasn't really a theory. It was the only way to see things; an elegant and untouchable passageway between microbe and disease. Virulent microbe gains access to host, multiplies in tissue, and causes sickness, sometimes lethal. As Dubos said, "the germ theory of disease has a quality of obviousness and lucidity which makes it equally satisfying to a schoolboy and to a trained physician." But, he went on: "In reality, however, this view of the relation between patient and microbe is so oversimplified that it rarely fits the facts of disease. Indeed it corresponds almost to a cult—generated by a few miracles, undisturbed by the inconsistencies and not too exacting about evidence."

The problem with germ theory is that it doesn't address why so many people harbor the microbes that cause disease and yet remain healthy and vibrant. The microbe that causes sickness and ulceration of the stomach and small intestine—*Helicobacter pylori*—is one example of many. This microbe has been following the human species around in our migrations over the globe for the last 100,000 years, diversifying its genes along the way. Less clear, however, is why about half of all humans are carrying it in the gut right now, but only a small percentage of humans succumb to infection from this little beast.

The presence, or actually, now, the *absence*, of *H. pylori* is also implicated in many non-communicable diseases that were at one time considered to be psychosomatic. Emerging research has connected the microbe to migraine headaches, diabetes, rheumatoid arthritis, autoimmune conditions, and dementia. But annihilation of the microbe isn't the answer either. Eradication of *H. pylori* with antibiotics and *lack* of exposure in early life may prevent proper training of the immune system. The Hygiene Hypothesis once again. This is a microbe we have carried with us for millennia. Removing it unnecessarily may strip away yet one more layer of protection against subsequent development of asthma and atopic disease.

But the same cloudy picture emerges when focusing only on stress and the psyche as well. Psyche-dominant theories posit that stress *is* the disease. But drilling down, researchers find it hard to explain why there are significant individual differences in the physiological responses to psychological stress, and the risk of diseases that are linked to stress. Even when researchers do their best to hone in on individuals who have shared somewhat similar life histories and previous stress exposures, differences in stress responses and disease risk emerge. Thus, along both lines of inquiry, specific microbe alone, or stress alone, neither can fully explain all disease by itself.

The germ theory for the 21st century is the probiotic theory of disease. Rather than focusing on one germ, or pathogenic microbe, we must now ask which microbes are preserving the health of the host?

It is evident that some collection of microbes must be probios, acting in the interest of life. The exciting part of the new probiotic theory of disease is that it will allow for personalized treatment.

Transferring Empathy

Parents and children in westernized nations are constantly bombarded with messages concerning the frightening ways in which germs are spread by human contact, including shaking hands. North American supermarkets, including ours, have enormous center-of-the-aisle displays emblazoned with "*Kills 99.9% of Bacteria*!" These chemical spray products surround a smiling baby depicted with food all over her little face. The message is clear: germs are to be feared. The current President of the United States refers to shaking hands as a barbaric practice. Maybe so. But the transfer of microbes always seems to be ominous. Does it have to be? There are indirect ways in which we can transfer a healthy microbiome to one another as well. Kindness. Empathy.

Research in humans and wild primates indicates that the characters we hang around with, not necessarily even our blood relatives, but the ones we are near, can shape our own microbiome. There are distinct similarities in the microbiome profile when we share space. With good reason this has been attributed to transfer via physical contact and similar dietary habits. But is it possible that the similarities are also a product of how we treat one another? Does altruism, emotional solidarity, nurturing, and love play a role?

Volumes of animal studies and a growing number of human studies show that stress is the provocateur of dysbiosis. Thus, if we want to protect the health of our microbiome, it hinges on more than just the critically important loop of healthy foods and spending time in nature. It also rests on stress management and doing our best to mitigate sources of stress, at least to the extent we can. There are countless books on stress management techniques for self care. Mindfulness in particular can help reduce stress and enhance the joy and benefits of nature experiences. But what about the way we treat one another within the system of stress?

Earlier we discussed our grave concerns for losses in societal empathy. In fact, it chills us to the bone. We have covered the many ways in which empathy and emotional intelligence can help assure personal success in life. But it is important to consider the ways in which it might impact the health of your human relatives and the microbes that live upon them. Remember, the folks we interact with every day—family, friends, co-workers, classmates, and strangers—are holobionts; that is, multispecies beings going about their business at home, work, school, and all around town. Teeming with little microbial pets. It's hard to think of it that way because we can't see them. But the primary *Secret* of the microbiome revolution is that the whole universe in which we live really isn't what it appears to be.

We can look to research to show just how far a little empathy and some heart can go. In 2014, Dr. Jorge Fuentes and colleagues published a study involving empathy in the journal *Physical Therapy*. The subjects in the study had chronic lower back pain and were split into four groups. All were hooked up to an electrical current device that is commonly used to treat back pain in medical offices. However, the researchers only switched the "on" button to the device for half of the subjects. The other half had all the trappings of a machine working its electrical magic but it wasn't actually on. To make things even more interesting, for half of all subjects the therapist engaged in warm communication and fostered a sense of collaboration and support. In other words, the subjects experienced empathy.

The results tell a tale not only about therapeutics and the art of medicine, but about daily life. The patients for whom the "on" button remained off, and the therapist was "off" (in the sense that they paid little overt attention to the subject) still experienced a respectable 25 percent reduction in pain. Button switched on and the pain decreased by 46 percent. Not too shabby. But when the electrotherapy power button was off and the therapist was empathic, the pain reduction was 55 percent. Which means that empathy alone outperformed a standard intervention! When both switches were on—current and empathy—there was a whopping 77 percent reduction in pain.

It's easy to pigeonhole this study into yet another call for so-called allopathic medicine to wise up and start taking the *art* of medicine seriously. But it's far more than that. It would be naive to think that our mobile mind doesn't respond in healing ways to love, kindness, and empathy in all of our experiences. We certainly know how our immune health suffers when we experience stress and hostility. The converse is also likely. So we can tend to one another's immune systems and microbiomes through the way in which we choose to interact. We can choose to keep the empathic amp switched off, or crank it up to 11.

To illustrate how interactions might impact the microbiome, we can look to a 2014 study published in the journal *Brain Behavior and Immunity*. For 16 weeks, subjects with depression met once weekly with a therapist and engaged in cognitive behavioral therapy. Each session lasted an hour and a half. At the conclusion of the block of sessions, many patients improved. Blood tests revealed that subjects had decreased inflammatory immune responses to the endotoxin derived from intestinal bacteria. As you know from earlier discussions, endotoxin finds its way through a permeable intestinal lining. The remarkable inference is that talk therapy might be able to settle the microbiome and patch up a leaky gut.

This study is less about cognitive behavioral therapy and more about care in general. As we have pointed out, the meta-analyses of Pim Cuijpers and others show the effect size of this particular brand of talk therapy is only marginally better than placebo, and there is no convincing evidence that it stands above other forms of talk therapy. The point is that the patients experienced professional attention and care. Lots of it. It is likely that 1.5 hours a week of professionally guided, personalized horticultural therapy for 16 weeks would also soothe the gut lining. Or 16 sessions of holistic medicine focused on diet, culinary preparation, and healthy eating.

The placebo effect is a source of frustration for scientists trying to sort out the true efficacy of an agent. But outside labs it's a thing of beauty. Why does the placebo have to be something penned in by medical constructs? Its lessons can be set free, reaching out to all of us in daily life. We can fire up our well-documented healing response pathways—placebo or whatever the anthropocentric label—in the ways in which we communicate with one another. *Vis medicatrix naturae.*

Taking Back the Microbiome

Keeping the conversation only within the health of our own species, one of the reasons the microbiome must be taken so seriously is because it involves our children and the next generation of kids. Recall, microbes have been passed down to us at birth since time immemorial. Well, at least traceable to a 15-million-year-old ancestor, anyway. The implications of loss of microbial diversity, and the vicious cycle of junk food adherence which ensures suppressed diversity, can literally determine how a young life unfolds and set the stage for health in later life.

Modernity and the forces of dysbiosis have clearly transformed the microbes that we pass on. Taking back the microbiome means addressing the lifestyle that causes dysbiosis. But it's also a road that must be paved by equal opportunity. Inequalities stand in the way of access to biodiversity.

Opportunity is defined as the conditions favorable for the attainment of a goal, advancement, or success. Options. Prospects. Opportunity is the canvas of your future. Often, "opportunity" is assumed to be a level playing field for all; it is mixed in with assumed social and environmental equity. This golden chalice of conjecture is filled with mythical waters. An elite elixir where the conditions favorable for success are viewed just like regional weather; that is, they apply equally to all. But the fact is we don't have equal access to the conditions favorable to success.

Many of our youth are now seeing through the facade. Once emboldened, no group is better equipped to create change than our youth. Seeing through marketing manipulation and a system rigged toward global dysbiosis, they are already taking steps to change the system. In a 2016 study, researchers reported on the ways in which young people might exert their influence in transforming the local food environment in a disadvantaged neighborhood. The researchers found that empowered youth act as strong agents for social change. The teens reached out in multiple directions to individuals, shopkeepers, and society as they attempted to brighten a touch of grey space. They weren't motivated by the one millionth message that it's important to eat healthy. Rather, when the topic of healthy nutrition and the food environment is presented to youth in its actual ecological context—presented, that is, as exactly what they are, which is food justice and power inequity issues—they get fired up.

In the Los Angeles study, neighborhood corner stores were reorganized, one might say reverse engineered from their norms. Healthy foods were brought up front and unhealthy foods sent to the back; community high school students transformed the overall physical appearance of the stores and flipped normal marketing inside-out. They engaged in community marketing that emphasized healthy eating and the benefits of shopping locally and learning to eat healthy foods.

Youthful eyes glaze over when another food pyramid—or whatever shape the lobbyist-inspired nutritional information is formed into—is presented. Not so when teens and young adults are informed on the relationship between food *environment* issues and poor dietary behaviors. Knowledge becomes power and motivation for change. Understanding that grey space is a manipulated, AstroFood-like environment—one that is causing hurt in the community—is science knowledge they need. Empowered, youth become community advocates for change, educating parents, teachers, and others. No one likes getting hoodwinked, most especially our youth.

In 1973 the founder of McDonald's, Ray Kroc, had this to say to *Time* magazine concerning his (then) 500 million dollar fortune "I expect money like you walk into a room and turn on a light switch or a faucet, it's not enough." Today, young people are learning what that statement means to them and society. What it might say about gluttony, greed, and the marketing that supports them. What it says concerning the establishments that set themselves up in their neighborhoods. What it might say about health and dysbiosis on the grand scale.

However, these studies involving youth also show that local community efforts must be supported at the larger societal and policy levels. It's very hard to swim upstream to stop the wellsprings of disease and the barriers to full health potential when the forces of dysbiosis are a raging torrent. Government policies subsidizing the commodity foods—corn, soybean, wheat, and others—that are linked to disease promotion, the presence of fast foods in disadvantaged neighborhoods (and often either in or adjacent to hospitals and schools), the heavy-duty marketing and associated environmental degradation are not easy to overcome. Neither is the message that all surfaces need to be 99.9% free from bacteria, and that soil and the bark of trees are dangerous and dirty. But as Nelson Mandela famously said, "It always seems impossible until it's done."

In the youth-inspired community food transformation projects, one little nugget of information could help fast track the passage to change. The corner store owners reported a 20 percent increase in sales and high levels of satisfaction with the store transformation. It works from a business perspective. Businesses can still be very profitable and help solve social and environmental problems. One measure of whether a company is biophilic is an audit by the independent group known as B Corp. They want to know if a company is only in it for cash, or do they have the interests of the entire biosphere in mind? Are there signs of responsibilities to stakeholders (that would be all of us) as well as shareholders? This represents the way things can be. Corporations can be probios, and B Corp challenges companies to meet the highest standards of social and environmental performance. There are now over 1,000 B Corp companies and kudos for Danone for being the first global multinational entity to engage with B Corp principles. The world's largest probiotic company looking at probios.

The B Corp Philosophy is all about ecosystems. B Corporations and leaders of this emerging economy believe:

- That all business ought to be conducted as if people and place mattered.
- That, through their products, practices, and profits, businesses should aspire to do no harm and benefit all.
- To do so requires that we act with the understanding that we are each dependent upon another and thus responsible for each other and future generations.

Since each of us is a multispecies, covered in microbes, being dependent upon another means looking out for our microbial mates and the ecosystems they are dependent upon. That won't happen if all of us are walking around looking for money where it's like turning on a tap or a light switch.

Lunar Landing in Your Gut

If there was ever an example of what a sanitized, lifeless ecosystem might look like, we have our prime example. The moon. Imagine that as the extreme end of biodiversity loss. Think about that as the polar opposite of the flourishing ecosystem that should be within your gastrointestinal tract, on your skin, out in your community, and throughout the Earth. In the years that follow, scientists will provide us with more details on the microbiome and its place within ecosystems. But in the meantime we can still tend to our inner and outer gardens.

Percy A. Campbell not only defined probiotics in 1943, he also made an important point about humans and the environment. He referred to all life—microbial, plant, animal—as bioprecipitates. That is, we are biological forms drawn from the elements of life, the solution, crystallizing into form in the biosphere around us. Crystallizing into beauty. All living creatures take from their environment, but usually as part of a balanced system.

But Campbell worried about the extraordinary propensity of the humans to self-aggrandize as they bioprecipitate. That is, taking more than we need in order to crystallize simply because we can. Human self-aggrandized behavior has progressed into taking, and taking, and ah, what the heck, take even more from the living Earth. Increasingly like thoughtless trophy hunters on a grand scale, only to discard our wasted prizes for the next conquest. Campbell had this to say about bioprecipitation in the self-aggrandized fast lane, writing about 25 years before astronauts looked back at the Earth from the surface of the moon: "Man's powers of self-aggrandizement have become so extreme that the Earth's treasure house is being rapidly rifled irreversibly. High-powered civilization is self-aggrandizement in hypertrophied form...a run-down Earth will be a poor welcomer of posterity. Yet man is hurrying on the day when nature will be badly 'used up.' Bioprecipitation is doing its part towards duplicating terrestrially the fully precipitated condition of the Earth's satellite, the moon."

So, Percy Campbell warned against taking more from the environment than was required for survival, and René Dubos said the same thing. But Dubos also said that mere survival is not enough. We should be taking less and striving for more. That is, more thriving and greater quality of life. For Dubos, survival meant having a faint pulse. Respect for and living in harmony with the environment, that is a strong heartbeat. Understanding that the giant sequoia and the ant, the fungi and the Lactobacilli, are all dependent upon us, and we on them, that's your pathway to burst into full consciousness. To awaken from the anesthesia of modernity.

In This Together

We have crossed the threshold. We have entered a new era. The microbiome revolution has shown us that everything is interconnected—our health, our environment, our planet. We are seeing the landscape with new eyes. Yes, we have done much to damage and destroy that landscape, but in seeing our global challenges in a new way, through the interconnectivity of all things, we can come closer to finding new solutions.

Microbes are the oldest form of life on Earth. They have adapted to every new condition, every new habitat, across millennia. This alone is testament to their resilience. They have partnered the evolution of every other living creature on this planet. Truly old friends. Guardian angels of life itself. Vital partners. Yet, unseen and unknown for most of human history. Finally discovered only to be demonized. For most of modern history we have seen our friends as adversaries. We immediately assumed the worst the moment we developed the technology to see into their microscopic world. In a world beset with infectious disease, that was the logical conclusion. And there is no doubt that the antibiotic revolution changed the face of modern medicine. Completely. Antibiotics saved countless lives from infectious disease, and still do. The importance of that is not to be understated, and antibiotics continue to have an important place in many conditions. But they are not the only answer. They are overused. And they have created many new problems. Yet even with our every attempt to exterminate them, our ancient guardians prevail. Ever quietly moving to restore natural balance by default. Like patient parents of unruly children, they are still here with us and we can learn much from them.

History has also shown us that we need to understand situations far more deeply, and far more holistically. Infections are essentially a specific bacteria overgrowing and causing damage and destruction in the process. But even bacteria that cause infections are not inherently bad. The distinction between what is considered harmless or pathogenic lies not only in the properties of the microbe, but in the health of the ecosystem, the integrity of the barrier at the site of infection, and many of general factors such as nutrition and stress, which are, you guessed it, all interrelated. In essence, we are learning to look at the disruptions in ecosystems that lead to the infection in the first place. The conditions that allow a microbe to become a pathogen. It becomes a matter of understanding balance and why it can become altered, rather than pointing the finger in one direction. Yes, we might need to call for yet more weed killers when things get dangerously out of balance, but also we need to think of better ways to restore the balance again afterwards. Or, better, prevent imbalance in the first place. The real point is that these insights provide new opportunities.

In our search for sentient life in the heavens, sending our rockets to Pluto, we have not been aware of the extent of life all around us. We live on a garden planet, and it is within this biosphere where we will find abundance in our quest for greater consciousness. It is imperative that we explore our own home and hunt without harm for solutions to all that ails us. And we are ailing. Our garden is showing signs of wilting and so many of our fellow creatures need help. Diversity of all life, microbial and otherwise, is under threat. Put simply, we are more likely to find what we are looking for here, rather than on the surface of Mars.

Even at this very early stage of the microbiome revolution, overthrowing stale thinking and smashing dogma as it goes, the evidence may not yet provide a cure for depression or heart disease, but it does support a central manifesto that can be applied to the human condition. Interdependence, mutualism, and interconnectivity underpin all forms of life, including *Homo sapiens*. Here in the garden, no forms of life are truly separate. Diversity and mutualism buffer us, stabilizing the environment, however that might be defined. From the minute tip of one finger-like projection of the intestinal lining to the houses of political power that govern us, diversity and mutualism allow life to be resilient to the winds of change and external threats. In every situation, a healthy ecosystem will be far more robust and more able to respond to challenges.

The same patterns are repeated on every level. The fractals of life recur at progressively larger and smaller scales alike, governing both biology and behaviour. What we might learn from nature, large and small, we might apply to society. The story of the microbes, which are everywhere, is such a good example of how everything is interconnected, in ways that were once beyond our awareness.

Now we are gaining more confidence in the rich terrain of the microscopic world. Learning that we are welcome explorers and that we don't need a Desert Storm campaign. We need less Shock and more Awe. This journey is already taking us to many new places. We are learning much. About balance. About connectivity of life. About the need to take a holistic vision. We have learned to challenge assumptions. That all is not as it seems. That we are all more interconnected than we ever imagined.

Our view is changing completely. And so is our quest. We are no longer preoccupied with microbes as demons. We are beginning to see they *might* even be our saviors. Now we delve into this wonderland with great hope of finding answers from the creatures who have survived for billions of years before us on this planet. Already they are showing their tremendous potential in breaking down man-made waste, plastics, and pollutants—even the sorts we never imagined could ever be biodegradable. But even as microbes offer us solutions to the problems we have created, we must not shy from our responsibility to this relationship, and to our planet.

We have begun to see that microbes are the foundation of all life, not just in evolutionary terms, but right *here* and right *now*. They are the glue that holds everything together. Without them whole ecosystems collapse. Virtually every aspect of human health depends on microbes for healthy functioning. And the health of "our" microbes depends on the health of the wider ecosystems in which we reside. It is only with this knowledge we can move forward *with* nature to restore ecosystems. More than anything, the story of microbes shows us that the key to any problem is collaboration. Working together, symbiotically. A simple unifying principle that is much needed across every domain of human society today.

Look Within

There are so many gifts provided to us by the study of the most minute details of life. But it is now time to start reassembling these parts into holistic progress. The study of probiotics is best regarded in the wider frame of biophilic science. And for this we must turn toward nature, not against it. For this is where the answers lie, but only if we keep the big picture in mind. We must turn toward nature relatedness.

Our ancestors lived intimately connected to nature. As *part* of it. They absorbed the finest details of plants, of rocks, of animals in motion. They experienced the seasonal tilt of the Earth and knew with great precision where the sun's rays and stars would be at any given point in time. Life was fully dependent upon knowing nature. They accumulated vast knowledge and deep understanding of balance. But they also felt nature's awe. Its wonder. This knowledge was not merely intellectual, it was also at the very heart of their purpose and spirituality. We know this from the most ancient surviving cultures that still walk the Earth today, the Australian Aboriginals. They have carried not only knowledge of the land for more than 50,000 years, but wisdom. The wisdom of balance. The wisdom of interdependence. There is much we can yet learn from them, before that too is lost.

Many other ancient cultures coalesced their knowledge of the natural world, their *only* world, into the wisdom that all things are interconnected. Their knowledge, spanning thousands upon thousands of years, has somehow been dismissed as from a prescientific era. Yet, their detailed observation and deep understanding are no less evidence-based than the modern cultures that displaced them, often with judgement, ignorance, and false assumptions. It is modern culture, not traditional cultures, which drives the risk of chronic, non-communicable disease. Tragically, as traditional cultures have been forced to rapidly transition to modern lifestyles, they have become the *most* vulnerable to diabetes, heart disease, and other modern diseases.

But this is not the time to lament past follies. This is time to look forward and not repeat the same mistakes. To work together. To use knowledge more wisely. To listen to age-old wisdom as we make new discoveries. To bring the heart back to science. To bring caring back to life. Now more than ever we need to heed the wisdom of the ancient sages. For thousands of years, from countless cultures, the gentle message has been the same—we need to look *within* to find the answers to life. So, we must all ask: What *kind* of life do we want?

The answers lie within all of us.

11

Marlies' Recipes for *The Secret Life*

Your digestive tract is home to around 100 trillion bacteria—more than all the stars in the Milky Way galaxy.

I'm Marlies Venier—health and fermentation coach. I'm also known as the Gut Girl, which makes me feel like a superhero. I study, teach, and prepare all things fermented to promote and maintain a happy, healthy life.

I focus on healing and balancing the microbiome, because the bacteria in your gut can affect the health of your digestive system, immune system, and mental wellbeing. Since making fermentation an integral part of my diet, I've been able to increase my energy levels, reduce anxiety, and decrease belly bloat. My skin has cleared and brightened, I sleep better, and I finally feel content.

But my life didn't change overnight. My journey started after losing my best friend to cancer at the age of 30. I became very motivated to learn as much as I could about nutrition. As a mother of three it was important for me to keep my family—and myself—as healthy as I could.

Although I knew a lot about clean eating, I still suffered from stress-related inflammation in my joints. With all the conflicting information in the media, on the internet, and in research books, I was left feeling even more frustrated and confused. That's when I decided to go to nutrition school to seek the "truth." I graduated as a Certified Holistic Health Coach from the Institute of Integrative Nutrition and became a member of the American Association of Drugless Practitioners.

While training, I had many incredible teachers, but one woman really got my attention. Donna Gates is the author and creator of *The Body Ecology Diet*, and I had read her book years before. She introduced me to the importance of the microbiome and how your health is affected by the condition of the gut. To say I was inspired is an understatement.

I began to experiment with cultured and fermented vegetables, started to make up my own recipes and tried a host of new and exotic concoctions—beet kvass, anyone? I found a wonderful community of fermenters and furthered my education with Summer Bock's Fermentationist Certification Program.

I was so amazed by the health benefits, I began coaching clients and teaching workshops. When you know something this important, you just want to let the world in on your secret.

I'm thrilled to have this opportunity to share these simple gut-healing recipes with you. I hope they help you look and feel fabulous, and assist you on your own wellness journey!

Breakfast

Good Morning Smoothie

Start your day off right with this refreshing and healthy smoothie. If you like iced coffee you'll love this frosty breakfast treat. The coffee will give you a boost of energy, and the protein powder will keep you full until lunch.

Ingredients

1 cup brewed and chilled coffee

1 scoop Genuine Health fermented vegan protein powder (unsweetened, unflavored)

½ large frozen banana

½ cup unsweetened vanilla almond milk

1 tbsp. Bulletproof Brain Octane (MCT) Oil

½ tsp. cinnamon

Directions

1. Combine all ingredients in a blender and blend on high until smooth.

Aloe Vera and Coconut Water Smoothie

Aloe vera juice is helpful for healing leaky gut. It aids in digestion and boosts the immune system. It really has no flavor, so it doesn't distract from the ginger, mint, and vanilla featured in this recipe.

Ingredients

½ cup coconut water
¼ cup aloe vera juice
1 scoop Genuine Health fermented vanilla protein powder
1 small knob of ginger
1 small handful of mint

Directions

1. Combine all ingredients in a blender and blend on high until smooth.

Minty Berries and Yogurt Dip

I brought this to a parent breakfast gathering and it was a big hit. Super easy and healthy, and all that mint is great for digestion. You can make as little or as much of this as you'd like.

Ingredients

1 cup blueberries
1 cup raspberries
1 cup strawberries
1 cup blackberries
1 big bunch of mint (save a sprig for garnish)
2 cups of Dannon plain yogurt
1 tbsp. vanilla
½ cup honey
1 tsp. cinnamon

Directions

1. Add berries to a pretty white bowl. Chop the mint and add to the bowl. Gently incorporate the mint into the berries.
2. For the yogurt dip, in a separate bowl mix the yogurt, vanilla, honey, and cinnamon. Garnish with a sprig of mint on top.

Coconut Yogurt

My coconut yogurt is delicious and simple to make—without all the added sugar you'll find in many store-bought varieties. Plus, the strains of bacteria in this recipe are very helpful in balancing the gut flora.

Ingredients

1 package frozen young Thai coconut flesh, defrosted, or the flesh of a fresh Thai coconut

½ cup (or more) coconut water to thin out the flesh of the coconut

¼ tsp. yogurt starter (I use the yogurt starter Formula 2 from Custom Probiotics—it contains *Lactobacillus bulgaricus, Streptococcus thermophilus, Lactobacillus acidophilus, Lactococcus lactis*, and *Bifidobacterium lactis*)

Directions

1. Add the coconut flesh to a Vitamix (or high-speed) blender and blend.
2. Add the coconut water and blend again to thin out the coconut flesh (you want a nice smooth consistency). The blender should warm up the coconut.
3. Pour the yogurt into a glass bowl or yogurt maker and stir in the yogurt starter.
4. Cover loosely and store in a warm place or in your yogurt maker for 8 hours.

Raw Chocolate Chia Pancakes

This recipe will get you out of bed in the morning! These are delicious topped with half a sliced banana, a spoonful of Dannon vanilla yogurt, and a drizzle of maple syrup.

Ingredients

1 cup chia seeds

½ cup goji berries

½ cup raw cacao powder

½ cup walnuts or macadamia nuts, finely ground

1 tsp. maca powder

1 tsp. cinnamon

2 cups water

Directions

1. Mix all dry ingredients in a large bowl.
2. Add water and mix well with a fork.
3. Form rounds using an ice cream scoop. Put on a baking tray. Set the oven to the lowest temperature. Gently bake overnight, or use a dehydrator set at 115°F for 7 hours.

Favorite Superfood Smoothie

I have been making this smoothie for myself, and teaching my clients how to make it, for years. It's delicious and full of so much nutritional goodness that it keeps you feeling full between meals. It's great to share or to enjoy throughout the day.

Ingredients

2 cups ice
2 cups water
1 cup almond milk
½ cup aloe vera juice
½ banana
4 Brazil nuts
3 pitted dates
2 tbsp. Bulletproof Brain Octane (MCT) Oil
1 scoop chocolate protein powder or cacao powder (I prefer Genuine Health natural chocolate fermented vegan protein)
1 tbsp. maca powder
1 tbsp. green powder
1 handful goji berries
2 tsp. organic vanilla or 1 vanilla pod
½ tsp. cinnamon
Pinch of sea salt

Directions

1. Combine all ingredients in a blender and blend on high until smooth.
2. Store leftovers in the fridge in a Mason jar.

Fermented Chai Low-Carb Granola

Adapted from Maria Emmerich's granola recipe

My Fermented Chai Low-Carb Granola recipe is the perfect gut-boosting Ketogenic Diet (or low-sugar) breakfast. A ¼-cup serving contains only about 2.3 grams of carbs, while traditional recipes and packaged varieties contain about 22 grams of carbs. That's a big savings! It helps keep my insulin levels low, but I don't miss out on my favorite flavors or textures.

Ingredients

1 cup chopped pecans
1 cup chopped walnuts
1 cup slivered almonds
1 cup sunflower seeds
1 cup raw pumpkin seeds
½ cup sesame seeds
½ cup vanilla whey protein powder (Genuine Health or any vanilla protein powder you like)
¼ cup Genuine Health Fermented Whole Body Nutrition with Greens+ Vanilla Chai Powder
½ cup erythritol (Organic Zero) or ½ cup xylitol or ½ cup Swerve
1 tsp. cinnamon
½ tsp. Celtic sea salt
1¼ cup melted coconut oil

Directions

1. Preheat the oven to 300°F.
2. In a large bowl, mix the nuts, seeds, protein powders, erythritol, cinnamon, and salt. Add the oil and combine.
3. Transfer the granola to a large cookie sheet and spread into a single layer. Place it in the oven for 20–25 minutes, until golden brown.
4. Remove and let cool. Break the granola into bite-sized pieces and store in an airtight container. Serve with almond or coconut milk.

Gut Girl Note: Opt for organic nuts and seeds if possible.

Coconut Water Kefir

Coconut water kefir is a fermented beverage made from coconut water and water kefir starter. The water kefir starter contains a combination of bacteria and yeast that promotes healthy digestion and an array of other health benefits. I recommend drinking it first thing in the morning to get your body ready for the day.

Ingredients

4–6 cups (or 1 liter) coconut water from young Thai coconuts or organic unpasteurised coconut water (I recommend the Feeding Change brand—find it in the frozen section at Whole Foods)

1 packet water kefir grains starter or 2 probiotic tablets, opened

Directions

1. Strain the coconut water into a saucepan.
2. Heat to 92°F. Be careful not to overheat it. If you don't have a thermometer, use your finger to check the temperature. It should feel neutral to the touch, not hot or cold.
3. Pour the warm coconut water into a blender. Add the kefir starter packet or the opened probiotic capsules.
4. Blend for a few seconds just to mix the ingredients.
5. Pour into a glass container and top with a tight-fitting lid.
6. Store at room temperature, ideally 70–75°F for 3 days.
7. Chill and enjoy. Stays fresh for about 3 weeks.

Protein-Packed Breakfast Shake

My Protein-Packed Breakfast Shake provides me with a super dose of my favorite superfoods. Plus, kale is brimming with vitamins, minerals, and antioxidants that cleanse your body of toxins and provide an abundance of plant-based energy.

Ingredients

1 scoop Genuine Health Fermented Whole Body Nutrition with Greens+ Vanilla Chai Powder
1–2 cups kale, roughly chopped
½ cup kefir
½ cup blueberries
1 tsp. Bulletproof Brain Octane (MCT) Oil
1 cup water or as needed

Directions

1. Combine all ingredients in a blender and blend on high until smooth.
2. Add more water if you desire a thinner consistency.

Vegan Brain-Boosting Coffee

Dave Asprey created Bulletproof Coffee after discovering that high-quality fat has a positive effect on brain function and cognition. Combined with coffee, it can provide a mental and physical boost without depleting your body or causing a post-caffeine crash.

Ingredients

1–2 cups regular or decaffeinated coffee (organic or high-quality is best)
2 tbsp. coconut milk
1–2 tbsp. Bulletproof or TruMarine Collagen
½–1 tbsp. Bulletproof Brain Octane (MCT) Oil
½ tsp. cinnamon

Directions

1. Place the coffee into a blender and add the remaining ingredients.
2. Blend until it's nice and frothy!

Almond Milk

Homemade almond milk is delicious, and far better than anything you can buy at the grocery store. I enjoy this poured on my cereal or just as a rich creamy drink. Bonus: Add 2 tbsp. cacao powder and blend to make chocolate almond milk. Kids love it!

Ingredients

1 cup raw almonds
4 cups filtered water + more for soaking
2–4 pitted Medjool dates (I use 3 depending on the size, but sweeten it to your liking)
1 whole vanilla bean, chopped (or ½–1 tsp. vanilla extract)
¼ tsp. cinnamon
Small pinch of fine-grain sea salt, to enhance the flavor

Directions

1. Place the almonds in a bowl and cover with water. Soak overnight (for about 8–12 hours). You can cut the time to 1–2 hours; however, the longer they soak the more plump and sprouted they will be.
2. Rinse and drain the almonds and place in a blender. Add the water, dates, and vanilla. Blend on the highest speed for 1 minute or more.
3. Place a nut milk bag over a large bowl and slowly pour the almond milk into the bag. Gently squeeze the bottom of the bag to release the milk.
4. Rinse the blender and pour the milk back in. Add the cinnamon and salt, and blend on low to combine.
5. Pour into a glass jar to store in the fridge for up to 3–5 days. The mixture will separate so give it a shake before using.

Gut Girl Note: You can replace the dates with other sweeteners, like maple syrup, if you prefer. If your dates or vanilla bean are dry, soak in warm water to soften before blending.

Light Meals, Salads, and Dressings

Nonna's Radicchio Salad

I love and miss my husband's Nonna. There are many delicious recipes she left us, but we serve her radicchio salad almost every night. I've introduced many friends to this lettuce and they've come to appreciate the bitter taste. It's worth trying, as radicchio aids in digestion and colon cleansing with its high fiber content.

Ingredients

Salad

1 head radicchio

1–2 tomatoes, sliced

1 avocado, sliced

½ small red onion, thinly sliced

Dressing

3–4 tbsp. organic cold-pressed olive oil

1–2 tbsp. organic red wine vinegar

Splash of balsamic vinegar

½ tsp. garlic powder

½ tsp. onion powder

Salt and pepper to taste

Directions

1. Tear lettuce into bite-size pieces. Wash and dry in a salad spinner. Place in a large bowl, then add the tomato, avocado, and onion.
2. In a small bowl, whisk all the dressing ingredients together.
3. Pour dressing over salad and toss to combine.

Gut Girl Note: I don't usually measure this dressing. I go by instinct and taste, and I encourage you to do the same. Play around with it until it tastes perfect to your palate.

Riced Cauliflower and Herb Salad

A best girlfriend made this for a luncheon, and I was excited to experiment with the recipe. This is a beautiful, nutrient-rich salad that can be served as a side or meal. I like to layer the ingredients (dressing on the bottom) in Mason jars and keep them in the fridge for grab-and-go lunches.

Ingredients

1 medium head cauliflower
2 garlic cloves
1 cup pitted and chopped Kalamata olives
1 English cucumber, diced
1 pint cherry tomatoes
2 green onions, sliced
1 bunch flat leaf parsley, finely chopped
½ cup mint, finely chopped
1 cup cilantro, finely chopped
Juice of 1 lemon
⅓ cup extra virgin olive oil
Black pepper to taste

Directions

1. Grate the cauliflower using a cheese grater or a food processor with the grater attachment.
2. Using a garlic press, crush the garlic in a large bowl. Add the olives, cucumber, tomatoes, onion, parsley, mint, and cilantro.
3. Add the lemon juice, oil and pepper. Toss well.
4. Add the cauliflower, and toss again. Adjust seasonings to taste.

Turmeric Tahini Dressing

I love this dressing from *Keepin' It Kind*. I added fresh ginger, which is known to ease indigestion. It's great on a Buddha bowl or a chicken and kale salad.

Ingredients

¼ cup tahini
¼ cup water
¼ cup lemon juice
1 tbsp. maple syrup
1½ tsp. organic soy sauce
½ tsp. turmeric
¼ tsp. grated ginger

Directions

1. Add all the ingredients to a blender and blend.
2. Store in the fridge in a Mason jar.

Mushroom Miso Soup

Full of probiotics, miso soup aids in digestion. It's also very alkalizing. I love it because it's warm and comforting, but still light.

Ingredients

1 tbsp. avocado oil
1 lb. mixed mushrooms (portobello, cremini, oyster, beech, and/or shiitake), sliced
6 scallions (white and light green parts only), sliced
6 cups water
¼ cup + 2 tbsp. miso paste

Directions

1. In a large soup pot, heat the oil. Add the mushrooms and cook for 5 minutes, stirring occasionally, to soften.
2. Add the scallions, water, and miso. Stir to combine.
3. Bring to a boil, reduce heat to low and cover. Simmer for 15 minutes.

Roasted Prebiotic Veggie Bowl with Quinoa

This recipe packs a lot of flavor and provides you with sustainable energy. It's a great way to get all your veggies in, and it's also a great way to round up any produce that's been sitting in the fridge.

Ingredients

3 beets
4 carrots
6 Jerusalem artichokes
3 tbsp. Bulletproof Brain Octane (MCT) Oil
2 tsp. thyme
1 tsp. rosemary
3 cloves garlic
1 cup uncooked quinoa, rinsed
2 cups bone broth
Salt and pepper to taste

Directions

1. Preheat the oven to 350°F.
2. Scrub and peel the root vegetables. Chop them into medium-sized, even chunks. Place them on a baking sheet in a single layer (use 2 baking sheets if necessary).
3. In a small bowl, mix the oil, thyme, rosemary, garlic, salt, and pepper. Pour over the veggies and toss to coat.
4. Roast for 35–45 minutes, until just slightly crunchy and beginning to lightly brown.
5. Meanwhile, combine the quinoa and broth in a pot. Bring to a boil, then reduce heat to low. Cover and simmer until tender and most of the liquid has been absorbed, about 15–20 minutes. Let sit, covered, for 5 minutes. Fluff with a fork.
6. Divide the quinoa into bowls and top with the roasted vegetables. Adjust seasonings to taste.

Salmon Chowder

I love chowders, and this one is quick and easy to make. Salmon is one of the best fish sources of anti-inflammatory omega-3 fat, an essential fat that supports gut health. The added bonus? Kids love it, too!

Ingredients

4 tbsp. coconut oil
1 large onion, sliced
3 large carrots, diced
3 celery stalks, diced
1 cup kale, torn or chopped
4 new potatoes, cut into quarters
2 cups coconut milk (look for cans that are BPA free if you can't find coconut milk in a glass container)
1 cup chicken broth
1 bay leaf
1 lb. wild salmon fillets
Salt and pepper to taste
Handful of parsley

Directions

1. Place a large pot over medium heat, and add 2 tbsp. oil. Add the onion, carrots, celery, kale, and potatoes. Sauté for 8 minutes.
2. Add the milk, broth, and bay leaf. Bring to a simmer and continue to cook.
3. In a large frying pan, add the remaining 2 tbsp. oil and cook the salmon, skin side down, until crisp and brown. Flip and cook the other side for about 2 minutes. Remove and discard the skin. Chop the salmon and add it to the pot.
4. Poach the salmon for about 10–12 minutes. Make sure the potatoes are tender before serving. Add salt and pepper to taste.
5. Serve in bowls garnished with parsley.

Swiss Chard Salad with Asian Dressing

Dark leafy greens are perfect as they are, but there are a few easy ways to dress them up so everyone at the table is actually excited to eat them! This dressing combines bold flavors to yield an Asian-inspired dish that's way better than takeout.

Ingredients

For the salad

1 bunch rainbow Swiss chard

½ package baked tofu, cut into bite-sized cubes

2 cups red cabbage, coarsely shredded

2–3 carrots, grated

1 cup toasted walnuts

For the dressing

2–3 garlic cloves, pressed or chopped

½ cup walnut oil

¼ cup rice vinegar

2 tbsp. toasted sesame oil

Directions

1. Wash the chard and remove the leaves from the stems. Chop the stems into small pieces.
2. Stack the leaves and roll them up as you would a sushi roll. Then slice crosswise to create long strips.
3. Place the chard stems and leaves into a large bowl, and add the remaining salad ingredients.
4. In a small bowl, whisk all of the dressing ingredients until well incorporated.
5. Pour the desired amount of dressing over the salad and toss well. Leftover dressing will last for a few days in the refrigerator.

Buddha Bowl

We're all super busy, but you still want to feed your family healthy, energy-packed dinners, right? I make this a couple times a week and eat the leftovers for lunch. Just remember that you can use whatever veggies you have and, of course, that changes with the seasons. I also don't use exact measurements, but invite you to go with the proportions that feel good to you.

Ingredients

2 tbsp. coconut oil
Onion
Garlic
Red or green cabbage, chopped
Chopped veggies (think broccoli, cauliflower, red peppers, zucchini, and grated carrot)
Brown rice
Cherry tomatoes (when in season), whole or cut in half
Cilantro, finely chopped
Parsley, finely chopped
1 jar of Naam Famous Miso Gravy or homemade miso gravy
Sesame seeds

Directions

1. Sauté onions and garlic in oil for 3–4 minutes.
2. Add the cabbage and cook for 3–4 minutes.
3. Add your chopped veggies of choice (except tomatoes, cilantro, and parsley) and a little water. Cover and steam until all vegetables are tender.
4. Scoop the rice into a bowl. Add the veggies and top with tomatoes, cilantro, and parsley.
5. Warm the miso gravy and pour over the Buddha Bowl. Sprinkle with the sesame seeds.

Dinners, Sides, and Greens

Slow-Cooker Golden Cauliflower and Chickpea Curry

Cauliflower is extremely good for you, so I love incorporating it into my weekly menus. And curry powder is known for preventing cancer, infections, and heart disease. If you're looking for an easy weeknight meal that packs a nutritional punch, this one's for you!

Ingredients

1 head cauliflower (6" diameter), cut into florets
1 cup organic homemade chicken broth
1 15-oz. can chickpeas, drained and rinsed
1 cup frozen peas
1 cup chopped yellow onion
1 tsp. turmeric powder
1 tbsp. curry powder
1 cup light coconut milk
1 bay leaf
1 lb. extra firm organic tofu, cut into cubes
1 cup fresh basil, shredded
Salt and pepper to taste

Directions

1. Place cauliflower in slow cooker. Add broth, chickpeas, peas, onion, turmeric, and curry. Cover and cook on high for 2 hours.
2. Stir in the milk and bay leaf. Cover and cook for 1 hour. Season with salt and pepper to taste.
3. Place the tofu in serving bowls. Ladle the hot curry over the tofu, add basil, and serve.

Ground Turkey Casserole

Do you love Thanksgiving—and dream of turkey dinners all year long? You're in luck, because this recipe is a quick and easy substitute. Not to mention, turkey is full of protein, and a great source of iron, potassium, and zinc.

Ingredients

1 cup canned navy beans, drained and rinsed
1 cup whole milk or milk substitute
1 tsp. turmeric
1 tsp. coriander
1½ tsp. salt (add more if needed to the puree)
1 cup organic chicken broth
1 lb. ground turkey
1 cup sweet onion, diced
1 cup baby carrots, thinly sliced
1 cup asparagus, cut into 1-inch pieces
2 cups green cabbage, grated
½ cup sliced almonds
½ cup organic cheddar, shredded

Directions

1. Preheat the oven to 350°F.
2. Add the navy beans, half of the milk, turmeric, coriander, and salt to a blender. Puree until smooth. Add the rest of the milk and the broth.
3. Place a non-stick frying pan over medium heat. Sauté the turkey in batches until lightly browned. Transfer to a large casserole dish.
4. Add the remaining ingredients to the casserole dish, including the pureed bean mixture. Stir well to incorporate and bake for 25–30 minutes.

Brook's Slow-Cooked Chicken Adobo

This is my daughter's favorite dish, so I make it often. She loves the sweet and savory combination, which is very balancing for the palate. To save time, I marinate the chicken the night before, and quickly cook it up for dinner.

Ingredients

2 lbs. chicken thighs
1 small onion, minced
7 garlic cloves, minced
3 tbsp. honey
¼ tsp. cayenne pepper
½ tsp. pepper
¾ cup organic Tamari (gluten-free soy sauce)
½ cup white vinegar
1 can coconut milk (can should be BPA free) or 2 cups chicken stock
2 bay leaves
Brown or basmati rice, for serving

Directions

1. For the marinade, place the chicken in a large Ziploc bag. Add the onion, garlic, honey, cayenne, pepper, Tamari, and vinegar. Gently mix to coat the chicken. Leave in the fridge overnight.
2. Place the chicken and seasonings in a slow cooker. Add the milk or stock and the bay leaves. Cover and cook on low for 5–6 hours.
3. Preheat the oven to broil. Place a cooling rack inside a baking sheet lined with foil.
4. Remove the chicken thighs from the slow cooker and place onto the prepared baking sheet.
5. Strain the remaining sauce in a pot and cook over medium high heat, stirring frequently, until reduced by half, about 8–10 minutes. Brush both sides of each chicken thigh with the sauce. Place in oven and broil for 2–3 minutes, or until slightly charred and crispy.
6. Serve with brown or basmati rice. Or, for extra gut power, serve over grated cabbage.

Organic Taco Seasoning

Ever since discovering this recipe from *The Heal Your Gut Cookbook*, I keep this taco spice mix on hand. Spices are full of phytonutrients, essential oils, antioxidants, minerals, and vitamins, and are essential for overall wellness and gut health. Preparing your own is much better for you than the packaged taco seasoning mix you find in the supermarket, and it's very simple to make. Just be sure to purchase organic seasonings.

Ingredients

2 tbsp. chili powder
1 tbsp. onion powder
1 tbsp. + 2 tsp. sweet paprika
4 tsp. ground cumin
2½ tsp. garlic powder
2 tsp. sea salt
⅛ tsp. cayenne pepper

Directions

1. Combine all the spices, mix to incorporate, and store in a glass jar for up to a year.

Paige's Slow-Cooked Beef Taco Salad

This is my other daughter's favorite dish. It's also a brilliant slow cooker meal, as it's easy to prepare during the busy school year. And it's especially handy during the winter months. Put it on in the morning, and when you get home all you need to do is add lettuce, tomatoes, avocados, and sauerkraut for a delicious taco salad.

Ingredients

2 tbsp. coconut oil
1 onion, diced
2 lb. grass-fed ground beef or bison
¼ cup Taco Seasoning (recipe above)
2 cups bone broth

Directions

1. Sauté onion in 1 tbsp. oil in a frying pan.
2. Add the beef or bison and lightly brown.
3. With the slow cooker on high, add the other 1 tbsp. oil and sweat the taco seasoning to release and intensify the flavors of the spices (about 1 minute).
4. Add the ground meat, onion, and broth to the slow cooker.
5. Cook on low for 5–6 hours.

Cherry Blossom Cauliflower Kvass

I recently had lunch at my favorite eatery and ordered the fermented cauliflower. To my surprise, the cauliflower was pink! It was so gorgeous (and flavorful) I had to talk to the chef about it. It turns out this colorful culture was inspired by the beauty of cherry blossom season. I was so enamored of the dish, I made my own recipe!

Ingredients

2 cups water
½ tsp. sea salt
3 garlic cloves
2 large beets, diced
½ small onion, diced
1 head cauliflower, cut into florets

Directions

For the Kvass

1. Create brine by combining the water and salt. Set aside.
2. Release the flavor of the garlic by crushing each clove with the flat side of your knife.
3. Layer the garlic, beets, and onion in a 1-liter Mason jar. Fill the jar with the brine, leaving one inch of space at the top of the jar.
4. Place the lid on tightly and let it ferment at room temperature, about 70–85°F, for 10 days.
5. Turn the jar upside down every day for the first week until the gas builds up enough to keep the floating vegetables covered.
6. After 10 days, the beet mixture (or kvass) can be strained out. Notice the rich deep purple color!

For the Cauliflower

1. Clean the jar and pour in the kvass. Add the cauliflower florets.
2. Put the lid on and let it sit on the counter for 2 days, then refrigerate.
3. Once chilled, serve as a side dish or use in rice bowls and salads.

Gut Girl Note: If you already have kvass, skip the "For the Kvass" section and go straight to the cauliflower instructions.

Broccoli Rabe

Broccoli rabe is packed with potassium, iron, and calcium. It also contains dietary fiber as well as vitamins A, C, and K, making it the perfect addition to your dinner plate. If you haven't heard of nutritional yeast flakes, give them a try. They have a cheesy flavor, but don't contain dairy. You can usually find them in the bulk bins.

Ingredients

1 bunch broccoli rabe
2 tbsp. olive oil
2 garlic cloves
2 tbsp. water
Pinch of sea salt
Parmesan cheese or nutritional yeast flakes, optional

Directions

1. Wash broccoli rabe and cut off the tough stem ends. Chop the rest into ½ inch pieces.
2. Warm the oil in a pan and add the garlic. Sauté for a few minutes.
3. Add the broccoli rabe and salt, then sauté for about 3 minutes.
4. Add water, cover, and steam for about 2 minutes. Check for desired tenderness.
5. If needed, add a bit more water and steam for a few more minutes.
6. Serve with freshly grated Parmesan cheese or sprinkle with nutritional yeast flakes, if desired.

Collards with Dill and Parsley

Adding fresh herbs, like dill and parsley, to your dark leafy greens can make them more palatable—especially for those who struggle to eat veggies. If you want to add a bit of heat, mix the juice of 1 lime and a dash of cayenne pepper. Toss with the greens and serve as a spicy side.

Ingredients

1 bunch collard greens
2 tbsp. olive oil
1 tsp. black pepper
2 tbsp. water
½ cup fresh chopped dill
½ cup fresh chopped parsley
Pinch of sea salt

Directions

1. Wash the collards and remove the leaves from the stems. Chop the stems into small pieces.
2. Stack the leaves and roll them up, as you would a sushi roll. Then, slice crosswise to create long strips.
3. Warm the oil and pepper in a pan, then add the stems. Sauté for a few minutes.
4. Add the leaves and salt. Sauté for about 3 minutes.
5. Add water, cover, and steam for about 3–4 minutes. Remove from heat.
6. Add the dill and parsley, toss well and let sit uncovered for a few minutes before serving.

Sautéed Greens with Pine Nuts and Raisins

Mustard, kale, and dandelion greens can be bitter. To balance the flavors, I cook them with chewy, sweet raisins and top them with toasted pine nuts. Serve with lemon wedges, and discover a whole new way to eat these intense, but super healthy greens.

Ingredients

¼ cup pine nuts
½ bunch mustard greens, chopped
2 tbsp. olive oil

½ bunch kale, chopped
½ bunch dandelion greens, chopped
½ tsp. sea salt
⅓ cup raisins
Lemon, cut into wedges

Directions

1. Preheat the oven to 325°F. Place the pine nuts on a cookie sheet in a single layer and toast in the oven for 5 minutes. Set aside.
2. Heat the oil in a pan. Add the greens, salt, and raisins. Stir and cook for 5 minutes.
3. Before serving, top with toasted pine nuts. Drizzle some fresh lemon juice on top or serve with lemon wedges.

Gayatri Greens

These Indian-style greens bear the name of a powerful Hindu Goddess. Gayatri is also a beautiful mantra (prayer) said to represent the divine awakening of the mind and soul. A favorite recipe of the Institute of Integrative Nutrition.

Ingredients

1 bunch Swiss chard
2 tbsp. coconut oil
1 tsp. black mustard seeds
1 tsp. ground cumin
1 tsp. ground coriander
½ tsp. sea salt
½ cup organic plain yogurt

Directions

1. Wash the chard and remove the leaves from the stems. Chop the stems into small pieces.
2. Chop the leaves into 1-inch pieces. Set aside.
3. Place a frying pan over medium-high heat, then add the oil.
4. When oil is hot, add the mustard seeds and stir for 1 minute.
5. Add the cumin and coriander and cook for another 30 seconds, stirring constantly.
6. Add the chard stems and salt. Sauté for 1 minute, then add the leaves. Mix well and cook for 3–5 minutes until chard is wilted.
7. Remove pan from heat and stir in the yogurt.

Snacks and Drinks

Almond Pesto

Pestos are delicious and great to have on hand to make a quick and easy gut-friendly lunch or dinner. The almonds in this pesto are a prebiotic food that the gut flora loves. Spread it on gluten-free or sprouted bread for sandwiches, or add a spoonful to a bowl of soup. To make a delicious salad dressing, I add a little red wine or balsamic vinegar to the pesto (my favorite is pesto on zucchini pasta).

Ingredients

1 cup almonds
1 bunch of basil
2 garlic cloves
Zest and juice of 1 lemon
½ cup cold-pressed olive oil
½ cup Parmesan (optional)

Directions

1. Cover the almonds with water and soak overnight. Strain and place them in a dehydrator or in an oven on low. Dry them for 12–24 hours. It's a good idea to do about 4 cups at a time to have extra on hand, so you can easily whip up a batch of pesto when needed.
2. Add all ingredients to a blender and blend until smooth.

Coconut Gelatin Squares

Gelatin is very beneficial for the gut and joints. These protein-packed squares are also a yummy way to decrease inflammation. Enjoy one or two a day to support gut health. They'll even naturally satisfy your sweet tooth if a craving strikes!

Ingredients

1 can, or about 2 cups (BPA free) coconut milk
½ cup honey
½ tsp. vanilla
6 tbsp. grass-fed gelatin

Directions

1. Mix 1 cup of milk with honey and vanilla.
2. Stir gelatin into the coconut milk mixture. Let it sit and thicken.
3. In a saucepan over low heat, warm the remaining cup of milk. Do not boil. Pour the hot milk over the gelatin and whisk.
4. Pour the mixture into a 9" × 13" Pyrex dish and refrigerate for 2 hours.
5. Cut into squares and store in an airtight container in the fridge.

Lemon Gut Healing Tonic

Feel sluggish when you wake up? Try this anti-inflammatory tonic that can help stimulate the digestive tract when taken first thing in the morning. Drink for a week and notice the difference in your skin, energy, and digestive health.

Ingredients

3 lemons
1 large knob ginger
1 large knob raw turmeric
1 tbsp. honey
2 cups spring water or mineral water

Directions

1. Put all ingredients into a high-powered blender (like a Vitamix) and blend well.
2. Store in a glass Mason jar. Drink 2 oz. in the morning and before eating.

Crunchy Roasted Chickpeas

Chickpeas, also called garbanzo beans, help boost digestion, keep blood sugar levels stable, increase protection against disease, and more. They offer a potent package of protein, vitamins, and minerals. Make a big batch and keep them in your purse for an anytime snack.

Ingredients

1 can (BPA free) chickpeas, drained
¼ tsp. Himalayan salt
1 tsp. extra virgin olive oil
¼ tsp. garlic powder
⅛ tsp. cayenne pepper

Directions

1. Preheat the oven to 425°F.
2. Blot the chickpeas with paper towels to remove as much moisture as possible.
3. Transfer chickpeas to large mixing bowl. Add salt, oil, garlic, and cayenne. Stir to coat.
4. Spread the chickpeas on a baking sheet in a single layer (use 2 sheets if necessary). Bake for 20 minutes.
5. Stir, then return to oven for 5–10 minutes or until crisp, but not burned. Check frequently after 20 minutes, as cooking times may vary.

Cruciferous Vegetables with White Beans

This dish features three cruciferous superfoods: Brussels sprouts, cauliflower, and kale. Combine the finished veggies with white beans to make it a well-rounded meal, or use it as a side dish to accompany chicken or beef. It's savory, salty, garlicky, hearty, and just plain delicious.

Ingredients

¼ cup chopped yellow onion
3 garlic cloves, minced
½ lb. Brussels sprouts, halved
½ head cauliflower, separated into florets
1 tbsp. avocado oil, divided
4 cups kale, chopped
1 15-oz. can cannellini beans, drained and rinsed
Salt to taste

Directions

1. Preheat the oven to 400°F.
2. In a large bowl, combine the onion, garlic, Brussels sprouts, and cauliflower. Add ½ tbsp. oil. Stir to coat.
3. Add the kale, remaining oil, and salt. Stir again.
4. Spread the vegetables on a baking sheet in a single layer (use 2 baking sheets if necessary) and place in the oven for 25 minutes.
5. Stir and return to oven for 10–15 minutes.
6. Transfer the vegetables to a serving dish and add the cannellini beans. Toss to combine.

Kale Chips

By now you've probably heard about how healthy kale is, and for good reason. It's full of fiber, which means it's great for digestion. Plus, it's filled with vitamins and nutrients. These chips are a fun way to get more greens into your diet—and satisfy your crunch craving.

Ingredients

1–2 bunches kale
2 tbsp. avocado oil or coconut oil
Sea salt

Directions

1. Preheat the oven to 425°F.
2. Remove the kale leaves from the stems, leaving them in large pieces. (Save the stems for one of the other sautéed greens recipes.)
3. Pour the oil into a bowl. Dip your fingers into the oil and massage a very light coat over the kale. Season well with salt.
4. Place the kale on baking sheet in a single layer (cook in batches if necessary) and bake for 5 minutes, or until they start to turn a bit brown. Keep an eye on them, as they can burn quickly!
5. Flip and bake until both sides are golden and crisp, about 5 more minutes. Remove and cool completely.

Gut Girl Notes: Try different kinds of leafy greens, like collards and purple kale. For added flavor, sprinkle the kale with spices, such as curry, cumin, nutritional yeast, and red pepper flakes after rubbing on the olive oil.

Spicy Korean Kimchi

What's so great about kimchi? It's full of beneficial microbes that aid in digestion, which is why Koreans eat it with nearly every meal. Kimchi is also known for its anti-aging and skin-clearing properties. In other words, it makes you look *and* feel good!

Ingredients

1 packet of culture starter
3½ cups filtered water
1 handful fresh parsley
1 tsp. chili flakes (or crushed red pepper)
1 large cabbage, core removed
5 carrots
1 zucchini
1 thumb-sized piece of turmeric (1–2 inches), peeled and minced
6 garlic cloves, minced
1 tbsp. sea salt

Directions

1. Add culture starter, water, parsley, and chili flakes to your blender. Liquefy.
2. Cut the cabbage, carrots, and zucchini into chunks that will fit down the top of your food processor. Attach the shredding blade and shred the vegetables.
3. Place the shredded vegetables, culture mixture, turmeric, garlic, and salt into a large bowl. Mix together and massage with your hands until tender and juicy.
4. Store in jars at 72°F for 10 days or longer (let it sit for up to 1 month for a really pungent product).
5. If the jar remains closed, the kimchi will last at room temperature for years! Once opened, refrigerate immediately.

Fermented Macadamia Nut Feta

If you're avoiding dairy, you'll love this recipe. Macadamia nuts support digestion and bone health and protect skin and hair. Combined with kefir, you have a probiotic-rich feta that can add lots of intrigue to salads and Mediterranean dishes.

Ingredients

2 cups raw macadamia nuts
1 cup coconut kefir, or 1 cup water with 2 capsules of probiotic powder
1 tbsp. finely chopped green onion
1–2 tsp. nutritional yeast flakes or miso
1 tsp. lemon juice
1 tsp. minced garlic
⅛–¼ tsp. salt, plus more for soaking the nuts

Directions

1. Soak the nuts overnight in salt water, then dry them in a warm oven or dehydrator.
2. Blend the nuts and kefir in a high-speed blender and continue to mix until smooth. Add more of your liquid if necessary to form a smooth, creamy texture.
3. Pour into a nut bag or cheesecloth-lined strainer. Allow to strain for 18–48 hours.
4. After the fermentation process, remove the cheese from the nut bag or cheesecloth. Combine with the green onion, nutritional yeast, lemon, garlic, and salt.
5. Refrigerate the cheese until ready to use. Season further or sweeten your cheese as desired before serving.

Gut Girl Notes: Be sure to store your cheese in an airtight container in the refrigerator; nut cheese will last for up to two weeks. When straining, you can place the nut bag over a bowl or a colander with a weight on top. This will apply pressure and push out the excess liquid over the process time. The longer it sits the more tart your nut cheese will be, so feel free to give it a little taste along the way. I like to let it sit for a couple of days.

Essential Sauerkraut

Yes, you can buy sauerkraut at the grocery store, but it's healthier, cheaper, and more fun to make at home. And all you really need are two ingredients: cabbage and salt. Once you become more confident in your fermenting skills, you can add in a variety of veggies and aromatics.

Ingredients

1 head cabbage
Sea salt

Directions

1. Chop or grate the cabbage, keeping the pieces relatively even in size.
2. Transfer to a large bowl and sprinkle with sea salt. I recommend about 3 tbsp. of salt per 5 lbs. of cabbage, but taste as you go. If the cabbage tastes like a potato chip, and you can't stop at just one, then you have added the perfect amount.
3. Using your hands, massage the salt into the cabbage, which will create liquid (aka brine). This can take about 15–20 minutes.
4. Place the cabbage into a sterilized wide-mouth glass jar or fermentation crock. Add in all of the brine.
5. Using your fist or tamper, push all the cabbage below the liquid, getting all the air bubbles out.
6. If the brine does not rise above the cabbage, you can add a little salt water (1 tsp. salt in 1 cup filtered water). However, I'd prefer to wait 24 hours to see if more natural brine is produced.
7. Seal the jar or follow manufacturer's instructions for using your fermentation crock. Leave the jar or crock in a cool, dry place and check the kraut every day or two. Don't be afraid to give it a taste. You can enjoy in 3 days or, for a stronger product, 2–3 weeks.
8. When it has the right amount of tang for you, store it in the fridge and enjoy as often as you'd like.

Dill Pickles

If you're like me and you love cucumber season, pickling is a great option. Not only do you get to continue enjoying the cucumbers, but your body also gets to enjoy the many health benefits of cultured foods.

Ingredients

4 lb. small cucumbers
2 garlic bulbs, peeled and chopped
1 handful dill weed
1 handful grape leaves (or other tannin-rich leaves)
1 tbsp. peppercorns
2 tsp. minced horseradish
1 tsp. hot chili pepper flakes
2½ quarts spring or filtered water
¼ cup sea salt

Directions

1. Gently wash the cucumbers. Place the garlic, dill weed, grape leaves, peppercorns, horseradish, and chili pepper flakes in a 1-gallon crock.
2. Tightly pack the cucumbers in the crock, placing larger cucumbers at the bottom. Combine the water and salt to make brine and pour over the cucumbers. Put a plate or other weight on top of the cucumbers and add brine to completely submerge all of the ingredients. Cover the crock with a towel held in place with a rubber band.
3. In a few days, fermentation will begin. Bubbling can last anywhere from 2–4 weeks, depending on the temperature. When bubbling has ceased, sample a cucumber. If it has not pickled through to the center, give them some more time. When they are fully pickled, transfer to fridge for storage. (A half-sour pickle will still be raw and crunchy in the center.)

Gut Girl Note: This recipe can also be made in two 2-liter jars—simply divide the seasonings between them.

Beet Kvass

Beet kvass is a fermented, healing tonic. It originated in Eastern Europe, where it was originally prepared by fermenting stale bread. The resulting liquid was taken to fight against illness and disease. Today, you'll find that kvass made from beets is just as healthy (if not more). Plus, it's a bit tastier than stale bread!

Ingredients

2 cups water
1½ tsp. sea salt
1 medium beet, diced

Directions

1. Combine the water and salt to create brine. Set aside.
2. Fill a 1-liter Mason jar with the diced beet.
3. Pour the brine over the beet pieces, leaving one inch of space at the top of the jar.
4. Place the lid on tightly and let it ferment at room temperature, about 70–85°F.
5. Let it sit on the counter for 2 weeks, turning it upside down every day for the first week. This will allow the gas to build up so that the beets stay covered in the brine.
6. When ready, strain the liquid into a clean jar and store it in the fridge. You can use the leftover beets in salads and rice bowls.

Simply Delicious Fermented Ginger Carrots

My Simply Delicious Fermented Ginger Carrots recipe was the first one I made when I started to teach myself how to ferment foods. If you're a beginner and have fear around starting, this is a great go-to, as it only takes a few days to ferment on your counter.

Ingredients

2 cups filtered water
1½ tsp. unrefined sea salt
4 cups grated carrots
1 tbsp. freshly grated ginger
1 probiotic capsule

Directions

1. Pour the water and salt in a small glass jar and mix until the salt is completely dissolved to create brine. Set aside.
2. Combine the carrots and ginger and place into a ½-gallon Mason jar or fermentation crock. Pack down firmly with your fist or a tamper and remove any air bubbles.
3. Pour the brine over the carrots until the water level is just above the carrots. Leave room in the jar, as the carrots will release more liquid.
4. Be sure to cover your jar with a tea towel fastened with a rubber band to block out any light. Leave the carrots on the counter for 3 days.
5. Transfer to fridge when ready.

Gluten-Free Focaccia Bread

I follow the Ketogenic Diet, so I'm always on the lookout for low-carb, low-sugar recipes. Made with ground flaxseeds or flax meal, this bread ticks all the right boxes. Not to mention, the herbs and spices make it taste just like traditional focaccia bread, but without all the refined ingredients.

Dry Ingredients

2 cups ground flaxseeds or flax meal
1 tbsp. Flavor Plus Probiotic Italian Seasoning
1 tbsp. dried onion flakes
1 tbsp. garlic powder
1 tbsp. baking powder
1 tbsp. baking soda
1 tsp. sea salt

Wet Ingredients

5 large eggs
½ cup water
¼ cup olive or coconut oil

Directions

1. Preheat the oven to 350°F. Line a loaf pan with parchment paper or use a lined muffin tin to make buns.
2. In a large bowl, combine the dry ingredients.

3. In a blender, combine the wet ingredients and blend until it yields a foamy texture.
4. Add the wet mixture to the dry mixture and gently mix.
5. Let the batter sit for a few minutes to thicken up.
6. Pour the batter into the loaf pan or muffin tins.
7. Bake for 20 minutes or until golden.
8. Remove from oven and put the bread or buns on a cooling rack.

Paleo Bone Broth

Bone broth is wonderfully comforting when simply sipped. But it can also be used as a base for soups, gravies, and sauces. You can even replace water with bone broth to make grain and pasta dishes like risotto. So go ahead and prepare a big batch to have on hand. You can always store leftovers in the freezer in ice cube trays.

Ingredients

5 lb. grass-fed beef bones
1 onion or 1–2 leeks, chopped
6 garlic cloves
2 carrots, chopped
2 celery stalks, chopped
2 tbsp. apple cider vinegar
1–2 dried bay leaves
2 tsp. salt

Directions

1. Place all ingredients in a slow cooker and cover with water (about 12 cups).
2. Cook for at least 8–10 hours on low (I like to cook mine for 48 hours on low).
3. After it has cooled, the fat will float to the top and you can easily remove it with a slotted spoon.
4. Pour the broth through a strainer and discard the veggies.
5. Taste the broth and add more salt if needed.
6. The broth will keep in the fridge for 3 days or you can freeze it for up to 3 months.

Prebiotic Veggie Plate and Dip

Ingredients

1 bunch of asparagus, lightly steamed and put into an ice-water bath

1 large jicama, cut into sticks

1 bunch of carrots, cut into sticks

1 bunch of radishes, quartered

Savory Cream Cheese Dip

Ingredients

½ cup organic cream cheese

1 tbsp Dannon plain yogurt

1–2 Tbsp kimchi or sauerkraut

Directions

1. Strain the asparagus and dry with paper towels.
2. Arrange vegetables on a plate.
3. In a bowl mix the cream cheese, yogurt, and kimchi or sauerkraut together and serve with the veggie plate.

Bio

Marlies Venier, also known as The Gut Girl, is a Fermentation Coach. A few years ago, she was faced with stress-related health issues and embarked on a healing journey. It was when she introduced fermented foods to her diet that she experienced an astounding transformation, from brighter skin and a flatter tummy to a happier disposition. Marlies is now on a mission to share the healing and beautifying benefits of probiotic-rich foods with busy women everywhere! For more details and delicious recipes, visit marliesvenier.com

References

Chapter 1: Savoring the Biosphere

Adams, K. M. et al. 2010. Nutrition education in US medical schools: Latest update of a national survey. *Academic Medicine* 85:1537–1542.

Cardinal, B. J. et al. 2015. If exercise is medicine, where is exercise in medicine? Review of US medical education curricula for physical activity-related content. *Journal of Physical Activity and Health* 12:1336–1343.

Castillo, M. et al. 2016. Nutrition knowledge of medical graduates. *Journal of Advanced Nutrition and Human Metabolism* 2:e1188.

Castillo, M. et al. 2016. Basic nutrition knowledge of recent medical graduates entering a pediatric residency program. *International Journal of Adolescent Medicine and Health* Nov. 1 28(4):357–361.

Ceballos, G. et al. 2015. Accelerated modern human-induced species losses: Entering the sixth mass extinction. *Science Advances* June 19;1(5):e1400253.

Cosmides, L. and J. Tooby. 2013. Evolutionary psychology: New perspectives on cognition and motivation. *Annual Review Psychology* 64:201–29.

Editors, "The British Medical Association: 82nd Annual Meeting, at Aberdeen." *Lancet* Aug 8, 1914:377–400.

Erlich, N. et al. 2013. Of hissing snakes and angry voices: Human infants are differentially responsive to evolutionary fear-relevant sounds. *Developmental Science* 16: 894–904.

Holtz, K. A. et al. 2013. Exercise behaviour and attitudes among fourth-year medical students at the University of British Columbia. *Canadian Family Physician* 59: e26–e32.

Huber, M. et al. 2012. Anthropogenic and natural warming inferred from changes in earth's energy balance. *Nature Geoscience* 5:31–36.

Huxley, Thomas H. *On the Origin of Species or the Causes of Phenomena of Organic Nature: A Course of Six Lectures to Working Men.* New York: D. Appleton, 1872.

Igarashi, M. et al. 2015. Effect of stimulation by foliage plant display images on prefrontal cortex activity: A comparison with stimulation using actual foliage plants. *Journal of Neuroimaging* 25(1):127–130.

Igarashi, M. et al. 2014. Effects of stimulation by three-dimensional natural images on prefrontal cortex and autonomic nerve activity: A comparison with stimulation using two-dimensional images. *Cognitive Processing* 15(4):551–6.

Johnson, G. E. 1909. Why teach a child to play? *Third Annual Proceedings of the Playground Association of America.* 3:357–65.

Kaplan, R. The Green Experience in *Humanscape: Environments for the People*, ed. Stephen Kaplan and Rachael Kaplan, 186–193. Anne Arbor: Ulrich's, 1982.

Kim, G. W. et al. 2011. Neuro-anatomical evaluation of human suitability for rural and urban environment by using fMRI. *Korean Journal of Medical Physics* 22:18–27.

Kim, G. W. et al. 2010. Functional neuroanatomy associated with natural and urban scenic views in the human brain: 3.0–T functional MR imaging. *Korean Journal of Radiology* 11:507–513.

Kim, G. W. et al. 2014. Brain activation patterns associated with the human comfortability of residential environments: 3.0–T functional MRI. *Neuroreport* 25(12):915–20.

Kim, T. H. et al. 2010. Human brain activation in response to visual stimulation and rural urban scenery pictures:A functional magnetic resonance imaging study. *Science of the Total Environment* 408:2600–2607.

LoBue, V. et al. 2010. Superior detection of threat-relevant stimuli in infancy. *Developmental Science* 13:221–8.

Meert, K. et al. 2014. Taking a shine to it: How the preference for glossy items stems from an innate need for water. *Journal of Consumer Psychology* 4:195–206.

Melander, B. Emylcamate, a potent tranquillizing relaxant. *J. Med. Chem.* 1 (5), (1958):443–457.

Muir, John. *My First Summer in the Sierra.* Boston: Houghton Mifflin, 1911.

New, J. et al. 2007. Category-specific attention for animals reflects ancestral priorities, not expertise. *Proceedings of the National Academy of Sciences of the United States of America* 104:16598–603.

New, J. et al. 2007. Spatial adaptations for plant foraging: Women excel and calories count. *Proceedings of the Royal Society B: Biological Science* 274:2679–84.

Prescott, S. L. et al. 2016. Transforming life: A broad view of the developmental origins of health and disease concept from an ecological justice perspective. *International Journal of Environmental Research and Public Health.* Nov. 3;13(11) pii:E1075.

Pye-Smith, P. H. "Opening of the Medical Schools: Introductory Addresses." *BMJ* 2 (Oct 9, 1919):977–1000.

Sarfaty, M. et al. 2016. Views of AAAAI members on climate change and health. *Journal of Allergy Clinical Immunology: In Practice* 4:333–335.

Sarfaty, M. et al. 2015. American thoracic society member survey on climate change and health. *Annals of the American Thoracic Society* 12:274–278.

Sarfaty, M. et al. 2016. Views of allergy specialists on health effects of climate change. *George Mason University Centre for Climate Change Communication.* (Accessed 29 August 2016). Available online: https://www.aaaai.org/Aaaai/media/Media Library/PDF%20Documents/Libraries/Climate-Change-Survey.pdf

Thomson, J. A. *Biology for Everyman. Volume Two.* London: J. M. Dent and Sons, 1934.

——. *Introduction to Science*, London: Henry Holt, 1911.

——. *The System of Animate Nature: The Gifford Lectures 1915–16*, London: Henry Holt, 1920.

——. The seasonal study of natural history. *Official Report of the Nature Study Exhibition and Conferences July 25th–August 2, 1902.* London: Blackie and Son, 1903:124–134.

———. *Toward Health*, London: Methuen, 1927.
Tinio, P. et al. Natural scenes are indeed preferred, but image quality might have the last word. *Psychology of Aesthetics, Creativity, and the Arts* 3 (2009):52–56.
Wang, S. et al. 2015. Preferential attention to animals and people is independent of the amygdala. *Social Cognitive and Affective Neuroscience* 10 (3):371–80.

Chapter 2: Surviving the Dysbiosphere

Babcock, S. et al. 2017. Two replications of an investigation on empathy and utilitarian judgment across socioeconomic status. *Scientific Data* Jan. 17; 4:160129.
Blanke, E. S. et al. 2016. Does being empathic pay off? Associations between performance-based measures of empathy and social adjustment in younger and older women. *Emotion* Aug; 16(5):671–83.
Blinkhom, V. et al. 2016. Drop the bad attitude! Narcissism predicts acceptance of violent behaviour. *Personality and Individual Differences* 98:157–61.
Campbell, W. K. et al. 2005. Understanding the social costs of narcissism: The case of the tragedy of the commons. *Personality and Social Psychology Bulletin* Oct. 31 (10):1358–68.
Ceballos, G. et al. 2015. Accelerated modern human-induced species losses: Entering the sixth mass extinction. *Science Advances* Jun 19; 1(5):e1400253.
Chang, A. M. et al. 2015. Evening use of light-emitting eReaders negatively affects sleep, circadian timing, and next-morning alertness. *Proceedings of the National Academy of Science* Jan 27; 112(4):1232–7.
Cheung, I. N. et al. Morning and evening blue-enriched light exposure alters metabolic function in normal weight adults. 2016. *PLOS ONE* May 18; 11(5):e0155601.
Cook, J. et al. Consensus on consensus: A synthesis of consensus estimates on human-caused global warming. *Environmental Research Letters* 11:048002.
Cox, D. et al. 2017. The rarity of direct experiences of nature in an urban population. *Landscape and Urban Planning* 160:79–84.
Dauchy, R. T. et al. 2015. Daytime blue light enhances the nighttime circadian melatonin inhibition of human prostate cancer growth. *Comparative Medicine* Dec; 65 (6):473–85.
Debnam, Dean. 2017. Huge rise in global employee depression, stress, and anxiety since 2012. *The Huffington Post*, Jan. 7, 2017. huffingtonpost.com/dean-debnam/huge-rise-in-global-employee_b_8923252
DeWall, C. N. et al. 2011. Tuning in to psychological change: Linguistic markers of psychological traits and emotions over time in popular U.S. song lyrics. *Psychology of Aesthetics, Creativity and the Arts* 5:200–207.
Dos Santos, C. F. et al. 2016. Queens become workers: Pesticides alter caste differentiation in bees. *Scientific Reports* Aug 17; 6:31605.
Dubos, René. 1968. Adapting to pollution. *Scientist and Citizen* 10:1–8.
———. 1982. Education for the celebration of life: Optimism despite it all. *Teachers College Record* 84 (1)255–76.
———. 1968. Environmental determinants of human life. In *Environmental Influences*, ed. D. C. Glass, 138–154. New York: Rockefeller University Press.
———. 1970. The human landscape. *Bulletin of the Atomic Scientists* 26:31–37.

——. 1966. Man and his environment: Scope, impact and nature. In *Environmental Improvement*, ed. R. W. Marquis, 3–21 Washington, DC: Graduate School Press, United States Department of Agriculture.

——. The spaceship earth. 1969. *Journal of Allergy*. July; 44(1):1–9.

Falchi, F. et al. 2016. The new world atlas of artificial night sky brightness. *Science Advances* June 10; 2(6):e1600377.

Ffrench-Constant, R. H. et al. 2016. Light pollution is associated with earlier tree budburst across the United Kingdom. *Proceedings of the Royal Society B: Biological Sciences* June 29; 283(1833). pii:2016.0813.

Goldney, R. D. et al. 2004. Subsyndromal depression: Prevalence, use of health services and quality of life in an Australian population. *Social Psychiatry and Psychiatric Epidemiology* 39:293–8.

Goulson, D., E. Nicholls. The canary in the coalmine: Bee declines as an indicator of environmental health. *Science Progress* 99:312–26.

Guadagni, V. et al. 2016. The relationship between quality of sleep and emotional empathy. *Journal of Psychophysiology.* In press.

Helander, H. F. et al. 2014. Surface area of the digestive tract—revisited. *Scandinavian Journal of Gastroenterology* June; 49(6):681–9.

Houtrow, A. J. 2014. Changing trends of childhood disability, 2001–2011. *Pediatrics* Sept.;134(3):530–8.

Jacobsen, M. B. et al. 2000. Relation between food provocation and systemic immune activation in patients with food intolerance. *Lancet* July 29; 356(9227):400–1.

Kesebir, S. 2017. Growing disconnect from nature is evident in cultural products. *Perspectives on Psychological Science.* In press.

Khajehzadeh, I. et al. 2016. How New Zealanders distribute their daily time between home indoors, home outdoors and out of home. *Kotuitui New Zealand Journal of Social Sciences Online.* In press.

Kolappa, K. et al. 2013. No physical health without mental health: Lessons unlearned? *Bulletin of the World Health Organization* 91:3–3A.

Konrath, S. H. et al. 2011. Changes in dispositional empathy in American college students over time: A meta-analysis. *Personality and Social Psychology Review* May; 15(2):180–98.

Lin, C. H. et al. 2012. Urbanization and prevalence of depression in diabetes. *Public Health* 126:104–11.

Lull, R. B. and T. Dickinson. 2016. Does television cultivate narcissism? Relationships between television exposure, preferences for specific genres, and subclinical narcissism. *Psychology of Popular Media Culture.* In press.

Marmot, M. et al. 2016. Social inequalities in health: a proper concern of epidemiology. *Annals of Epidemiology* April; 26(4):238–40.

Marmot, M. et al. Empowering Communities. *American Journal of Public Health* Feb.; 106(2):230–1.

Matz, C. J. et al. 2014. Effects of age, season, gender and urban-rural status on time-activity: Canadian Human Activity Pattern Survey 2 (CHAPS 2). *International Journal of Environmental Research and Public Health*. Feb. 19; 11(2):2108–24.

Nechita, F. et al. 2015. Circadian malfunctions in depression: Neurobiological and psychosocial approaches. *Romanian Journal of Morphology and Embryology* 56(3):949–55.

Patty, Anna. 2016. Stress-related absence from work on the rise. *Sydney Morning Herald*, March 5, 2016. smh.com.au/business/workplace-relations/stressrelated-absence-from-work-on-the-rise-20160302-gn96a1

Pergams, O. R. and P. A. Zaradic. 2008. Evidence for a fundamental and pervasive shift away from nature-based recreation. *Proceedings of the National Academy of Science*. Feb. 19; 105(7):2295–300.

Pergams, O. R. and P. A. Zaradic. 2006. Is love of nature in the US becoming love of electronic media? 16-year downtrend in national park visits explained by watching movies, playing video games, internet use, and oil prices. *Journal of Environmental Management* Sept.; 80(4):387–93.

Persson. B. N., P. J. Kajonius. 2016. Empathy and universal values explicated by the empathy-altruism hypothesis. *The Journal of Social Psychology* Nov-Dec; 156(6): 610–619.

Pietrzak, R. H. et al. 2013. Subsyndromal depression in the United States: Prevalence, course, and risk for incident psychiatric outcomes. *Psychological Medicine* 43:1401–14.

Plutzer, E. et al. 2016. Climate confusion among US teachers. *Science* 351:664–665.

Rodríguez, M. R. et al. 2012. Definitions and factors associated with subthreshold depressive conditions: A systematic review. *BMC Psychiatry* 12:181.

Rucci, P. et al. 2003. Subthreshold psychiatric disorders in primary care: Prevalence and associated characteristics. *Journal of Affective Disorders* 76:171–81.

Rybnikova, N. A., A Haim, B. A. Portnov. 2016. Is prostate cancer incidence worldwide linked to artificial light at night exposures? Review of earlier findings and analysis of current trends. *Archives of Environmental and Occupational Health* March 30:1–12.

Sartorius, N. and L. Cimino. 2012. The co-occurrence of diabetes and depression: An example of the worldwide epidemic of comorbidity of mental and physical illness. *Annals of the Academy of Medicine, Singapore* 41:430–1.

Swanson, G. R. et al. 2015. Decreased melatonin secretion is associated with increased intestinal permeability and marker of endotoxemia in alcoholics. *American Journal of Physiology. Gastrointestinal Liver Physiology* June 15; 308(12): G1004–11.

Thomson, J. A. A biologist's philosophy. In *Contemporary British Philosophy, vol* 2, ed. J. J. Muirheas, 307–334. London: Unwin, 1925.

Twenge, J. M. 2013. Time period and birth cohort differences in depressive symptoms in the US, 1982–2013. *Social Indicators Research* 121:437–54.

Westerman, J. W. et al. 2016. Ecological values, narcissism, and materialism: A comparison of business students in the USA and the Netherlands. *International Journal of Innovative Sustainable Development* 8:92–104.

www.rita.dot.gov/bts/sites/rita.dot.gov.bts/files/data_and_statistics/by_subject/freight/freight_facts_2015/chapter2/table2_4

www.newsmaker.com.au/release/pdf/id/164063
Zhou, H. et al. 2016. Smaller gray matter volume of hippocampus/parahippocampus in elderly people with subthreshold depression: a cross-sectional study. *BMC Psychiatry* Jul 7;16:219.

Chapter 3: Meet Your Maestro: The Immune System

Atterby, C. et al. 2016. Increased prevalence of antibiotic-resistant *E. coli* in gulls sampled in Southcentral Alaska is associated with urban environments. *Infection Ecology & Epidemiology* Sept. 19; 6:32334.
Bailly, J. et al. 2016. Negative impact of urban habitat on immunity in the great tit Parus major. *Oecologia*. Dec; 182(4):1053–1062.
Belsky, D. W. et al. 2014. Is chronic asthma associated with shorter leukocyte telomere length at midlife? *American Journal of Respiratory and Critical Care Medicine* Aug. 15; 190(4):384–91.
Brown, E. S. et al. 2015. Hippocampal volume in healthy controls given 3-day stress doses of hydrocortisone. *Neuropsychopharmacology* Mar. 13; 40(5):1216–21.
Clark, S. M. et al. 2015. Dissociation between sickness behavior and emotionality during lipopolysaccharide challenge in lymphocyte deficient Rag2(-/-) mice. *Behavioural Brain Research* Feb. 1;278:74–82.
Daruna, J. H. 2012. *Introduction to Psychosomatic Medicine, 2nd ed.* London: Elsevier.
Das, A. 2016. Psychosocial distress and inflammation: Which way does causality flow? *Social Science & Medicine* Oct. 5; 170:1–8.
Denda. M. 2016. Sensing environmental factors: The emerging role of receptors in epidermal homeostasis and whole-body health, 403–414, in *Skin Stress Response Pathways*, ed., G. T. Wondrak. Cham, Switzerland: Springer International Publishing.
de Sousa Rodrigues, M. E. et al. 2016. Chronic psychological stress and high-fat high-fructose diet disrupt metabolic and inflammatory gene networks in the brain, liver, and gut and promote behavioral deficits in mice. *Brain, Behavior, and Immunity* Sept. 1. pii:S0889-1591(16)30409-3.
Dubos, René. Man and his environment: Scope, impact and nature. In *Environmental Improvement* ed. R. W. Marquis, 3–21. Washington, DC: Graduate School Press, United States Department of Agriculture.
Filiano, A. J. et al. 2016. Unexpected role of interferon-γ in regulating neuronal connectivity and social behaviour. *Nature* July 21; 535(7612):425–9.
Freitas-Simoes, T. M. et al. 2016. Nutrients, foods, dietary patterns and telomere length: Update of epidemiological studies and randomized trials. *Metabolism* April; 65(4):406–15.
Gidlow, C. J. et al. 2016. Natural environments and chronic stress measured by hair cortisol. *Landscape and Urban Planning* 148:61–67.
Goodwin, R. D. et al. 2016. Childhood atopy and mental health: A prospective, longitudinal investigation. *Psychological Medicine* Oct. 20:1–9.
Grosse, L. et al. 2016. Deficiencies of the T and natural killer cell system in major

depressive disorder: T regulatory cell defects are associated with inflammatory monocyte activation. *Brain, Behavior, and Immununity* May; 54:38–44.

Hales, D. 2012. Special reports: Psycho-Immunity. *Science Digest* 1981; 89:12–14.

Hoffman, M. C. et al. 2016. Measures of maternal stress and mood in relation to preterm birth. *Obstetrics & Gynecology* March; 127(3):545–52.

Koelsch, S. et al. 2016. The impact of acute stress on hormones and cytokines, and how their recovery is affected by music-evoked positive mood. *Scientific Reports* Mar. 29; 6:23008.

Lattey, R. M. 1969. Dr. Sigmund Freud, Pseudoscientist. *Canadian Family Physician* Feb.; 15(2):59, [61], 63.

Melody, G. F. 1950. The doctor-patient relationship. Gynecological aspects. *Bulletin University Califronia Medical Center* 2:25–54.

Mischkowski, D. et al. 2016. From painkiller to empathy killer: Acetaminophen (paracetamol) reduces empathy for pain. *Social Cognitive & Affective Neuroscience* Sept.; 11(9):1345–53.

Pirrie, William. 1867. *On Hay Asthma, and the Affection Known as Hay Fever.* London: John Churchill and Sons.

Puterman, E. et al. 2016. Lifespan adversity and later adulthood telomere length in the nationally representative US Health and Retirement Study. *Proceedings of the National Academy of Science* Oct. 18; 113(42):E6335-E6342.

Rather, L. J. 1971. Disturbance of function (functio laesa): The legendary fifth cardinal sign of inflammation, added by Galen to the four cardinal signs of Celsus. *Bulletin of the New York Academy of Medicine* March; 47(3):303–322.

Ronaldson, A. et al. 2016. Increased percentages of regulatory T cells are associated with inflammatory and neuroendocrine responses to acute psychological stress and poorer health status in older men and women. *Psychopharmacology* May; 233(9):1661–8.

Sandiego, C. M. et al. 2015. Imaging robust microglial activation after lipopolysaccharide administration in humans with PET. *Proceedings of the National Academy of Science* Oct. 6; 112(40):12468–73.

Sato, Y. et al. 2016. Asthma and atopic diseases in adolescence and antidepressant medication in middle age. *Journal of Health Psychology* July 26. pii:1359105316660181.

Schutte, N. S. et al. 2016. The relationship between positive psychological characteristics and longer telomeres. *Psychology & Health* Dec; 31(12):1466–1480.

Spiers, J. G. et al. 2015. Activation of the hypothalamic-pituitary-adrenal stress axis induces cellular oxidative stress. *Frontiers in Neuroscience* Jan. 19; 8:456.

Szentivanyi, A. et al. 2003. The discovery of immune-neuroendocrine circuitry: A generation of progress, in *The Immune-Neuroendocrine Circuitry*, ed. A. Szentivanyi and I. Berczi, 15–18. London: Elsevier Science.

Talmage, D. W. 1969. The nature of immunological response, in *Immunology and Development*, ed. Adinolfi M. Suffolk: Lavenham Press.

Tayama, J. et al. 2012. Effects of personality traits on the manifestations of irritable bowel syndrome. *BioPsychoSocial Medicine* 6:20.

Tian, R. et al. 2014. A possible change process of inflammatory cytokines in the prolonged chronic stress and its ultimate implications for health. *The Scientific World Journal* 2014:780616.

Wickware, F. S. 1945. Psychosomatic medicine: Upset emotion can cause illness, obesity. *Life* 18:49–56.

Chapter 4: Biodiversity: Getting It Under Your Skin

Aas, R. et al. 2016. Patients' recovery experiences of indoor plants and views of nature in a rehabilitation center. *Work* 53:45–55.

Albrecht, G. et al. 2007. Solastalgia: The distress caused by environmental change. *Australasian Psychiatry* 15 Suppl 1:S95–8.

Annerstedt, M. et al. 2013. Inducing physiological stress recovery with sounds of nature in a virtual reality forest—Results from a pilot study. *Physiology & Behavior* 118:240–250.

Bai, Z. Y. et al. 2008. A preliminary study on interaction of negative air ion with plant aromatic substance. *Journal of Chinese Urban Forestry* 6:56–8.

Beyer, K. M. et al. 2014. Exposure to neighborhood green space and mental health: Evidence from the survey of the health of Wisconsin. *International Journal of Environmental Research and Public Health* March 21; 11(3):3453–72.

Bowen, D. J. et al. 2016. Wilderness adventure therapy effects on the mental health of youth participants. *Evaluation and Program Planning* Oct.; 58:49–59.

Branas, C. C. et al. 2016. Urban blight remediation as a cost-beneficial solution to firearm violence. *American Journal of Public Health* e1–e7.

Brown, S. C. et al. 2016. Neighborhood greenness and chronic health conditions in medicare beneficiaries. *American Journal of Preventive Medicine* July; 51(1):78–89

Capaldi, C. et al. 2015. Flourishing in nature: A review of connecting with nature and its application as a well-being intervention. *Intenational Journal of Wellbeing* 5:1–6.

Carrus, G. et al. 2015. Go greener, feel better? The positive effects of biodiversity on the well-being of individuals visiting urban and peri-urban green areas. *Landscape and Urban Planning* 134:221–228.

Clash, J. 2015. Astronaut Bill Anders recalls famous "Earthrise" photo he took from moon. *Forbes*, April 17.

Cohen-Cline, H. et al. 2015. Access to green space, physical activity and mental health: A twin study. *Journal of Epidemiology and Community Health* June; 69(6): 523–9.

Cracknell, D. et al. 2016. Marine biota and psychological well-being: A preliminary examination of dose–response effects in an aquarium setting. *Environment and Behavior* 48:1242–69.

Dadvand, P. et al. 2015. Green spaces and cognitive development in primary schoolchildren. *Proceedings of the National Academy of Science* June 30; 112(26):7937–42.

Dadvand, P. et al. 2016. Green spaces and spectacles use in schoolchildren in Barcelona. *Environmental Research* Nov. 2; 152:256–262.

Dalton, A. M. et al. 2016. Residential neighbourhood greenspace is associated with reduced risk of incident diabetes in older people: A prospective cohort study. *BMC Public Health* Nov. 18; 16(1):1171.

Donovan, G. H. et al. 2013. The relationship between trees and human health: Evidence from the spread of the emerald ash borer. *American Journal of Preventive Medicine* 44, 139–145.

Donovan, G. H. et al. 2015. Is tree loss associated with cardiovascular-disease risk in the women's health initiative? A natural experiment. *Health & Place* 36, 1–7.

Duarte-Tagles, H. et al. 2015. Biodiversity and depressive symptoms in Mexican adults: Exploration of beneficial environmental effects. *Biomedica* Aug.; 35 Spec: 46–57.

Dubos, René. 1965. *Man Adapting*. New Haven: Yale University Press.

——. 1971. Civilizing technology. In *Essays in Honor of David Lyall Patrick*. 3–17. Tuscon: University of Arizona.

——. 1974. The despairing optimist. *The American Scholar* 43:188–93.

Fuller, R. A. et al. 2007. Psychological benefits of greenspace increase with biodiversity. *Biology Letters* 3:390–94.

Gardiner, M. M. et al. 2013. The value of urban vacant land to support arthropod biodiversity and ecosystem services. *Environmental Entomology* 42, 1123–1136.

Garvin, E. C. et al. 2013. Greening vacant lots to reduce violent crime: A randomised controlled trial. *Injury Prevention* 19, 198–203.

Gao, L. et al. 2016. Negative oxygen ions production by superamphiphobic and antibacterial TiO_2/Cu_2O composite film anchored on wooden substrates. *Scientific Reports* May 27; 6:26055.

Geniole, S. N. et al. 2016. Restoring land and mind: The benefits of an outdoor walk on mood are enhanced in a naturalized landfill area relative to its neighboring urban area. *Ecopsychology* 8:107–20.

Grafetstätter, C. et al. 2017. Does waterfall aerosol influence mucosal immunity and chronic stress? A randomized controlled clinical trial. *Journal of Physiological Anthropology* Jan. 13; 36(1):10.

Gueguen, N. 2012. Dead indoor plants strengthen beliefs in global warming. *Journal of Environmental Psychology* 32:173–77.

Gunnarsson, B. et al. 2016. Effects of biodiversity and environment-related attitude on perception of urban green space. *Urban Ecosystems*. In press.

Haack, R. A. et al. 2002. The emerald ash borer: A new exotic pest in North America. *Michigan Entomological Society* 47:1–5.

Haahtela, T. 2009. Allergy is rare where butterflies flourish in a biodiverse environment. *Allergy* Dec.; 64(12):1799–803.

Hamann, G. A. et al. 2016. 30 minutes in nature a day can increase mood, well-being, meaning in life and mindfulness: Effects of a pilot programme. *Social Inquiry into Well-Being* 2:34–46.

Hanski, I. et al. 2012. Environmental biodiversity, human microbiota, and allergy are interrelated. *Proceedings of the National Academy of Science* May 22; 109(21): 8334–9.

Hedblom, M. et al. 2014. Bird song diversity influences young people's appreciation of urban landscapes. *Urban Forestry & Urban Greening* 13:469–74.

Hendrryx, M et al. 2013. Increased risk of depression for people living in coal mining areas of central Appalachia. *Ecopsychology* 5, 179–187.

Ikei, H. et al. 2017. Physiological effects of wood on humans: A review. *Journal of Wood Science.* In press.

Jones, B. A. 2016. Work more and play less? Time use impacts of changing ecosystem services: The case of the invasive emerald ash borer. *Ecological Economics* 124, 49–58.

Li, Q. et al. 2016. Effects of forest bathing on cardiovascular and metabolic parameters in middle-aged males. *Evidence-Based Complementary and Alternative Medicine* 2016:2587381.

Luck, G. W. et al. 2011. Relations between urban bird and plant communities and human well-being and connection to nature. *Conservation Biology* 25:816–826.

Maas, J. et al. 2009. Morbidity is related to a green living environment. *Journal of Epidemiology and Community Health* 63:967–73.

Maas, J. et al. 2006. Green space, urbanity, and health: How strong is the relation? *Journal of Epidemiology and Community Health* 60:587–92.

McEachan, R. R. et al. 2016. The association between green space and depressive symptoms in pregnant women: Moderating roles of socioeconomic status and physical activity. *Journal of Epidemiology and Community Health* March 70(3): 253–9.

Mitchell, R. et al. 2015. Neighborhood environments and socioeconomic inequalities in mental well-being. *American Journal of Preventive Medicine* 49:80–84.

Mitchell, R. and F. Popham. 2008. Effect of exposure to natural environment on health inequalities: An observational population study. *Lancet* 372:1655–60.

Muratet, A. et al. 2015. Perception and knowledge of plant diversity among urban park users. *Landscape and Urban Planning* 137:95–106.

Perez, V. et al. 2013. Air ions and mood outcomes: A review and meta-analysis. *BMC Psychiatry* Jan. 15; 13:29.

Pereira, G. et al. 2013. The association between neighborhood greenness and weight status: An observational study in Perth, Western Australia. *Environmental Health* 12:49.

Richardson, M. et al. 2016. 30 days wild: Development and evaluation of a large-scale nature engagement campaign to improve well-being. *PLoS One* Feb. 18; 11(2):e0149777.

Schwartz, A et al. 2014. Enhancing urban biodiversity and its influence on city-dwellers: An experiment. *Biological Conservation* 171:82–90.

Song, C. et al. Physiological effects of nature therapy: A review of the research in Japan. *International Journal of Environmental Research and Public Health* Aug. 3; 13(8). pii:E781.

South, E. C. et al. 2015. Neighborhood blight, stress, and health: A walking trial of urban greening and ambulatory heart rate. *American Journal of Public Health* 105, 909–913.

Southon, G. E. et al. 2017. Biodiverse perennial meadows have aesthetic value and increase residents' perceptions of site quality in urban green-space. *Landscape and Urban Planning* 158:105–18.

Speldewinde, P. C. et al. 2009. A relationship between environmental degradation and mental health in rural western Australia. *Health & Place* 15, 880–887.

Tikhonov, V. P. 2004. Complex therapeutical effect of ionized air: Stimulation of the immune system and decrease in excessive serotonin. H_2O_2 as a link between the two counterparts. *IEEE Transactions on Plasma Science* 32:1661–67.

Ulrich, R. 1981. Natural versus urban scenes—some psychophysiological differences. *Environment and Behavior* 13:523–56.

Ulrich, R. 1984. View through a window may influence recovery from surgery. *Science* 224:420–1.

Vincent, E. et al. 2010. The effects of presence and influence in nature images in a simulated hospital patient room. *HERD* Spring; 3(3):56–69.

Wang, D. et al. 2016. Neighboring green space and all-cause mortality in elderly people in Hong Kong: A retrospective cohort study. *Lancet.* In press.

Warber, S. L. et al. 2015. Addressing "nature-deficit disorder": A mixed methods pilot study of young adults attending a wilderness camp. *Evidence-Based Complementary and Alternative Medicine* 2015:651827.

Wheeler, B. W. et al. 2015. Beyond greenspace: An ecological study of population general health and indicators of natural environment type and quality. *International Journal of Health Geographics* 14:17.

Willox, A. C. et al. 2013. Climate change and mental health: An exploratory case study from Rigolet, Nunatsiavut, Canada. *Climatic Change* 121, 255–270.

Chapter 5: Your Microbial Orchestra

Anwesh, M. et al. 2016. Elucidating the richness of bacterial groups in the gut of Nicobarese tribal community: Perspective on their lifestyle transition. *Anaerobe* June; 39:68–76.

Avershina, E. et al. 2015. Potential association of vacuum cleaning frequency with an altered gut microbiota in pregnant women and their 2-year-old children. *Microbiome* Dec. 21; 3:65.

Bested, A. C. et al. 2013. Intestinal microbiota, probiotics and mental health: From Metchnikoff to modern advances: Part II—contemporary contextual research. *Gut Pathogens* 5:3.

Blaser, M. J. et al. 2013. Distinct cutaneous bacterial assemblages in a sampling of South American Amerindians and US residents. *The ISME Journal* Jan.; 7(1):85–95.

Center for Disease Dynamics, Economics & Policy. 2015. *The State of the World's Antibiotics 2015*. Washington, DC: CDDEP.

Chassaing, B. et al. 2017. Dietary emulsifiers directly impact the human gut microbiota increasing its pro-inflammatory potential and ability to induce intestinal inflammation. *Inflammatory Bowel Diseases* Feb. 23 Supplement 1:S5.

Clement, J. C. et al. 2015. The microbiome of uncontacted Amerindians. *Science Advances* April 3; 1(3). pii:e1500183.

Conrad, M. L. et al. 2009. Maternal TLR signaling is required for prenatal asthma protection by the nonpathogenic microbe *Acinetobacter lwoffii* F78. *Journal of Experimental Medicine* Dec. 21; 206(13):2869–77.

Cui, M. et al. Circadian rhythm shapes the gut microbiota affecting host radiosensitivity. *International Journal of Molecular Sciences* 17(11), 1786.

Dahl, W. J. et al. 2017. Health benefits of fiber fermentation. *Journal of the American College of Nutrition* Jan. 9:1–10.

Desai, M. S. et al. 2016. A dietary fiber-deprived gut microbiota degrades the colonic mucus barrier and enhances pathogen susceptibility. *Cell* Nov. 17; 167(5):1339–1353.e21.

Dubos, René. 1962. *The Unseen World.* New York: The Rockefeller Institute Press.

Dubos R. et al. 1966 Biological Freudianism: Lasting effects of early environmental influences. *Pediatrics* Nov.; 38(5):789–800.

Fyhrquist, N. et al. 2014. Acinetobacter species in the skin microbiota protect against allergic sensitization and inflammation. *Journal of Allergy and Clinical Immunology* Dec.; 134(6):1301–1309.e11.

Gasperotti, M. et al. 2015. Fate of microbial metabolites of dietary polyphenols in rats: Is the brain their target destination? *ACS Chemical Neuroscience* Aug. 19; 6(8):1341–52.

Gazzaniga, F. S. et al. 2016. Veggies and intact grains a day keep the pathogens away. *Cell* 167:1161–2.

Gomez, A. et al. 2016. Gut microbiome of coexisting BaAka pygmies and Bantu reflects gradients of traditional subsistence patterns. *Cell Reports* March 8; 14(9): 2142–53.

Hanski, I. et al. 2012. Environmental biodiversity, human microbiota, and allergy are interrelated. *Proceedings of the National Academy of Science* May 22; 109(21): 8334–8339.

Hegstrand, L. R. et al. 1986. Variations of brain histamine levels in germ-free and nephrectomized rats. *Neurochemical Research* 11:185–91.

Jain, A. K. et al. 1998. Biodiversity of angiospermic taxa in the air of central India. In *Perspectives in Environment* ed. S. K. Agarwal SK et al. 371–376. New Delhi: APH Publishing.

Joly Condette, C. et al. 2015. Chlorpyrifos exposure during perinatal period affects intestinal microbiota associated with delay of maturation of digestive tract in rats. *Journal of Pediatric Gastroenterology and Nutrition* July; 61(1):30–40.

Hu, J. et al. 2016. Effect of postnatal low-dose exposure to environmental chemicals on the gut microbiome in a rodent model. *Microbiome.* June 14; 4(1):26.

Imhann, F. et al. 2017. The influence of proton pump inhibitors and other commonly used medication on the gut microbiota. *Gut Microbes.* In press.

Kaliannan, K. et al. 2015. A host-microbiome interaction mediates the opposing effects of omega-6 and omega-3 fatty acids on metabolic endotoxemia. *Scientific Reports* June 11; 5:11276.

Kiraly, D. D. et al. 2016. Alterations of the host microbiome affect behavioral responses to cocaine. *Scientific Reports* Oct. 18; 6:35455.

Leclercq, S. et al. 2014. Intestinal permeability, gut-bacterial dysbiosis, and behavioral markers of alcohol-dependence severity. *Proceedings of the National Academy of Science* Oct. 21; 111(42):E4485–93.

Lerner, A. et al. 2016. Multiple food additives enhance human chronic diseases. *SOJ Microbiology & Infectious Diseases* 4:1–2.

Li, C. et al. 2016. *Mycobacterium vaccae* nebulization can protect against asthma in balb/c mice by regulating Th9 expression. *PLoS One* Aug. 12; 11(8):e0161164.

Logan, A. C. 2015. Dysbiotic drift: Mental health, environmental grey space, and microbiota. *Journal of Physiological Anthropology* 34:23.

Logan A. C. and M. Katzman. 2005. Major depressive disorder: Probiotics may be an adjuvant therapy. *Medical Hypotheses* 64:533–8.

Logan, A. C. et al. 2003. Chronic fatigue syndrome: Lactic acid bacteria may be of therapeutic value. *Medical Hypotheses* 60:915–23.

Logan, A. C. et al. 2014. Nutritional psychiatry research: An emerging discipline and its intersection with global urbanization, environmental challenges, and the evolutionary mismatch. *Journal of Physiological Anthropology* 33:22.

Logan, A. C. et al. 2016. The microbiome and mental health: Looking back, moving forward with lessons from allergic diseases. *Clinical Psychopharmacology and Neuroscience* May 31; 14(2):131–47.

Logan, A. C. et al. 2016. Immune-microbiota interactions: Dysbiosis as a global health issue. *Current Allergy and Asthma Reports* Feb.; 16(2):13.

Lymperopoulou, D. S. et al. 2016. Contribution of vegetation to the microbial composition of nearby outdoor air. *Applied and Environmental Microbiology* 82: 3822–3833.

Mahajan, R. et al. 2016. Microbe-bio-chemical Insights: Reviewing interactions between dietary polyphenols and gut microbiota. *Mini-Reviews in Medicinal Chemistry.* In press.

Mariotti Lippi, M. et al. 2015. Multistep food plant processing at Grotta Paglicci (Southern Italy) around 32,600 cal BP. *Proceedings of the National Academy of Science* Sept. 29; 112(39):12075–80.

Melamed, Y. et al. 2016. The plant component of an Acheulian diet at Gesher Benot Ya'aqov, Israel. *Proceedings of the National Academy of Science.* In press.

Mhuireach, G. et al. 2016. Urban greenness influences airborne bacterial community composition. *Science of the Total Environment* 571:680–687.

Moeller, A. H. et al. 2016. Cospeciation of gut microbiota with hominids. *Science.* July 22; 353(6297):380–2.

Norman, H. J. 1909. Lactic acid bacilli in the treatment of melancholia. *British Medical Journal* 1:1234–5.

Obersteiner, A. et al. 2016. Pollen-associated microbiome correlates with pollution parameters and the allergenicity of pollen. *PLoS ONE* 11:e0149545.

Ounnas, F. et al. 2017. Rye polyphenols and the metabolism of n-3 fatty acids in rats: A dose dependent fatty fish-like effect. *Scientific Reports* Jan. 10; 7:40162.

Ozdemir, P. G. et al. 2014. Assessment of the effects of antihistamine drugs on mood, sleep quality, sleepiness, and dream anxiety. *International Journal of Psychiatry in Clinical Practice* Aug; 18(3):161–8.

Parkar, S. G. et al. 2013. Fecal microbial metabolism of polyphenols and its effects on human gut microbiota. *Anaerobe* Oct; 23:12–9.

Parker-Pope, T. 1998. Dial soap campaign aims to soothe fear of germs. *Wall Street Journal* Jan. 20, 1998.

Patterson, E. et al. 2014. Impact of dietary fatty acids on metabolic activity and host intestinal microbiota composition in C57BL/6J mice. *British Journal of Nutrition* June 14; 111(11):1905–17.

Phillips, J. G. P. 1910. The treatment of melancholia by the lactic acid bacillus. *Journal of Mental Science* 56:422–31.

Prescott, S. L. et al. 1999. Development of allergen-specific T-cell memory in atopic and normal children. *Lancet* Jan. 16; 353(9148):196–200.

Prescott, S. L. et al. 2016. Biodiversity, the human microbiome and mental health: Moving toward a new clinical ecology for the 21st century? *International Journal of Biodiversity* 2016:2718275.

Reber, S. O. et al. 2016. Immunization with a heat-killed preparation of the environmental bacterium *Mycobacterium vaccae* promotes stress resilience in mice. *Proceedings of the National Academy of Science* May 31; 113(22):E3130–9.

Rees, T. et al. 2016. Waking up from antibiotic sleep. *Perspectives in Public Health* 136:202–204.

Reygner, J. et al. 2016. Changes in composition and function of human intestinal microbiota exposed to chlorpyrifos in oil as assessed by the SHIME® model. *International Journal of Environmental Research and Public Health* Nov. 4; 13(11). pii:E1088.

Salim, S. Y. et al. 2014. Air pollution effects on the gut microbiota: A link between exposure and inflammatory disease. *Gut Microbes* March-April; 5(2):215–9.

Searle, F. W. 1897. The land impoverished by the sea; a plea for the return of sewage to the soil. *Journal of Medical Science* 1897; 4:6–10.

Segata, N. 2015. Gut microbiome: Westernization and the disappearance of intestinal diversity. *Current Biology* July 20; 25(14):R611–3.

Selhub, E. M. et al. 2014. Fermented foods, microbiota, and mental health: Ancient practice meets nutritional psychiatry. *Journal of Physiological Anthropology* 33:2.

Serafim, K. R. et al. 2013. H_1 but not H_2 histamine antagonist receptors mediate anxiety-related behaviors and emotional memory deficit in mice subjected to elevated plus-maze testing. *Brazilian Journal of Medical and Biological Research* May; 46(5):440–6.

Sheflin, A. M. et al. 2016. Linking dietary patterns with gut microbial composition and function. *Gut Microbes* Dec. 14:1–17.

Street, R. A. et al. 2013. *Cichorium intybus*: Traditional uses, phytochemistry, pharmacology, and toxicology. *Evidence-Based Complementary and Alternative Medicine* 2013:579319.

Sudo, N. et al. 2004. Postnatal microbial colonization programs the hypothalamic-pituitary-adrenal system for stress response in mice. *The Journal of Physiology* July 1; 558(Pt 1):263–75.

Thaiss, C. A. et al. 2016. Persistent microbiome alterations modulate the rate of post-dieting weight regain. *Nature*. In press.

Thoreau, Henry David. 1864. *The Maine Woods*. Boston: Ticknor and Fields.

Tillisch, K. et al. 2013. Consumption of fermented milk product with probiotic modulates brain activity. *Gastroenterology* 144:1394–401.

Tochitani, S. et al. 2016. Administration of non-absorbable antibiotics to pregnant mice to perturb the maternal gut microbiota is associated with alterations in offspring behavior. *PLoS ONE* 11:e0138293.

Tito, R. Y. et al. 2012. Insights from characterizing extinct human gut microbiomes. *PLoS One* 7:e51146.

Van Vleck Pereira, R. et al. 2016. Ingestion of milk containing very low concentration of antimicrobials: Longitudinal effect on fecal microbiota composition in preweaned calves. *PLoS One* Jan. 25; 11(1):e0147525.

Velmurugan, G. et al. 2017. Gut microbial degradation of organophosphate insecticides-induces glucose intolerance via gluconeogenesis. *Genome Biology* Jan. 24; 18(1):8.

Viennois, E. et al. 2016. Dietary emulsifier-induced low-grade inflammation promotes colon carcinogenesis. *Cancer Research.* In press.

Yee, A. L. et al. 2016. Is triclosan harming your microbiome? *Science* July 22; 353(6297):348–9.

Zota, A. R. et al. 2016. Recent fast food consumption and bisphenol A and phthalates exposures among the U.S. population in NHANES, 2003–2010. *Environmental Health Perspectives* Oct.; 124(10):1521–1528.

Chapter 6: Traditional Nutrition

Akter, S. et al. 2016. High dietary acid load score is associated with increased risk of type 2 diabetes in Japanese men: The Japan public health center-based prospective study. *Journal of Nutrition* May; 146(5):1076–83.

Alves, N. E. et al. 2014. Meal replacement based on Human Ration modulates metabolic risk factors during body weight loss: A randomized controlled trial. *European Journal of Nutrition* Apr.; 53(3):939–50.

Angoorani, P. et al. 2016. Dietary consumption of advanced glycation end products and risk of metabolic syndrome. *International Journal of Food Sciences and Nutrition* 67(2):170–6.

Azzini, E. et al. 2011. Mediterranean diet effect: An Italian picture. *Nutrition Journal* Nov. 16; 10:125.

Bell, L. et al. 2015. A review of the cognitive effects observed in humans following acute supplementation with flavonoids, and their associated mechanisms of action. *Nutrients* Dec. 9; 7(12):10290–306.

Berardi, J. M. et al. 2008. Plant based dietary supplement increases urinary pH. *Journal of the International Society of Sports Nutrition* Nov. 6; 5:20.

Black, J. L. et al. 2013. Do Canadians meet Canada's Food Guide's recommendations for fruits and vegetables? *Applied Physiology, Nutrition, and Metabolism* Mar; 38(3):234–42.

Boon H., J. Clitheroe J, and T. Forte, 2004. Effects of greens+: A randomized, controlled trial. *Canadian Journal of Dietetic Practice and Research* Summer; 65(2): 66–71.

Caciano, S. L. et al. 2015. Effects of dietary acid load on exercise metabolism and anaerobic exercise performance. *Journal of Sports Science and Medicine* May 8; 14(2):364–71.

Casas, R. et al. 2016. Long-term immunomodulatory effects of a Mediterranean diet in adults at high risk of cardiovascular disease in the PREvención con DIeta MEDiterránea (PREDIMED) randomized controlled trial. *Nutrition* Sept.; 146(9): 1684–93.

Chilton, S. N. et al. 2015. Inclusion of fermented foods in food guides around the world. *Nutrients* Jan. 8; 7(1):390–404.

Cotton, R. H. 1971. Engineering in relation to nutrition: A parameter of the educational environment. Summary Report and Recommendations Based on a Symposium and Workshops Held at the 8th Annual Meeting of the National Academy of Engineering. Nov 1–2; 205–217.

Cotton, R. H. et al. 1971. Astrofood: A fortified baked product with creamed filling. *Cereal Science Today* 16:188–89.

Davis, D. 2009. Declining fruit and vegetable nutrient composition: What is the evidence? *Horticultural Science* 44:15–19.

Di Pino, A. et al. 2016. Low advanced glycation end product diet improves the lipid and inflammatory profiles of prediabetic subjects. *Journal of Clinical Lipidology* Sept.–Oct. 10(5):1098–108.

Dumas, J. A. et al. 2016. Dietary saturated fat and monounsaturated fat have reversible effects on brain function and the secretion of pro-inflammatory cytokines in young women. *Metabolism* Oct.; 65(10):1582–8.

Dunne, J. et al. 2012. First dairying in green Saharan Africa in the fifth millennium BC. *Nature* June 20; 486(7403):390–4.

Epel, E. 2009. Psychological and metabolic stress: A recipe for accelerated cellular aging? *Hormones* 8:7–22.

Esche, J. et al. 2016. Higher diet-dependent renal acid load associates with higher glucocorticoid secretion and potentially bioactive free glucocorticoids in healthy children. *Kidney International* Aug.; 90(2):325–33.

Guo, X, et al. 2016. Effects of polyphenol, measured by a biomarker of total polyphenols in urine, on cardiovascular risk factors after a long-term follow-up in the PREDIMED study. *Oxidative Medicine and Cellular Longevity* 2016:2572606.

Hadrup, N. et al. 2016. Juvenile male rats exposed to a low-dose mixture of twenty-seven environmental chemicals display adverse health effects. *PLoS One* Sept. 6; 11(9):e0162027.

Haro, C. et al. 2016. The gut microbial community in metabolic syndrome patients is modified by diet. *The Journal of Nutritional Biochemistry* Jan.; 27:27–31.

Hendaus, M. A. et al. 2016. Allergic diseases among children: Nutritional prevention and intervention. *Journal of Therapeutics and Clinical Risk Management* March 7; 12:361–72,

Horrobin, D. F. 2002. Food, micronutrients, and psychiatry. *International Psychogeriatrics* Dec.; 14(4):331–4.

Howes, M. J. et al. 2014. The role of phytochemicals as micronutrients in health and disease. *Current Opinion in Clinical Nutrition and Metabolic Care* Nov.; 17(6): 558–66.

Jacobson, M. F. Food Ingredients and additives. Quoted in: *Canner and Packer Yearbook 1972–73*. pp 41–42.

Jacques, P. F. et al. 2015. Dietary flavonoid intakes and CVD incidence in the Framingham Offspring Cohort. *British Journal of Nutrition* Nov. 14; 114(9):1496–503.

Jonvik, K. L. et al. 2016. Nitrate-rich vegetables increase plasma nitrate and nitrite concentrations and lower blood pressure in healthy adults. *Journal of Nutrition* May; 146(5):986–93.

Kaliannan, K. et al. 2015. A host-microbiome interaction mediates the opposing effects of omega-6 and omega-3 fatty acids on metabolic endotoxemia. *Scientific Reports* June 11; 5:11276.

Kiefte-de Jong, J. C. et al. 2016. Diet-dependent acid load and type 2 diabetes: Pooled results from three prospective cohort studies. *Diabetologia.* In press.

Knekt, P. et al. 1996. Flavonoid intake and coronary mortality in Finland: A cohort study. *British Journal of Medicine* 312:478–81.

Knowledge Center of the American Dietetic Association. 2001. Nutrition Fact Sheet. Straight facts About Beverage Choices. Chicago: American Dietetic Association.

Lanham-New, S. A. The balance of bone health: Tipping the scales in favor of potassium-rich, bicarbonate-rich foods. *Journal of Nutrition* 138:172S–177S.

Lee, J. et al. 2015. Switching to a 10-day Mediterranean-style diet improves mood and cardiovascular function in a controlled crossover study. *Nutrition* May; 31(5): 647–52.

Lieberman, H. M. et al. 1976. Evaluation of a ghetto school breakfast program. *Journal of the American Dietetic Association* Feb.; 68(2):132–8.

Logan, A. C. et al. 2014. Nutritional psychiatry research: An emerging discipline and its intersection with global urbanization, environmental challenges and the evolutionary mismatch. *Journal of Physiological Anthropology* July 24; 33:22.

Lopez-Moreno, J. et al. 2016. Mediterranean diet reduces serum advanced glycation end products and increases antioxidant defenses in elderly adults: A randomized controlled trial. *Journal of the American Geriatrics Society* April; 64(4):901–4.

Marco, J. L. et al. Health benefits of fermented foods: Microbiota and beyond. *Current Opinion in Biotechnology* Dec. 17; 44:94–102.

Marungruang, N. et al. 2016. Heat-treated high-fat diet modifies gut microbiota and metabolic markers in apoe-/- mice. *Nutrition & Metabolism* March 12; 13:22.

Medina-Remón, A. et al. 2016. Polyphenol intake from a Mediterranean diet decreases inflammatory biomarkers related to atherosclerosis: A sub-study of the PREDIMED trial. *British Journal of Clinical Pharmacology.* In press.

Medina-Remón, A. et al. 2017. Polyphenol intake from a Mediterranean diet decreases inflammatory biomarkers related to atherosclerosis: A substudy of the PREDIMED trial. *British Journal of Clinical Pharmacology* Jan.; 83(1):114–128.

Moubarac, J. C. et al. 2014. Processed and ultra-processed food products: Consumption trends in Canada from 1938 to 2011. *Canadian Journal of Dietetic Practice and Research* Spring; 75(1):15–21.

Mahajan, R. et al. 2016. Microbe-bio-chemical insight: Reviewing interactions between dietary polyphenols and gut microbiota. *Mini-Reviews in Medicinal Chemistry* Oct. 24.

Mayer, A. M. 1997. Historical changes in the mineral content of fruits and vegetables. *British Food Journal* 99:207–11.

McGill, C. R. et al. 2015. Ten-year trends in fiber and whole grain intakes and food sources for the United States population: National Health and Nutrition Examination Survey 2001–2010. *Nutrients* Feb. 9; 7(2):1119–30.

Miller, M. G. et al. 2016. Role of fruits, nuts, and vegetables in maintaining cognitive health. *Experimental Gerontology* Dec. 20. pii:S0531-5565(16)30606-4.

Moghadam, S. K. et al. 2016. Association between dietary acid load and insulin resistance: Tehran lipid and glucose study. *Preventive Nutrition and Food Science* June; 21(2):104–9.

Murakami, S. et al. 2015. The consumption of bicarbonate-rich mineral water improves glycemic control. *Evidence-Based Complementary and Alternative Medicine* 2015:824395.

Murphy, M. M. et al. 2012. Phytonutrient intake by adults in the United States in relation to fruit and vegetable consumption. *Journal of the Academy of Nutrition and Dietetics* Feb.; 112(2):222–9.

Ni, Y. et al. 2015. A molecular-level landscape of diet-gut microbiome interactions: Toward dietary interventions targeting bacterial genes. *mBio* Oct. 27; 6(6). pii:e01263-15.

Norman, J. et al. 2016. The impact of marketing and advertising on food behaviours: Evaluating the evidence for a causal relationship. *Current Nutrition Reports* Sept.; 5:139–49.

Papanikolaou, Y. et al. 2014. U.S. adults are not meeting recommended levels for fish and omega-3 fatty acid intake: Results of an analysis using observational data from NHANES 2003–2008. *Nutrition Journal* April 2; 13:31.

Phan, M. et al. 2017. Interactions between phytochemicals from fruits and vegetables: Effects on bioactivities and bioavailability. *Critical Reviews in Food Science Nutrition*. In press.

Pim, S. L. et al. 2015. How many plant species are there, where are they, and at what rate are they going extinct? *Annals of the Missouri Botanical Garden* 100:170176.

Pusceddu, M. M. et al. 2015. N-3 polyunsaturated fatty acids (PUFAs) reverse the impact of early-life stress on the gut microbiota. *PLoS One* Oct. 1; 10(10):e0139721.

Rao, V. et al. 2011. In vitro and in vivo antioxidant properties of the plant-based supplement greens+. *International Journal of Molecular Science* 12(8):4896–908.

Rebholz, C. M. et al. 2016. Dietary acid load and incident chronic kidney disease: Results from the ARIC study. *American Journal of Nephrology* Jan. 21; 42(6):427–435.

Remer, T. et al. 2016. Increased protein intake and corresponding renal acid load under a concurrent alkalizing diet regime. *Physiological Reports* July; 4(13). pii:e12851.

Rodriguez-Casado, A. 2016. The health potential of fruits and vegetables phytochemicals: Notable examples. *Critical Reviews in Food Science and Nutrition* May 18; 56(7):1097–107.

Sadeghirad, B. et al. 2016. Influence of unhealthy food and beverage marketing on children's dietary intake and preference: A systematic review and meta-analysis of randomized trials. *Obesity Reviews* Oct.; 17(10):945–59.

Sánchez-Villegas, A. et al. 2015. A longitudinal analysis of diet quality scores and the risk of incident depression in the SUN Project. *BMC Medicine* 13:197.

Sheflin, A. M. et al. 2016. Linking dietary patterns with gut microbial composition and function. *Gut Microbes*. In press.
Tan, Z. X., R. Lal R and K. D. Wiebe. 2005. Global soil nutrient depletion and yield reduction. *Journal of Sustainable Agriculture* 26:123–46.
Tan, X. et al. 2016. Effect of six-month diet intervention on sleep among overweight and obese men with chronic insomnia symptoms: A randomized controlled trial. *Nutrients* Nov. 23; 8(11). pii:E751.
Thomas, D. 2006. Meat and dairy: Where have all the minerals gone? *Food* 72:10.
Tresserra-Rimbau, A. 2014. Polyphenol intake and mortality risk: A re-analysis of the PREDIMED trial. *BMC Medicine* May 13; 12:77.
United Nations. 2004. Training Manual of the Food and Agriculture Organization of the United Nations. Building on Gender, Agrobiodiversity and Local Knowledge. FAO.
U.S. Congress. 1971. Senate Select Committee on Nutrition and Human Needs. Statement of Robert H. Cotton, PhD, Vice-President, Research, ITT-Continental Bakery, Co. 163–167. Wednesday, February 24th, 1971.
Valls-Pedret, C. et al. 2015. Mediterranean diet and age-related cognitive decline: A randomized clinical trial. *JAMA Internal Medicine* July; 175(7):1094–103.
van Dooren, F. E. et al. 2016. Advanced glycation end product (AGE) accumulation in the skin is associated with depression: The Maastricht study. *Depression and Anxiety*. In press.
Williams, R. S. et al. 2015. The role of dietary acid load and mild metabolic acidosis in insulin resistance in humans. *Biochimie* Sept. 10. pii:S0300–9084(15)00287–4.
Zaraska, M. 2015. Bitter truth: how we're making fruit and veg less healthy. *New Scientist* 227:2630.
Zwerdling, D. 1972. USDA decrees: Let them eat cake for breakfast. *Lansing State Journal* Oct. 12, 1972.
Zwerdling, D. 1972. If kids don't have bread for breakfast, let them eat cake. *The Morning News*. Sept. 25, 1972.

Chapter 7: On Being a Biophilist: An Injection of Nature Relatedness

Abdollahi, A. et al. 2015. Emotional intelligence and depressive symptoms as predictors of happiness among adolescents. *Iranian Journal of Psychiatry and Behavioral Sciences* 9:e2268.
Adams, C. J. 1907. Biophilism. The *Phrenological Journal of Science and Health* 120: 14–6.
Amel, E. L. 2009. Mindfulness and sustainable behavior: Pondering attention and awareness as means for increasing green behavior. *Ecopsychology* 1:14–25.
Allan, B. A. et al. 2015. Connecting mindfulness and meaning in life: Exploring the role of authenticity. *Mindfulness* 6:996–1003.
Allemand, M. et al. 2015. Empathy development in adolescence predicts social competencies in adulthood. *Journal of Personality* 83:229–241.
Aminabadi, N. A. et al. The impact of maternal emotional intelligence and parenting style on child anxiety and behavior in the dental setting. *Medicina Oral, Patologia Oral y Cirugia Bucal* 17:E1089–E1095.

Andersen, T. E. et al. 2016. A 13-weeks mindfulness based pain management program improves psychological distress in patients with chronic pain compared with waiting list controls. *Clinical Practice and Epidemiology in Mental Health* June 30; 12:49–58.

Aspy, D. J. et al. 2017. Mindfulness and loving-kindness meditation: Effects on connectedness to humanity and to the natural world. *Psychological Reports.* In press.

Barton, J. et al. 2016. The wilderness expedition: An effective life course intervention to improve young people's well-being and connectedness to nature. *Journal of Experiential Education* 39 (1)59–72.

Batson, C. et al. 1997. Is empathy-induced helping due to self–other merging? *Journal of Personality and Social Psychology* 73:495–509.

Bourgault, P. et al. 2015. Relationship between empathy and well-being among emergency nurses. *Journal of Emergency Nursing* 41:323–328.

Bruni, C. M. et al. 2017. Getting to know nature: Evaluating the effects of the Get to Know Program on children's connectedness with nature. *Environmental Education Research* 23:43–62.

Capaldi, C. A. et al. 2014. The relationship between nature connectedness and happiness: A meta-analysis. *Frontiers in Psychology* Sept. 8; 5:976.

Castillo, R. et al. 2013. Effects of an emotional intelligence intervention on aggression and empathy among adolescents. *Journal of Adolescence* 36:883–892.

Cho, Y. et al. 2017. 'Love honey, hate honey bees': Reviving biophilia of elementary school students through environmental education program. *Environmental Education Research.* In press.

Collado, S. et al. 2015. Effect of frequency and mode of contact with nature on children's self-reported ecological behaviors. *Journal of Environmental Psychology* 41:65–73.

Costa, A. et al. 2015. The impact of emotional intelligence on academic achievement: A longitudinal study in Portuguese secondary school. *Learning and Individual Differences* 37:38–47.

Creswell, J. D. et al. 2016. Alterations in resting-state functional connectivity link mindfulness meditation with reduced interleukin-6: A randomized controlled trial. *Biological Psychiatry* July 1; 80(1):53–61.

Curci, A. et al. 2014. Emotions in the classroom: The role of teachers' emotional intelligence ability in predicting students' achievement. *American Journal of Psychology* 127:431–445.

Di Fabio, A. et al. 2016. Promoting well-being: The contribution of emotional intelligence. *Frontiers in Psychology* Aug 17; 7:1182.

Di Fabio, A. et al. 2016. Green positive guidance and green positive life counseling for decent work and decent lives: Some empirical results. *Frontiers in Psychology.* In press.

Dubos, R. 1970. *Reason Awake: Science for Man*. New York: Columbia University Press.

——. 1974. *Beast or Angel?* New York: Scribner's.

Extremera, N. et al. 2016. Ability, emotional intelligence and life satisfaction: Positive and negative affect as mediators. *Personality and Individual Differences* 102:98–101.

Galante, J. et al. 2016. Loving-kindness meditation effects on well-being and altruism: A mixed-methods online RCT. *Applied Psychology: Health and Well Being* Nov.; 8(3):322–350.

Goodman, G. et al. 2014. Mars can wait: Facing the challenges of our civilization. *Israel Medical Association Journal* Dec; 16(12):744–7.

Gotink, R. A. et al. 2016. Mindfulness and mood stimulate each other in an upward spiral: A mindful walking intervention using experience sampling. *Mindfulness* 7(5):1114–1122.

Gueguen, N. et al. 2015. Carrying flowers on a city street increases others' spontaneous helping behavior. *Ecopsychology* 7:153–159.

Gueguen, N. et al. 2016. "Green Altruism": Short immersion in natural green environments and helping behavior. *Environment and Behavior* 48:324–342.

Harrod, N. R. et al. 2005. An exploration of adolescent emotional intelligence in relation to demographic characteristics. *Adolescence* 40:503–512.

Hiraoka, D. et al. 2016. The influence of cognitive load on empathy and intention in response to infant crying. *Scientific Reports.* In press.

Hojat, M. 2016. Empathy as related to personal qualities, career choice, acquisition of knowledge, and clinical competence. In *Empathy in Health Professions Education and Patient Care*, ed. M. Hojat. Cham, Switzerland: Springer International.

Howell, A. J. et al. 2011. Nature connectedness: Associations with well-being and mindfulness. *Personality and Individual Difference* 51:166–171.

Howell, A. J. et al. 2013. Meaning in nature: Meaning in life as a mediator of the relationship between nature connectedness and well-being. *Journal of Happiness Studies* 14:1681–96.

Kelly, B. 2014. Student's connectedness to nature in relation to academic major and learning style. Environmental Studies Undergraduate Student Theses. Paper 136. http://digitalcommons.unl.edu/envstudtheses/136

Killgore, W. D. et al. 2008. Sleep deprivation reduces perceived emotional intelligence and constructive thinking skills. *Sleep Medicine* 9:517–526.

Kim, W. et al. The effect of cognitive behavior therapy-based psychotherapy applied in a forest environment on physiological changes and remission of major depressive disorder. *Psychiatry Investigation* 6(4): 245–254.

Lekies, K. S. et al. 2015. Urban youth's experiences of nature: Implications for outdoor adventure recreation. *Journal of Outdoor Recreation and Tourism* 9:1–10.

Lenzi, D. et al. 2016. Mothers with depressive symptoms display differential brain activations when empathizing with infant faces. *Psychiatry Research: Neuroimaging* 249, 1–11.

Lin, Y. H. et al. 2014. Does awareness effect the restorative function and perception of street trees? *Frontiers in Psychology* 5:906.

Loucks, E. B. et al. 2015. Mindfulness and cardiovascular disease risk: State of the evidence, plausible mechanisms, and theoretical framework. *Current Cardiology Reports* Dec.; 17(12):112.

Maio, M. et al. 2016. Meaning in life promotes proactive coping via positive affect: A daily diary study. *Journal of Happiness Studies.* In press.

Ma-Kellams, C. et al. 2016. Trust your gut or think carefully? Examining whether an intuitive, versus a systematic, mode of thought produces greater empathic accuracy. *Journal of Personality and Social Psychology.* In press.

Martyn, P. and E. Brymer. 2016. The relationship between nature relatedness and anxiety. *Journal of Health Psychology* 21:1436–45.

Mavroveli, S. et al. 2011. Trait emotional intelligence influences on academic achievement and school behaviour. *British Journal of Educational Psychology* 81:112–134.

McGrath, M. 1969. Fromm, ethics, and education. Bulletin of the Bureau of School Service. 42:5–31.

Metz, A. L. Back to nature: The impact of nature relatedness on empathy and narcissism in the millennial generation. *Educational Specialist.* JMU Scholarly Commons Paper 65.

Nisbet, E. et al. 2011. Underestimating nearby nature: Affective forecasting errors obscure the happy path to sustainability. *Psychological Science* 22:1101–06.

Nyklicek, I. et al. 2015. The role of emotional intelligence in symptom reduction after psychotherapy in a heterogeneous psychiatric sample. *Comprehensive Psychiatry* 57:65–72.

Obery, A. et al. 2017. Exploring the influence of nature relatedness and perceived science knowledge on proenvironmental behavior. *Education Sciences* 7:17.

Okonofua, J. A. et al. 2016. Brief intervention to encourage empathic discipline cuts suspension rates in half among adolescents. *Proceedings of the National Academy of Science* 113:5221–5226.

Richardson, M. et al. 2017. Three good things in nature: Noticing nearby nature brings sustained increases in connection with nature. *Psyecology.* In press.

Rudd, M. et al. 2012. Awe expands people's perception of time, alters decision making, and enhances well-being. *Psychological Science* 23:1130–6.

Sanchez-Alvarez, N. et al. 2016. The relation between emotional intelligence and subjective well-being: A meta-analytic investigation. *Journal of Positive Psychology* 11:276–285.

Schutte, N. S. et al. 2007. A meta-analytic investigation of the relationship between emotional intelligence and health. *Personality and Individual Differences* 42: 921–933.

Schutte, N. S. et al. 2014. A meta-analytic review of the effects of mindfulness meditation on telomerase activity. *Psychoneuroendocrinology* April; 42:45–8.

Schutte, N. S. et al. 2016. The relationship between positive psychological characteristics and longer telomeres. *Psychology & Health* Dec.; 31(12):1466–1480.

Schreier, H. M. et al. 2013. Effect of volunteering on risk factors for cardiovascular disease in adolescents: A randomized controlled trial. *JAMA Pediatrics* 167: 327–332.

Shanahan, D. F. et al. 2017. Variation in experiences of nature across gradients of tree cover in compact and sprawling cities. *Landscape and Urban Planning* 157:231–8.

Shiota, M. N. et al. 2007. The nature of awe: Elicitors, appraisals, and effects on self-concept. *Cognition and Emotion* 21:944–63.

Shrira, A. et al. 2015. How do meaning in life and positive affect relate to adaptation to stress? The case of firefighters following the Mount Carmel forest fire. *Israel Journal of Psychiatry and Related Sciences* 52(3):68–70.

Slaski, M. et al. 2003. Emotional intelligence training and its implications for stress, health and performance. *Stress Health* 19:233–239.

Snell, T. L. et al. 2016. Contact with nature in childhood and adult depression. *Children, Youths and Environment* 26:111–124.

Soga, M. et al. 2016. Both direct and vicarious experiences of nature affect children's willingness to conserve biodiversity. *International Journal of Environmental Research and Public Health* May 25; 13(6). pii:E529.

Tillisch, K. et al. 2103. Consumption of fermented milk product with probiotic modulates brain activity. *Gastroenterology* June; 144(7):1394–401, 1401.e1–4.

Torkar, G. 2015. Creating meaningful life for a responsible and sustainable future. In *Responsible Living. Concepts, Education and Future Perspectives*, ed. V. W. Thoresen et al, 73–82. London: Springer.

Tousignant, B. et al. 2016. A developmental perspective on the neural bases of human empathy. *Infant Behavior and Development.* In press.

Tseng, T. A. et al. 2016. The health benefits of children by different natural landscape contacting level. *Environment-Behaviour Proceedings Journal* 1:168–79.

Tuke, D. H. 1892. *A Dictionary of Psychological Medicine*. London: J. A. Churchill. 39:59–72.

Ulutas, I. et al. 2012. Maternal attitudes, emotional intelligence and home environment and their relations with emotional intelligence of sixth years old children. In *Emotional Intelligence—New Perspectives and Applications*, ed. Annamaria di Fabio. InTech,

Unsworth, S. et al. 2016. The impact of mindful meditation in nature on self-nature interconnectedness. *Mindfulness* 7:1052–1060.

Walsh, E. et al. 2016. Brief mindfulness training reduces salivary IL-6 and TNF-α in young women with depressive symptomatology. *Journal of Consulting and Clinical Psychology* Oct.; 84(10):887–97.

Weinstein, N. et al. 2009. Can nature make us more caring? Effects of immersion in nature on intrinsic aspirations and generosity. *Personality and Social Psychology Bulletin* 35:1315–29.

Williams, A. et al. 2014. The influence of empathic concern on prosocial behavior in children. *Frontiers in Psychology* 5:425.

Zeidner, M. et al. 2012. The emotional intelligence, health, and well-being nexus: What have we learned and what have we missed? *Applied Psychology: Health and Well-Being* 4:1–30.

Zelenski, J. M. et al. 2015. Cooperation is in our nature: Nature exposure may promote cooperative and environmentally sustainable behavior. *Journal of Environmental Psychology* 42:24–31.

Zeng, X. et al. 2015. The effect of loving-kindness meditation on positive emotions: A meta-analytic review. *Frontiers in Psychology* Nov. 3:6:1693.

Zhang, J. W. et al. 2014. An occasion for unselfing: Beautiful nature leads to prosociality. *Journal of Environmental Psychology* 37:61–72.

Zhang, J. W. et al. 2014. Engagement with natural beauty moderates the positive relation between connectedness with nature and psychological well-being. *Journal of Environmental Psychology* 38:55–63.

Zilioli, S. et al. 2015. Purpose in life predicts allostatic load ten years later. *Journal of Psychosomatic Research* 79:451–457.

Chapter 8: Encephalobiotics Turn the Mind

Allen, A. P. et al. 2016. *Bifidobacterium longum* 1714 as a translational psychobiotic: Modulation of stress, electrophysiology and neurocognition in healthy volunteers. *Translational Psychiatry* Nov. 1; 6(11):e939.

Altmaier, E. et al. 2013. Metabolomic profiles in individuals with negative affectivity and social inhibition: A population-based study of Type D personality. *Psychoneuroendocrinology* Aug.; 38(8):1299–309.

Andersson, H. et al. 2016. Oral administration of *Lactobacillus plantarum* 299v reduces cortisol levels in human saliva during examination induced stress: A randomized, double-blind controlled trial. *International Journal of Microbiology* 2016:8469018.

Campbell, P. A. 1943. *Why the Universe? Or, Cosmopoietic Space*. San Francisco: G. Fields.

Chisholm, A. 1972. *Philosophers of the Earth: Conversations with Ecologists.* New York: E. P. Dutton.

Daland, J. 1909. Indicanuria and its significance. *JAMA* 53:14461449.

Dinan, T. G. et al. 2013. Psychobiotics: A novel class of psychotropic. *Biological Psychiatry* Nov. 15; 74(10):720–6.

Dolnick, E. 1998. *Madness on the Couch: Blaming the Victim in the Heyday of Psychoanalysis*. New York: Simon & Schuster.

Fiocchi, A. et al. 2016. Probiotics, prebiotics & food allergy prevention: Clinical data in children. *Journal of Pediatric Gastroenterology and Nutrition* Jul; 63 Suppl 1:S14–7.

Fijan, S. 2014. Microorganisms with claimed probiotic properties: An overview of recent literature. *International Journal of Environmental Research and Public Health* 11:4745–67.

Gomes, A. C. et al. 2017. The additional effects of a probiotic mix on abdominal adiposity and antioxidant status: A double-blind, randomized trial. *Obesity* Jan.; 25(1):30–38.

Greenblum, S. et al. 2015. Extensive strain-level copy-number variation across human gut microbiome species. *Cell* 160(4):583–94.

Gunsalus, K. T. et al. 2015. Manipulation of host diet to reduce gastrointestinal colonization by the opportunistic pathogen *Candida albicans*. *mSphere* Nov. 18; 1(1). pii: e00020-15.

Harima-Mizusawa, N. et al. 2016. Citrus juice fermented with *Lactobacillus plantarum* YIT 0132 alleviates symptoms of perennial allergic rhinitis in a double-blind, placebo-controlled trial. *Beneficial Microbes* Nov. 30; 7(5):649–658.

Hirano, S. et al. 2016. Effect of *Lactobacillus plantarum* Tennozu-SU2 on *Salmonella typhimurium* infection in human enterocyte-like HT-29-Luc cells and sALB/c mice. *Probiotics and Antimicrobial Proteins*. In press.

Huang, R. et al. 2016. Effect of probiotics on depression: A systematic review and meta-analysis of randomized controlled trials. *Nutrients* Aug. 6; 8(8). pii: E483.

Joint FAO/WHO Working Group Report on Drafting Guidelines for the Evaluation of Probiotics in Food. London, Ontario, Canada, April 30 and May 1, 2002.

Jukes, T.H. and W.L. Williams. 1953. Nutritional effects of antibiotics. *Pharmacological Review*. Dec; 5(4):381–420.

Jukes, T.H. 1973. Public health significance of feeding low levels of antibiotics to animals. *Advances in Applied Microbiology* 16:1–54.

Kalliomäki, M. et al. 2001. Probiotics in primary prevention of atopic disease: A randomised placebo-controlled trial. *Lancet* April 7; 357(9262):1076–9.

Kalliomäki, M. et al. 2003. Probiotics and prevention of atopic disease: 4-year follow-up of a randomised placebo-controlled trial. *Lancet* May 31; 361(9372):1869–71.

Kashyap, P.C. et al. 2013. Complex interactions among diet, gastrointestinal transit, and gut microbiota in humanized mice. *Gastroenterology* May; 144(5):967–77.

Kato-Kataoka, A. et al. 2016. Fermented milk containing *Lactobacillus casei* strain Shirota preserves the diversity of the gut microbiota and relieves abdominal dysfunction in healthy medical students exposed to academic stress. *Applied and Environmental Microbiology* May 31; 82(12):3649–58.

Kim, M. et al. 2017. Effects of weight loss using supplementation with *Lactobacillus* strains on body fat and medium-chain acylcarnitines in overweight individuals. *Food & Function* Jan. 25; 8(1):250–261.

Kirsch, I. 2014. Antidepressants and the placebo effect. *Zeitschrift für Psychologie* 222(3):128–134.

Kligler, B. et al. 2016. Complementary/Integrative therapies that work: A review of the evidence. *American Family Physician* Sept. 1; 94(5):369–74.

Lenoir-Wijnkoop, I. et al. 2016. The clinical and economic impact of probiotics consumption on respiratory tract infections: Projections for Canada. *PLoS One* Nov. 10; 11(11):e0166232.

Lovene, M.R. et al. 2016. Intestinal dysbiosis and yeast isolation in stool of subjects with autism spectrum disorders. *Mycopathologia*. In press.

Lundelin, K. et al. 2016. Long-term safety and efficacy of perinatal probiotic intervention: Evidence from a follow-up study of four randomized, double-blind, placebo-controlled trials. *Pediatric Allergy and Immunology*. In press.

Majlesi, M. et al. 2017. Effect of probiotic *Bacillus coagulans* and *Lactobacillus plantarum* on alleviation of mercury toxicity in rat. *Probiotics and Antimicrobial Proteins*. In press.

Martin, F.P. et al. 2009. Metabolic effects of dark chocolate consumption on energy, gut microbiota, and stress-related metabolism in free-living subjects. *Journal of Proteome Research* 8 (12):5568–5579.

McKean, J. et al. 2016. Probiotics and subclinical psychological symptoms in healthy participants: A systematic review and meta-analysis. *Journal of Alternative and Complementary Medicine*. In press.

Miyazaki, K. et al. 2014. Dietary heat-killed *Lactobacillus brevis* SBC8803 promotes voluntary wheel-running and affects sleep rhythms in mice. *Life Sciences* Aug. 28; 111(1-2):47–52.
Moncrieff, J. et al. 2015. Empirically derived criteria cast doubt on the clinical significance of antidepressant-placebo differences. *Contemporary Clinical Trials* Jul; 43:60–2.
Monoi, N. et al. 2016. Japanese sake yeast supplementation improves the quality of sleep: A double-blind randomised controlled clinical trial. *Journal of Sleep Research* Feb.; 25(1):116–23.
Nakaita, Y. et al. 2013. Heat-killed *Lactobacillus brevis* SBC8803 induces serotonin release from intestinal cells. *Food and Nutrition Sciences* 4:767–771.
Nakakita, Y. et al. 2016. Effect of dietary heat-killed *Lactobacillus brevis* SBC8803 (SBL88™) on sleep: A non-randomised, double-blind, placebo-controlled, and crossover pilot study. *Beneficial Microbes* Sept.; 7(4):501–9.
Nakamura, Y. et al. 2016. Oral administration of Japanese sake yeast (*Saccharomyces cerevisiae* sake) promotes non-rapid eye movement sleep in mice via adenosine A2A receptors. *Journal of Sleep Research* Dec.; 25(6):746–753.
Naseribafrouei, A. et. al. 2014. Correlation between the human fecal microbiota and depression. *Journal of Neurogastroenterology and Motility* 26:1155–1162.
Nielsen, S. J. et al. 2015. Trends in yogurt consumption, US adults, 1999–2012. *The FASEB Journal* 29 Supplement 1:587.17.
Ogata, T. et al. 1999. Effect of *Bifidobacterium longum* BB536 yogurt administration on the intestinal environment of healthy adults. *Microbial Ecology in Health and Disease* 11:41–46.
Orwell, George. 1949. *Nineteen Eighty-Four*. London: Secker & Warburg.
Parker, R. B. 1974. Probiotics, the other half of the antibiotics story. *Animal Nutrition and Health* 29:4–8.
Pärtty, A. et al. 2015. A possible link between early probiotic intervention and the risk of neuropsychiatric disorders later in childhood: A randomized trial. *Pediatric Research* June; 77(6):823–8.
Pirbaglou, M. et al. 2016. Probiotic supplementation can positively affect anxiety and depressive symptoms: A systematic review of randomized controlled trials. *Nutrition Research* Sept.; 36(9):889–98.
Proceedings of the 15th International Veterinary Congress. *Probiotic Epoch*. 1953. Proceedings Part I, Volume II. Stockholm, Sweden. p. 914.
Santelmann, H. et al. Effectiveness of nystatin in polysymptomatic patients. A randomized, double-blind trial with nystatin versus placebo in general practice. *Family Practice* June; 18(3):258–65.
Seleem, D. et al. Review of flavonoids: A diverse group of natural compounds with anti-*Candida albicans* activity in vitro. *Archives of Oral Biology* Aug. 27. pii: S0003-9969(16)30227-8.
Severance, E. G. et al. 2016. *Candida albicans* exposures, sex specificity and cognitive deficits in schizophrenia and bipolar disorder. *npj Schizophrenia* May 4; 2:16018.
Severance, E. G. et al. 2016. Probiotic normalization of *Candida albicans* in schizo-

phrenia: A randomized, placebo-controlled, longitudinal pilot study. *Brain, Behavior and Immunity* Nov. 18. pii: S0889-1591(16)30521-9.

Stenman, L. K. et al. 2016. Probiotic with or without fiber controls body fat mass, associated with serum zonulin, in overweight and obese adults-randomized controlled trial. *EBioMedicine* Nov.; 13:190–200.

Sugawara, T. et al. 2016. Regulatory effect of paraprobiotic *Lactobacillus gasseri* CP2305 on gut environment and function. *Microbial Ecology in Health and Disease* Mar. 14; 27:30259.

Sweileh, W. M. et al. 2016. Assessing worldwide research activity on probiotics in pediatrics using Scopus database: 1994–2014. *World Allergy Organization Journal* July 25; 9:25.

Taibi, A. et al. 2014. Practical approaches to probiotics use. *Applied Physiology, Nutrition, and Metabolism* Aug.;3 9(8):980–6.

Tohyama, K. et al. 1981. Effect of *Lactobacilli* on urinary indican excretion in gnotobiotic rats and in man. *Microbiology and Immunology* 25:101–112.

Tomasik, J. et al. 2015. Immunomodulatory effects of probiotic supplementation in schizophrenia patients: A randomized, placebo-controlled trial. *Biomark Insights* June 1; 10:47–54.

Turnbaugh, P. J. et al. 2010. Organismal, genetic, and transcriptional variation in the deeply sequenced gut microbiomes of identical twins. *Proceedings of the National Academy of Science* April 20; 107(16):7503–8.

Vergin, F. 1954. Anti- und Probiotika. *Hippokrates* 25:116–119.

Wang, F. et al. 2016. Gut homeostasis, microbial dysbiosis, and opioids. *Toxicologic Pathology.* In press.

Zelaya, H. et al. 2015. Nasal priming with immunobiotic *Lactobacillus rhamnosus* modulates inflammation-coagulation interactions and reduces influenza virus-associated pulmonary damage. *Inflammation Research* Aug.; 64(8):589–602.

Zhang, Y. et al. 2016. Effects of probiotic type, dose and treatment duration on irritable bowel syndrome diagnosed by Rome III criteria: A meta-analysis. *BMC Gastroenterology* June 13; 16(1):62.

Zheng, M. Y. et al. 2016. Mechanisms of inflammation-driven bacterial dysbiosis in the gut. *Mucosal Immunology.* In press.

Chapter 9: Biophilic Science: A Brighter Future

Ashton, J. R. 2006. Virchow misquoted, part-quoted, and the real McCoy. *Journal of Epidemiology and Community Health* Aug.; 60(8):671.

Bapayeva, G. et al. 2016. Organochlorine pesticides and female puberty in South Kazakhstan. *Reproductive Toxicology* Oct.;65:67–75.

Barmeyer, C. et al. 2017. Long-term response to gluten-free diet as evidence for non-celiac wheat sensitivity in one third of patients with diarrhea-dominant and mixed-type irritable bowel syndrome. *International Journal of Colorectal Disease* 32:29–39.

Beasley, R. T. et al. 2008. Association between paracetamol use in infancy and childhood, and risk of asthma, rhinoconjunctivitis, and eczema in children aged 6–7 years: Analysis from Phase Three of the ISAAC programme. *Lancet* 372(9643): 1039–1048.

Bill Moyers and Company. Jonas Salk on searching for the next medical miracle. Feb, 1990. http://billmoyers.com/content/jonas-salk/

Burmaster, K. B. et al. 2015. Impact of a private sector living wage intervention on depressive symptoms among apparel workers in the Dominican Republic: A quasi-experimental study. *BMJ Open* 5:e007336.

Century of Progress International Administration. 1933. *A Century of Progress: The Official Guidebook of the 1933 World's Fair*. Chicago: Century of Progress Administration Publication.

Cleary, D. W. et al. 2016. Long-term antibiotic exposure in soil is associated with changes in microbial community structure and prevalence of class 1 integrons. *FEMS Microbiology Ecology* Oct.;92(10). pii: fiw159.

Cristea, I. A., et al. 2017. Sponsorship bias in the comparative efficacy of psychotherapy and pharmacotherapy for adult depression: Meta-analysis. *British Journal of Psychiatry* 210:16–23.

Di Sabatino, A. et al. 2015. Small amounts of gluten in subjects with suspected nonceliac gluten sensitivity: A randomized, double-blind, placebo-controlled, crossover trial. *Clinical Gastroenterology and Hepatology* Sept.; 13(9):1604–12.e3.

de Gavelle, E. et al. 2016. Chronic dietary exposure to pesticide residues and associated risk in the French ELFE cohort of pregnant women. *Environment International* 92–93: 533–42.

Dubos, R. 1959. *The Mirage of Health: Utopias, Progress and Biological Change*. Brunswick, NJ: Rutgers University Press.

———. 1968. *So Human an Animal*. New York: Scribner's.

———. 1970. The human landscape. *Bulletin of the Atomic Scientists* 26:31–37.

Eswaran, S. L. et al. 2016. A randomized controlled trial comparing the low FODMAP diet vs. modified NICE guidelines in US adults with IBS-D. *American Journal of Gastroenterology* Dec.; 111(12):1824–1832.

Furukawa, T. K. et al. 2017. Initial severity of depression and efficacy of cognitive-behavioural therapy: Individual-participant data meta-analysis of pill-placebo-controlled trials. *British Journal of Psychiatry* Jan 19. pii: bjp.bp.116.187773.

Hall, N. 2014. The Kardashian index: A measure of discrepant social media profile for scientists. *Genome Biology* July 30; 15(7):424.

Hand, K. L. et al. 2016. A novel method for fine-scale biodiversity assessment and prediction across diverse urban landscapes reveals social deprivation-related inequalities in private, not public spaces. *Landscape and Urban Planning* 151:33–44.

Hard, A. D. 1911. Honesty in medical practice is not a handicap. *Med World* 29:521.

Herman, P. M. et al. 2014. A naturopathic approach to the prevention of cardiovascular disease: Cost-effectiveness analysis of a pragmatic multi-worksite randomized clinical trial. *Journal of Occupational and Environmental Medicine* Feb; 56(2)171–6.

Hu, Y. et al. 2016. Health risk from veterinary antimicrobial use in China's food animal production and its reduction. *Environmental Pollution* 219:993–97.

Kim Y. S. et al. High-throughput 16S rRNA gene sequencing reveals alterations of mouse intestinal microbiota after radiotherapy. *Anaerobe*. 2015 Jan 16;33:1–7.

Komro, K. A. et al. The effect of an increased minimum wage on infant mortality and birth weight. *American Journal of Public Health* 106(8):1514–1516.

Lust, B. 1924. Blood washing treatments and the fountain of youth. *Strength Magazine* p 13.

Maccoby, M. 1964. Government, scientists, and the priorities of science. *Dissent* 6: 55–67.

Moodie, R. 2013. Profits and pandemics: Prevention of harmful effects of tobacco, alcohol, and ultra-processed food and drink industries. *Lancet* Feb. 23; 381(9867): 670–9.

Nelis, D. et al. 2011. Increasing emotional competence improves psychological and physical well-being, social relationships, and employability. *Emotion* April; 11(2):354–66.

Ohira, H. et al. 2013. Pro-inflammatory cytokine predicts reduced rejection of unfair financial offers. *Neuroendocrinology Letters* 34(1):47–51.

Peters, S. L. et al. 2014. Randomised clinical trial: Gluten may cause depression in subjects with non-coeliac gluten sensitivity - an exploratory clinical study. *Alimentary Pharmacology & Therapeutics*May; 39(10):1104–12.

Reeve, A. B. 1908. Back to nature. *Scrap Book* 6:343–46.

Reeves, A. et al. 2016. Introduction of a national minimum wage reduced depressive symptoms in low-wage workers: A quasi-natural experiment in the UK. *Health Economics*. In press.

Richert, R. A. et al. 2010. Word learning from baby videos. *Archives of Pediatric and Adolescent Medicine* May; 164(5):432–7.

Sager, I. et al. 1998. Back to the future at Apple. *Business Week*, May 25:46–9.

Salk J. 1979. Lemmings or fruit flies? *The Age*, April 30, p 21.

Seely, D. et al. 2013. Naturopathic medicine for the prevention of cardiovascular disease: A randomized clinical trial. *Canadian Medical Association Journal* June 11; 185(9):E409–16.

Steif, W. 1974. Salk switches medicine for morality. Polio fighter takes on larger scourge—mankind. *The Pittsburgh Press*, October 26th.

Taubman, H. 1966. Unconquered darkness beyond. Salk faces west to new horizons. *Pittsburgh Post-Gazette*. Nov 24th.

Vermes, K. 2016. The latest and greatest fast food apps for 2016. *Tech Times*, April 19th.

Salk, J. 1973. *The Survival of the Wisest*. San Francisco: Harper and Row.

Salk, J. 1992. Nobel Gathering. Influencing Evolution. *Asbury Park Press*, October 9.

Stassinopoulos, A. 1984. Dr. Jonas Salk's formula for the future. Courage. Love. Forgiveness. *Parade Magazine*, Nov 4th.

Taubman, H. 1996. Unconquered darkness beyond: Salk faces west to new horizons. *Pittsburgh Post-Gazette*, Nov. 24.

Tessaro, L. W. et al. 2015. Bacterial growth rates are influenced by cellular characteristics of individual species when immersed in electromagnetic fields. *Microbiological Research* March; 172:26–33.

Tsao, T-Y. et al. 2016. Estimating potential reductions in premature mortality in New York City from raising the minimum wage to $15. *American Journal of Public Health* 106(6):1036–41.

Uhde, M. et al. 2016. Intestinal cell damage and systemic immune activation in individuals reporting sensitivity to wheat in the absence of coeliac disease. *Gut* July 25. pii: gutjnl–2016–311964.

Vitetta, L. et al. 2013. The gastrointestinal microbiome and musculoskeletal diseases: A beneficial role for probiotics and prebiotics. *Pathogens* 2, 606–626.

Wagner, R. F. 1958. How US labor aids science, health. *AFL-CIO Free Trade Union News* 13:2.

Wampold, B. E. et al. 2017. In pursuit of truth: A critical examination of meta-analyses of cognitive behavior therapy. *Psychotherapy Research* 27:14–32.

Xu, C. et al. 2016. Environmental exposure to DDT and its metabolites in cord serum: Distribution, enantiomeric patterns, and effects on infant birth outcomes. *Science of the Total Environment* Dec. 15. pii: S0048–9697(16)32662–6.

Chapter 10: The Concert of Life: Holistic Harmony

Achtman, M. 2016. How old are bacterial pathogens? *Proceedings of the Royal Society B: Biological Sciences* Aug. 17; 283(1836). pii: 20160990.

Amato, K. R. et al. 2017. Patterns in gut microbiota similarity associated with degree of sociality among sex classes of a neotropical primate. *Microbial Ecology.* In press.

Bahrndorff, S. et al. 2017. Bacterial communities associated with houseflies (*Musca domestica* L.) sampled within and between farms. *PLoS ONE* 12:e0169753.

Balmori, A. 2016. Radiotelemetry and wildlife: Highlighting a gap in the knowledge on radiofrequency radiation effects. *Science of the Total Environment* Feb. 1; 543(Pt A):662–9.

Baluška, F. et al. 2016. Vision in plants via plant-specific ocelli? *Trends in Plant Science* Sept.; 21(9):727–30.

Baluška, F. et al. 2016. Understanding of anesthesia - Why consciousness is essential for life and not based on genes. *Communicative and Integrative Biology* Nov. 4; 9(6):e1238118.

Bond-Lamberty, B. et al. 2016. Soil respiration and bacterial structure and function after 17 years of a reciprocal soil transplant experiment. *PLoS One* March 2; 11(3): e0150599.

Bose, J. C. 1929. Are plants sentient beings? Recent discoveries have revealed the unity of all life. *The Century* 117:385–93.

Bryan, C. J. et al. 2016. Harnessing adolescent values to motivate healthier eating. *Proceedings of the National Academy of Science* 113:10830–10835.

Cammaerts, M. C. et al. 2013. Food collection and response to pheromones in an ant species exposed to electromagnetic radiation. *Electromagnetic Biology and Medicine* Sept.; 32(3):315–32.

Cammaerts, M. C. et al. 2016. Ants can expect the time of an event on basis of previous experiences. *International Scholarly Research Notices* June 14; 2016:9473128.

Chopan, M. et al. 2017. The association of hot red chili pepper consumption and mortality: A large population-based cohort study. *PloS ONE* 12: e0169876.

Ciszak, M. et al. 2016. Plant shoots exhibit synchronized oscillatory motions. *Communicative & Integrative Biology* Oct. 7; 9(5):e1238117.

Cleveland, D. A. 2016. Prioritizing good diets. *Science.* Dec. 16; 354(6318):1385.

Darwin, C. 1880. *The Power of Movement in Plants.* London: John Murray.

Figuerola, E. L. et al. 2015. Crop monoculture rather than agriculture reduces the spatial turnover of soil bacterial communities at a regional scale. *Environmental Microbiology* March; 17(3):678–88.

Fuentes, J. et al. 2014. Enhanced therapeutic alliance modulates pain intensity and muscle pain sensitivity in patients with chronic low back pain: An experimental controlled study. *Physical Therapy* Apr.; 94(4):477–89.

Gorzelak, M. A. et al. 2015. Inter-plant communication through mycorrhizal networks mediates adaptive behavior in plant communities. *AoB Plants* 7:plv050.

Hand, K. L. et al. 2016. A novel method for fine-scale biodiversity assessment and prediction across diverse urban landscapes reveals social deprivation-related inequalities in private, not public spaces. *Landscape and Urban Planning* 151:33–44.

Hariri, A. R., et al. 2015. Finding translation in stress research. *Nature Neuroscience* Oct.; 18(10):1347–52.

Hu, J. et al. 2016. Probiotic diversity enhances rhizosphere microbiome function and plant disease suppression. *MBio* Dec. 13; 7(6). pii: e01790–16.

Hutchinson, R. F. 1885. Are plants sentient beings? *Knowledge* 207:337–38.

Waldmann-Selsam, C. et al. Radiofrequency radiation injures trees around mobile phone base stations. *Science of the Total Environment* Dec. 1; 572:554–569.

Kang, C. et al. 2016. Healthy subjects differentially respond to dietary capsaicin correlating with specific gut enterotypes. *Journal of Clinical Endocrinology and Metabolism* Dec.; 101(12):4681–4689.

Keri, S. et al. 2014. Expression of toll-like receptors in peripheral blood mononuclear cells and response to cognitive-behavioral therapy in major depressive disorder. *Brain, Behavior, and Immunity* Aug.; 40:235–43.

Khamechian, T. et al. 2015. Evaluation of the correlation between childhood asthma and *Helicobacter pylori* in Kashan. *Jundishapur Journal of Microbiology* June 1; 8(6):e17842.

Kim, Y. S. et al. 2015. High-throughput 16S rRNA gene sequencing reveals alterations of mouse intestinal microbiota after radiotherapy. *Anaerobe* Jan. 16; 33:1–7.

Kong, H. G. et al. 2016. Aboveground whitefly infestation-mediated reshaping of the root microbiota. *Frontiers in Microbiology* Sept. 7; 7:1314.

Korkama, T. et al. 2007. Interactions between extraradical ectomycorrhizal mycelia, microbes associated with the mycelia and growth rate of Norway spruce (*Picea abies*) clones. *New Phytologist* 173(4):798–807.

Kucukazman, M. et al. 2015. *Helicobacter pylori* and cardiovascular disease. *European Review for Medical and Pharmacological Science* Oct.;19 (19):3731–41.

Lareen, A. et al. 2016. Plant root-microbe communication in shaping root microbiomes. *Plant Molecular Biology* 90:575–587.

Lakshmanan, V. et al. 2016. Killing two birds with one stone: Natural rice rhizospheric microbes reduce arsenic uptake and blast infections in rice. *Frontiers in Plant Science* Oct 13; 7:1514.

Lazaro, A. et al. 2016. Electromagnetic radiation of mobile telecommunication antennas affects the abundance and composition of wild pollinators. *Journal of Insect Conservation* 20:315–24.

Lionetti, E. et al. 2014. *Helicobacter pylori* infection and atopic diseases: Is there a relationship? A systematic review and meta-analysis. *World Journal of Gastroenterology* Dec. 14; 20(46):17635–47.

Lupatini, M. et al. 2017. Soil microbiome is more heterogeneous in organic than in conventional farming system. *Frontiers in Microbiology* Jan. 4; 7:2.

Mancuso, S. and A. Viola. 2015. *Brilliant Green.* Washington, DC: Island Press.

Menezes, C. et al. 2015. A Brazilian social bee must cultivate fungus to survive. *Current Biology* Nov. 2;25(21):2851–5.

Margulis, L. and D. Sagan. 1995. *What is Life?* Berkeley: University of California Press.

Park, S. Y. et al. 2006. Reconstruction of the chemotaxis receptor–kinase assembly. *Nature Structural & Molecular Biology* 13:400–7.

Pereira, C. 2016. Is it quantum sentience or quantum consciousness? A review of social behaviours observed in primitive and present-day microorganisms. *NeuroQuantology* 14:16–27.

Radić, M. 2014. Role of *Helicobacter pylori* infection in autoimmune systemic rheumatic diseases. *World Journal of Gastroenterology* Sept. 28; 20(36):12839–46.

Schwarze, S. et al. Weak Broadband electromagnetic fields are more disruptive to magnetic compass orientation in a night-migratory songbird (*Erithacus rubecula*) than strong narrow-band fields. *Frontiers in Behavioral Neuroscience* March 22;10:55.

Shapiro, J. A. 2007. Bacteria are small but not stupid: Cognition, natural genetic engineering and socio-bacteriology. *Studies in History and Philosophy of Biological and Biomedical Sciences* Dec.; 38(4):807–19.

Shindler-Itskovitch, T. et al. A systematic review and meta-analysis of the association between *Helicobacter pylori* infection and dementia. *Journal of Alzheimer's Disease* April 15; 52(4):1431–42.

Song, Y. Y. et al. 2015. Defoliation of interior Douglas-fir elicits carbon transfer and stress signaling to ponderosa pine neighbors through ectomycorrhizal networks. *Scientific Reports* 5:8495.

Su, J. et al. 2014. Association between *Helicobacter pylori* infection and migraine: A meta-analysis. *World Journal of Gastroenterology* Oct. 28; 20(40):14965–72.

Tessaro, L. W. et al. 2015. Bacterial growth rates are influenced by cellular characteristics of individual species when immersed in electromagnetic fields. *Microbiological Research* March; 172:26–33.

The burger that conquered the country. 1973. *Time* Vol 102 September 17.

Thompson, B. et al. Strategies to empower communities to reduce health disparities. *Health Affairs* 35:1424–1428.

Wu, L. et al. 2016. Effects of consecutive monoculture of *Pseudostellaria heterophylla* on soil fungal community as determined by pyrosequencing. *Scientific Reports* May 24; 6:26601.

Xu, Y. et al. 2015. Is *Helicobacter pylori* infection a critical risk factor for vascular dementia? *International Journal of Neuroscience* Sept. 15:1–5.

Index

About the Authors

DR. SUSAN L. PRESCOTT MD, PHD, is an internationally acclaimed immunologist and pediatrician whose early work led to a paradigm shift in understanding the importance of the early environment in immune programming for the risk of subsequent disease. She is a former Director of the World Allergy Organisation and former Head of the School of Paediatrics and Child Health at the University of Western Australia. Prescott is the author of more than 250 scientific papers and several books including *The Allergy Epidemic*, *The Calling*, and *Origins*. She is married to Alan Logan and splits her time between Perth, Australia and New York.

DR. ALAN C. LOGAN ND, graduated *magna cum laude* from the State University of New York at Purchase, and was valedictorian at the Canadian College of Naturopathic Medicine in Toronto. He taught Mind-Body Medicine courses in Harvard's School of Continuing Medical Education, and his research and commentaries have appeared in over two dozen medical journals. He is a trusted source for many popular magazines including *Cosmopolitan*, *Elle*, *W*, *Health*, and *Life & Style*, and he is a commentator on US and Canadian television. Logan is the co-author of several books including *Your Brain on Nature* and *The Clear Skin Diet*, and is a contributor to the forthcoming *Oxford Textbook of Nature and Public Health*. He is married to Susan Prescott and splits his time between Perth, Australia and New York.

ABOUT NEW SOCIETY PUBLISHERS

New Society Publishers is an activist, solutions-oriented publisher focused on publishing books for a world of change. Our books offer tips, tools, and insights from leading experts in sustainable building, homesteading, climate change, environment, conscientious commerce, renewable energy, and more—positive solutions for troubled times.

DON'T EAT THIS BOOK (but you could)

We're proud to hold to the highest environmental and social standards of any publisher in North America. This is why some of our books might cost a little more. We think it's worth it!

- We print all our books in North America, never overseas
- All our books are printed on **100% post-consumer recycled paper**, processed chlorine-free, with low-VOC vegetable-based inks (since 2002)
- Our corporate structure is an innovative employee shareholder agreement, so we're one-third employee-owned (since 2015)
- We're carbon-neutral (since 2006)
- We're certified as a B Corporation (since 2016)

At New Society Publishers, we care deeply about *what* we publish—but also about *how* we do business.

New Society Publishers

ENVIRONMENTAL BENEFITS STATEMENT

For every 5,000 books printed, New Society saves the following resources:[1]

38	Trees
3,436	Pounds of Solid Waste
3,781	Gallons of Water
4,931	Kilowatt Hours of Electricity
6,246	Pounds of Greenhouse Gases
27	Pounds of HAPs, VOCs, and AOX Combined
9	Cubic Yards of Landfill Space

[1] Environmental benefits are calculated based on research done by the Environmental Defense Fund and other members of the Paper Task Force who study the environmental impacts of the paper industry.